No. 256

Cell–Cell Interactions
Second Edition
A Practical Approach

Edited by

Tom P. Fleming
School of Biological Sciences, University of
Southampton, Bassett Crescent East,
Southampton SO16 7PX, UK

OXFORD
UNIVERSITY PRESS

OXFORD
UNIVERSITY PRESS

Great Clarendon Street, Oxford OX2 6DP

Oxford University Press is a department of the University of Oxford.
It furthers the University's objective of excellence in research, scholarship,
and education by publishing worldwide in

Oxford New York

Athens Auckland Bangkok Bogotá Bombay Buenos Aires Calcutta
Cape Town Dar es Salaam Delhi Florence Hong Kong Istanbul
Karachi Kuala Lumpur Madrid Melbourne Mexico City Mumbai
Nairobi Paris São Paulo Singapore Taipei Tokyo Toronto Warsaw

and associated companies in Berlin Ibadan

Oxford is a registered trade mark of Oxford University Press
in the UK and in certain other countries

Published in the United States
by Oxford University Press Inc., New York

© Oxford University Press, 2002

British Library Cataloguing in Publication Data
Data available

Library of Congress Cataloging in Publication Data

Cell-cell interactions : a practical approach.—2nd ed. / edited by
Tom P. Fleming.
p. ; cm.—(Practical approach series ; 256)
Includes bibliographical references and index.
1. Cell interaction—Laboratory manuals. I. Fleming, Tom P. II. Series.
[DNLM: 1. Cell Communication. 2. Cell Adhesion Molecules. 3. Cytological
Techniques. 4. Intercellular Junctions. QH 604.2 C3922 2001]
QH604.2 .C4415 2001   571'.6—dc21   2001036903

ISBN 0 19 963864 0 (Hbk.)
ISBN 0 19 963863 2 (Pbk.)

10 9 8 7 6 5 4 3 2 1

Typeset in Swift by Footnote Graphics, Warminster, Wilts
Printed in Great Britain on acid-free paper
by The Bath Press Ltd, Avon

Preface

Cell-cell interactions are ubiquitous throughout biology. Virtually all cells in multicellular organisms will experience these interactions for a myriad of different purposes such as tissue construction, intercellular communication, information transfer, spatial awareness, control of differentiation and in inflammation response. As a consequence, it is very difficult, and probably unproductive, to try to bring together within a single text a comprehensive synthesis of all new advances in methodologies for investigating cell-cell interactions. I have therefore been selective, picking out those key systems and cell types where cell-cell interactions have a predominant role in biological and medical function. The chapters in this text cover keratinocytes, leukocytes, neurons and endothelial cells and, in addition, specialised areas of cell-cell interaction such as adhesion, tight and gap junction complexes. There are also chapters on cell-cell interactions during development, using Drosophila, Xenopus and mammalian models. Above all, the range of techniques detailed in the 105 Protocols supplied in the 10 Chapters are state-of-the-art and written in sufficient detail for investigators of cell-cell interactions to follow and repeat for their own system.

This book is a second edition of an earlier volume in the Practical Approach Series published some ten years ago on cell-cell interactions and edited by Bruce Stevenson and colleagues. Additional techniques associated with this fascinating and important area can be found in that excellent earlier text; the current edition does not overlap although recent developments in classical procedures are included.

I would like to thank all those who have contributed to the production of this volume. In particular, I am grateful to all the 44 authors who have given their time and considerable expertise to make this text so valuable to researchers interested in the ways cells interact with their neighbours. Thanks also to the OUP production staff and, of course, to all the granting agencies world-wide who have supplied the necessary funding for our research without which we would still be in the starting blocks.

Tom P. Fleming

Contents

Preface *page v*
Protocol list *xiii*

1 Cadherin adhesion regulation in keratinocytes *1*
Vania Braga

1 Introduction *1*
2 Keratinocytes and cadherins *1*
    Background *1*
    Model *3*
    Advantages of the model *4*
    Disadvantages of the model *4*
    Keratinocyte cultures *5*
3 Microinjection *7*
    Tools to modulate the activity of the small GTPases *7*
    Methodology *8*
    Strategy *8*
    Preparation of material to inject *10*
    Microinjection procedure to study keratinocyte junctions *11*
    Analysis of microinjection experiments *13*
4 Immunostaining and confocal analysis *14*
    Detergent extraction *14*
    Immunolabelling *16*
    Confocal microscocopy *19*
5 Cadherin functional assays *20*
    Inhibition by antibodies against E- and P-cadherin *20*
    Clustering cadherin receptors *21*
6 Biochemical analysis of cadherin complexes *24*
    Cadherin and catenins *24*
    Protein extracts *26*
    Immunoprecipitation and Western blotting *27*
7 Phosphorylation of cadherin complexes *30*
    Orthophosphate labelling *30*
    Isolation of phosphorylated proteins *31*
    Phosphopeptide mapping *33*
    Determination of the type of phosphorylated amino acid *34*

2 Analysis of intercellular adhesion molecules in endothelial cells 37
Gianfranco Bazzoni, Maria G. Lampugnani, Ofelia M. Martinez-Estrada, and Elisabetta Dejana

1 Introduction 37

2 Endothelial cell culture 38
   Human umbilical vein endothelial cells (HUVECs) 38

3 Vascular permeability and morphological changes associated with increased permeability 40
   Vascular permeability 42
   Immunofluorescence microscopy analysis of ECs 43

Acknowledgements 44

3 Gap junction-mediated interactions between cells 47
David Becker and Colin Green

1 Introduction 47

2 Establishing the pattern and extent of communication between cells 47
   Dye transfer 48

3 Gap junction density and connexin isoform expression 54
   Immunohistochemistry: connexin-specific antibodies 54
   Quantification of connexin plaque size and number 58
   Expression analysis of mRNA by quantitative RT–PCR 59
   Whole-mount *in situ* hybridization for connexins using digoxigenin 61

4 Manipulating gap junctional communication 63
   Pharmacological agents 64
   Antibody blockers 64
   Extracellular loop peptides 65
   Manipulation of gap junction gene expression 66

4 Functional analysis of the tight junction 71
Marcellino Cereijido, Liora Shoshani, and Ruben G. Contreras

1 Introduction 71

2 Model systems 72
   Monolayers of epithelial cells 72
   Co-cultures 74

3 Electron microscopy 77
   Electron-dense tracers 78
   Freeze-fracture 80

4 Information derived from the physical properties of permeating molecules 82
   Solute size and charge assay for paracellular transport 82

5 The tight junction also functions as a fence 85

6 Analysis of second messenger regulation of tight junction sealing and activity 87
   Ca^{2+} signalling 87
   Changes in phosphorylation of TJ-associated molecules 87

Acknowledgements 89

5 Leukocyte cell adhesion interactions *93*

Carl G. Gahmberg, Leena Valmu, Tiina J. Hilden, Eveliina Ihanus, Li Tian,
Henrietta Nyman, Ulla Lehto, Asko Uppala, Tanja-Maria Ranta,
and Erkki Koivunen

1   Introduction *93*
       General *93*
       The leukocyte β_2-integrins *94*
       The ICAMs *94*
       The selectins *96*
       Other leukocyte adhesion molecules *96*

2   Activation of leukocyte adhesion *97*
       Methods to activate integrins *97*

3   Identification and preliminary characterization of leukocyte adhesion
       molecules *98*

4   Phosphorylation of adhesion molecules *101*

5   Isolation of active β_2-integrins *104*

6   Preparation of I-domains *105*

7   Construction and purification of Fc chimeric proteins *108*

8   Cell adhesion assay *111*

9   Detection of ICAMs *112*

10 Use of phage display libraries for isolation of peptide ligands to
       β_2-integrins *113*

       Acknowledgements *117*

**6 Studying cell interactions during development of the nervous
system in *Drosophila*** *119*

David Shepherd, Louise Block, James Folwell, and Darren Williams

1   Introduction *119*
       Why study *Drosophila*? *119*
       For the absolute beginner *120*

2   Preparing tissues for study *120*
       Equipment required *121*
       Handling tissues *121*
       Dissections *121*

3   Revealing neurons and glia *123*
       Reporter genes *123*
       Dye injections *126*
       Using antibodies to reveal specific neurons *130*
       Lineage tracing and birthdating of cells *135*

4   Manipulation of cells *137*
       Methods of ectopic gene activation *137*
       The GAL4 enhancer-trap technique *139*
       Laser gene activation in single cells *140*
       Mosaic analysis *143*
       Methods of targeted cellular ablation *148*

       Acknowledgements *149*

CONTENTS

7 Tight junction protein expression in early *Xenopus* development and protein interaction studies *153*

Sandra Citi, Fabio D'Atri, Michelangelo Cordenonsi, and Pietro Cardellini

1 Introduction *153*

2 Tight junction formation during *Xenopus laevis* early development *153*
  Preparation of *Xenopus laevis* eggs and embryos *154*
  Whole-mount immunolocalization of tight junction proteins in *Xenopus laevis* eggs and early embryos *156*
  Scanning electron microscopic analysis of junctions in *Xenopus* embryos *159*
  Biochemical analysis of *Xenopus* eggs and embryos *160*
  Microinjection of *Xenopus* oocytes, eggs, and embryos *162*

3 Assays to study protein–protein interactions *in vitro* *164*
  Bacterial and insect cell expression of recombinant proteins *165*
  Preparation of vertebrate cell lysates containing tight junction proteins *168*
  GST pull-down *169*
  Determination of the dissociation constant for protein–protein interaction (K_d) *171*
  Immunoprecipitation *173*

  Websites for specialized suppliers *175*

  Acknowledgements *175*

8 Oocyte–granulosa cell interactions *177*

Barbara C. Vanderhyden

1 Introduction *177*

2 Mammalian oocyte growth and development *in vivo* *177*

3 Isolation and culture of pre-antral follicles and oocyte–granulosa cell complexes *179*
  Overview *179*
  Isolation and culture of rodent pre-antral follicles *180*
  Growth of pre-antral follicles from commercially important species *182*
  Growth of primate pre-antral follicles *183*
  Parameters indicative of successful follicle development *183*

4 *In vitro* maturation techniques *183*
  Meiotic (nuclear) and cytoplasmic maturation of oocytes *183*
  Factors to consider when designing oocyte maturation experiments *184*
  In vitro maturation of rodent oocytes *185*
  In vitro maturation of oocytes from domestic species *186*
  In vitro maturation of primate oocytes *187*
  Evaluation of *in vitro* maturation *187*

5 Evaluating granulosa cell effects on oocyte development *188*

6 Evaluating oocyte effects on granulosa cell development *188*
  Isolation and culture of cumulus granulosa cells *189*
  Isolation and culture of mural granulosa cells *192*
  Granulosa cell differentiation in culture *192*

7 Analysis of oocyte–granulosa cell gap junctional communication *193*

8 Antisense and other strategies for interfering with oocyte–granulosa cell interactions *194*
  Manipulation of gene expression in oocytes *194*
  Manipulation of gene expression in granulosa cells *198*

9 *In vivo* experimental systems for studying oocyte–granulosa interactions *198*
   Acknowledgements *198*

9 Cell–cell interactions in early mammalian development *203*
*Tom P. Fleming, Judith J. Eckert, Wing Yee Kwong, Fay C. Thomas,
Daniel J. Miller, Irina Fesenko, Andrew Mears, and Bhav Sheth*

1 Introduction *203*
   Cell–cell interactions in blastocyst morphogenesis *203*
2 Generation of synchronous embryos and cell clusters *206*
3 Analyses of gene and protein expression *210*
   Gene expression analysis *210*
   Protein localization at cell contact sites in pre-implantation embryos *214*
4 Analysis of cell lineage segregation *218*
   Acknowledgements *227*

10 Model systems for the investigation of implantation *229*
*John D. Aplin, Carlos Simón, Susan J. Kimber, Melanie J. Blissett, Helen Lacey,
Chie-Pein Chen, and Carolyn J. P. Jones*

1 Embryo–endometrial interactions *229*
   Co-culture of human embryos with human endometrial epithelial cells
      as a model for implantation studies *229*
   Murine uterine epithelial cell cultures *235*
2 Human placental development *243*
   Introduction: placental cell lineages *243*
   Cell culture methods *243*
   Extravillous trophoblast *244*
   Isolation and culture of placental fibroblasts *253*

A1 List of suppliers *259*

Index *265*

Protocol list

Cadherin adhesion regulation in keratinocytes

Chelation of divalent ions from serum 6

Preparation of samples for microinjection 10

Detergent extraction prior to immunolabelling 15

Clustering of cadherin receptors with latex beads 22

Protein extraction of keratinocytes 26

E-cadherin immunoprecipitation and Western blot from soluble and insoluble protein extracts 28

Orthophosphate labelling and cell lysis 31

Identification of phosphorylated proteins 32

Methionine and cysteine oxidation 33

Analysis of intercellular adhesion molecules in endothelial cells

Isolation of human umbilical vein endothelial cells (HUVECs) 38

Method for the analysis of vascular permeability 42

Staining of HUVECs for analysis of junctional molecules 43

Gap junction-mediated interactions between cells

Injection of neurobiotin and Lucifer yellow into avian retinal ganglion cells using sharp electrodes 49

Patch injection of neurobiotin and fluorescein isothiocyanate (FITC)–dextran into avian retinal ganglion cells 51

Scrape loading 52

Assessing coupling via cell settlement 53

Immunohistochemistry of frozen tissue sections 55

Immunohistochemistry of brain slices or whole-mount retina 56

Zamboni's method for immunohistochemistry on wax sections 57

Methanol freeze substitution 58

Quantification of connexin plaque size and number 58

Quantitative RT–PCR for connexins 60

Whole-mount *in situ* hybridization for connexins in embryos using digoxigenin 61

Bulk loading of blocking antibodies using DMSO 65

Pluronic gel and antisense application to chick embryos *in ovo* 67

Functional analysis of the tight junction

Cell culture and calcium switching 73

Co-culture of epithelial cells 75

Immunofluorescence of epithelial cell cultures *76*
Use of electron-dense tracers to evaluate tight junction sealing *79*
Freeze-fracture imaging of the tight junction *81*
Paracellular flux of FITC–dextran (J_{DEX}) *83*
Measurement of the transepithelial electrical resistance (TER) *84*
Measurement of diffusion of membrane lipids *86*
Measurement of cytosolic Ca^{2+} activity *88*

Leukocyte cell adhesion interactions
Isolation and activation of T cells *97*
Identification of potential leukocyte adhesion molecules *99*
Immunoprecipitation of leukocyte extracts and analysis by SDS–PAGE *99*
^{32}P-labelling and analysis of T lymphocytes *101*
Isolation of leukocyte integrins in an active form *105*
Expression and purification of the CD11a I-domain *105*
Construction of Fc chimeric plasmids and preparation of Fc-containing protein *109*
Cell adhesion assays *111*
Detection of soluble ICAM-5 (sICAM-5) in the cerebrospinal fluid using sandwich enzyme-linked immunosorbent assay (ELISA) *112*
Identification of integrin-binding sequences by phage display *114*

Studying cell interactions during development of the nervous system in *Drosophila*
Histochemical detection of *lacZ* gene expression in isolated CNS *125*
Labelling adult sensory neurons with DiI *128*
Photoconversion of DiI *129*
Antibody staining of embryos *131*
Antibody staining of the larval, pupal, and adult nervous system *133*
BrdU labelling of neurons *135*
Lineage tracing with BDA labelling *136*
Embryonic heat shock protocol *138*
Laser gene activation in embryos *141*
Laser gene activation in larvae *142*
MARCM *147*
Laser ablation in embryos *148*

Tight junction protein expression in early *Xenopus* development and protein interaction studies
Induction of egg production and normal fertilization in *Xenopus* *155*
Artificial fertilization of *Xenopus* eggs *155*
Removal of jelly coat and vitelline membrane from *Xenopus* eggs/embryos *157*
Whole-mount immunostaining of *Xenopus* eggs/embryos *157*
Preparation of *Xenopus* samples for scanning electron microscopy *159*
Preparation of lysates of *Xenopus* eggs or embryos *160*
Biochemical fractionation of *Xenopus* eggs/embryos *161*
Microinjection of *Xenopus laevis* *163*
Preparation of bacterial lysates containing recombinant GST fusion proteins *165*
Preparation of baculovirus-infected insect cell lysates containing recombinant proteins *166*
Preparation of lysates from cultured vertebrate cells *168*
Preparation of ^{35}S-labelled proteins in TNT reticulocyte lysates *169*
GST pull-down assay *170*

Determination of the K_d of protein–protein interaction using GST pull-down assay *172*

Immunoprecipitation from Triton-soluble and SDS-soluble fractions of cultured epithelial cells *174*

Oocyte–granulosa cell interactions

Isolation of rodent oocyte–granulosa cell complexes using enzymatic and mechanical means *180*

Collection and *in vitro* maturation of bovine oocytes *186*

Procedure for oocytectomy *189*

Preparation of plexiglass slides for micromanipulation *190*

Microinjection of oligonucleotides into oocytes *195*

Immunofluorescent detection of proteins expressed in oocytes *196*

Cell–cell interactions in early mammalian development

Generation of synchronized cell clusters from mouse pre-implantation embryos *207*

Long-term storage of embryos for mRNA extraction and RT–PCR *210*

Isolation of mRNA for multiple semi-quantitative RT–PCR analysis of embryos *211*

Semi-quantitative RT–PCR of pre-implantation embryos *212*

Preparation of immunolabelling chambers for embryos *215*

Fixation and attachment of embryos to immunolabelling chambers *216*

Immunolabelling of embryos within chambers *218*

Immunosurgery of mouse blastocysts *219*

Whole-mount fluorescence *in situ* hybridization of pre-implantation embryos *220*

Differential nuclear labelling of mouse or rat blastocysts *223*

Detection of apoptotic cells in blastocysts by terminal deoxynucleotidyl transferase (TdT)-mediated dUTP nick end labelling (TUNEL) *224*

Determination of pre-implantation embryo gender by PCR *225*

Model systems for the investigation of implantation

Isolation and culture of human non-polarized and polarized endometrial epithelial cells (EECs) *230*

Co-culture of human epithelium with embryos *235*

Establishment of mouse uterine luminal epithelium (LE) cell cultures *236*

Measurement of transepithelial resistance in uterine epithelial cell cultures *239*

In vitro culture of mouse embryos *240*

Assay of the rate of attachment of hatched blastocysts to cultured uterine epithelial cell monolayers *241*

Assessment of the effect of test reagents in attachment assays *242*

Preparation of villous tissue from first trimester villous placenta *244*

Explant culture of first trimester placental tissue *246*

Whole-mount staining of placental explant cultures in multiwell plates *250*

Whole-mount staining of placental explant cultures in tubes *251*

Cryosectioning of placental explant cultures *251*

Resin embedding of placental explant cultures *252*

Isolation and culture of placental fibroblasts from direct explants *255*

Isolation and culture of fibroblasts from enzymatic dispersal of placental tissue *255*

Isolation and culture of fibroblasts obtained via explant in collagen gel *256*

Abbreviations

AAM	antibiotic–antimycotic mix
AMV	avian myeloblastosis virus
BCIP	5-bromo-4-chloro-3-indoyl-phosphate
BDA	biotin-labelled 70 kDa dextran
BrdU	bromodeoxyuridine
BSA	bovine serum albumin
CD	cluster of differentiation antigens
CDMEM	complete Dulbecco's modified Eagle's medium
CHAPS	3-[(3-cholamidopropyl)-dimethylammonio]-1-propanesulphonate
CLSM	confocal laser scanning microscopy
CNF-1	cytotoxic necrotizing factor 1
CNS	central nervous system
CSK	cytoskeleton extraction buffer
Cx	connexin
DAB	diaminobenzidine
DAPI	4',6-diamidino-2-phenylindole dihydrochloride
DEAE	diethylaminoethyl
DEPC	diethylpyrocarbonate
DES	diethylstilbestrol
DIG	digoxigenin
DMEM	Dulbecco's modified Eagles's medium
DMSO	dimethylsulphoxide
DSC	desmocollin
DTT	dithiothreitol
EC	endothelial cell
ECL	enhanced chemiluminescence
ECM	extracellular matrix
EDTA	ethylenediamine tetraacetic acid
EEC	endometrial endothelial cell
ELISA	enzyme-linked immunosorbent assay
EGTA	ethylene glycol-bis(2-aminoethyl) tetraacetic acid
EM	electron microscopy

ESC	endometrial stromal cell
F	fluorescence intensity
F_{max}	maximum fluorescence intensity
F_{min}	minimum fluorescence intensity
F_{Mn}	manganese fluorescence intensity
FACS	fluorescence-activated cell sorting
FAD	DMEM: F12 medium 3:1 v/v
FBS	fetal bovine serum
FCS	fetal calf serum
FITC	fluorescein isothiocyanate
FLP	flippase
FRT	flippase recombinase target
FSH	follicle-stimulating hormone
GAL4	galactose-4
GAPDH	glyceraldehyde-3-phosphate dehydrogenase
GFP	green fluorescent protein
GST	glutathione S-transferase
GTP	guanosine triphosphate
GV	germinal vesicle
GVBD	germinal vesicle breakdown
HBSS	Hank's balanced salt solution; Mg^{2+}- and Ca^{2+}-free
hCG	human chorionic gonadotrophin
HEPES	N-(2-hydroxyethyl) piperazine-N'-(2-ethanesulphonic acid)
HGF	hepatocyte growth factor
HIFCS	heat-inactivated fetal calf serum
HNPP	2-hydroxy-3-naphthoic acid-2-phenylanilide phosphate
HPRT	hypoxanthine phosphoribosyltransferase
HRP	horseradish peroxidase
hsp	heat shock promoter
HUVEC	human umbilical vein endothelial cell
I	current
ICAM	intercellular cell adhesion molecule
ICM	inner cell mass
IMPTNT	impotent tetanus toxin
IFN	interferon
IgG	immunoglobulin G
IPTG	isopropyl-β-D-thiogalactopyranoside
IU	International Units
IVF	*in vitro* fertilization
J_{Dex}	dextran flux
JAM	junction adhesion molecule
K_d	dissociation constant
LA	leuprolide acetate
LAD	leukocyte adhesion deficiency
LB	Luria broth

LBT	lysis buffer Triton
LBX	lysis buffer *Xenopus*
LDL	low-density lipoprotein
LE	luminal epithelium (of uterus)
LH	luteinizing hormone
LLC-PK$_1$	pig kidney epithelial cell line
LS	large scale
LW	Landsteiner–Wiener antigen
MARCM	mosaic analysis with a repressible cell marker
MDCK	Madin–Darby canine kidney cell line
MEM	minimum essential medium
NA	numerical aperture
NBT	4-nitro blue tetrazolium chloride
NHS	normal horse serum
NP-40	Nonidet P-40
OCC	oocyte–cumulus cell complex
OD	optical density
ODN	deoxyoligonucleotide
OGP	*n*-octyl-β-D-glucopyranoside
P	permeability
P$_{Cl}$	Cl$^-$ permeability
P$_{Na}$	Na$^+$ permeability
PA	performic acid
PAGE	polyacrylamide gel electrophoresis
PBS	phosphate-buffered saline
PCR	polymerase chain reaction
PDBu	phorbol dibutyrate
PECAM	platelet–endothelial cell adhesion molecule
PIPES	piperazine-*N,N'*-bis-2-ethane sulphonic acid
PK	proteinase K
PLL	poly-L-lysine hydrobromide
PMSF	phenylmethylsulphonyl fluoride
PMSG	pregnant mares' serum gonadotrophin
PO	phosphodiester oligomer
PUFA	polyunsaturated fatty acid
PVDF	polyvinylidene difluoride
PVP	polyvinyl pyrrolidone
RASA	rabbit anti-mouse/rat spleen antiserum
RR	ruthenium red
RT-PCR	reverse transcriptase–PCR
SDS	sodium dodecyl sulphate
SEM	scanning electron microscopy
SF	scatter factor
SS	small scale
SSC	standard saline citrate

TBS	Tris-buffered saline
TBST	Tris-buffered saline with Tween-20
TEM	transmission electron microscopy
TER	transepithelial electrical resistance
TGF-β	transforming growth factor-β
TJ	tight junction
TLC	thin-layer chromatography
TLN	telencephalin
TLCK	*N-p*-tosyl-lysine chloromethyl ketone
TNBS	trinitrobenzene sulphonic acid
TNF	tumour necrosis factor
TPCK	*N-p*-tosyl-phenylalanyl chloromethyl ketone
TR	transepithelial resistance
TRITC	tetramethylrhodamine isothiocyanate
tu	transducing unit
TUNEL	terminal deoxynucleotidyl transferase-mediated dUTP nick end labelling
UAS	upstream activating sequence
VCAM	vascular cell adhesion molecule
VEGF	vascular endothelial growth factor
XPBS	*Xenopus* phosphate-buffered saline
ZO-1 (2 or 3)	zonulae occludens 1 (2 or 3)
ZP3	zona pellucida protein 3
ZPCK	*N*-benzyloxycarbonyl-L-phenylalanine chloromethyl ketone
$\Delta\Psi$	electrochemical potential
ΔV	voltage deflection

Chapter 1

Cadherin adhesion regulation in keratinocytes

Vania Braga

MRC Laboratory for Molecular Cell Biology, University College London,
Gower Street, London WC1E 6BT, UK
and
Division of Biomedical Sciences, Imperial College School of Medicine,
Sir Alexander Fleming Building, London SW7 2AZ, UK (current address)

1 Introduction

This chapter will describe the keratinocyte system as a model to study the regulation of cadherin-mediated cell–cell adhesion. However, some of the principles discussed here can be applied to the analyses of different adhesive systems in other cell types. The chapter describes the potentials and pitfalls of the keratinocyte culture system and the various techniques used to investigate the regulation of cell–cell adhesiveness. Among these, a detailed description is given of microinjection in epithelial cells as a technique to investigate regulation of cadherin function by members of the small GTPase family. The analysis of the microinjection results by immunostaining and confocal microscopy is also presented. In addition, this chapter illustrates biochemical methods to determine the composition of cadherin complexes and their phosphorylation status (orthophosphate labelling and phosphopeptide mapping). Finally, two functional assays are discussed: the use of inhibitory antibodies to block cadherin-dependent adhesion and a technique to cluster cadherin receptors artificially.

2 Keratinocytes and cadherins

2.1 Background

Cadherins are calcium-dependent cell–cell adhesion receptors that are expressed in many different cell types. In epithelia, cadherin-mediated adhesion is necessary to maintain three interlinked processes: tight adhesion between neighbouring cells, the organization of other adhesive structures at intercellular contacts, and the polarized cell morphology.

Cadherins bind to the same type of molecule on adjacent cells (homophilic binding, reviewed in ref. 1). Homophilic binding results in the clustering of the receptors as the two apposing membranes zip up together. The clustered receptors are held together at cell–cell contact sites by the association with the cortical cytoskeleton, and this process provides strength to the adhesion. In addition to

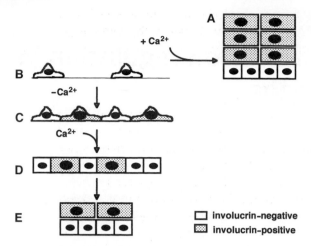

Figure 1 Keratinocyte culture system. (**a**) Under standard culture conditions (1.8 mM calcium ions), keratinocytes are organized as a stratified epithelium (multiple layers of cells), in which the suprabasal layers are formed by the differentiating cells (i.e. expressing the differentiation marker involucrin). (**b**) Cells are seeded in standard culture medium to encourage attachment and changed to low calcium medium when small colonies are visible. (**c**) When grown in medium containing a low concentration of calcium ions (0.1 mM), keratinocytes grow as a monolayer. (**d**) Calcium-dependent cell–cell contacts are quickly initiated following addition of calcium ions to the medium, in a process called calcium switch. (**e**) A few hours later, differentiating cells migrate upwards as stratification begins.

calcium ions, for full adhesiveness the receptors also require an association with the actin cytoskeleton and the activity of the RHO family of small GTPases, Rho, Rac, and Cdc42 (2). These molecules play a role in the actin reorganization in many cellular processes (3). The function of small GTPase family members is necessary for the stability of cadherin receptors at junctions, but the mechanisms involved in the regulation of junctions and epithelial polarity are not known.

The model we use to investigate cadherin-dependent adhesion is normal keratinocytes, isolated from human epidermis (see Section 2.2). They are the main cell type responsible for the protective barrier of the skin. The epidermis is composed of multiple layers of keratinocytes (stratified epithelium), organized in a three-dimensional structure according to their differentiation status (*Figure 1*). When isolated and cultured *in vitro*, keratinocytes can reproduce the same spatial organization shown *in vivo* and express most of the late differentiation markers.

To sustain mechanical stress in the skin, keratinocytes have well-developed cell–cell adhesive systems. Keratinocytes express classical cadherins (E- and P-cadherin) and different isoforms of desmosomal cadherins (desmoglein and desmocollins). These receptors assemble into the main keratinocyte adhesive contacts: adherens junctions and desmosomes, respectively. Contrary to simple epithelial and endothelial cells, keratinocytes do not assemble functional tight junctions. The fluid permeability barrier in the epidermis is conferred by the keratinization process that occurs during stratification and differentiation of the cells. Many other receptors such as integrins, CD44, junction adhesion molecule

(JAM), etc. are also found at cell–cell contacts. Although the presence of other receptors may help to stabilize junctions, their contribution to the keratinocyte adhesion or stratification is not clear.

2.2 Model

The keratinocyte model is well established and one of the best available to study cadherin-dependent cell–cell adhesion (4). In standard medium, normal keratinocytes grow as stratified cultures similar to the organization found *in vivo* (1.8 mM calcium ion concentration, *Figure 1a*). When placed in medium containing a reduced concentration of calcium (0.1 mM), keratinocytes are unable to stratify and grow as a monolayer (*Figure 1c*). This inability to form multiple layers and stratify is because adherens junctions cannot assemble in low calcium medium, as they require an extracellular concentration of calcium ions above 1 mM. However, in low calcium medium, the differentiation process is initiated, as early differentiation markers are expressed by a proportion of the keratinocytes (i.e. involucrin; ref. 5).

When grown in low calcium medium, cadherin receptors are expressed at the keratinocyte surface, but localize diffusely (*Figure 2*). Calcium-dependent cell–cell adhesion can be induced by transferring the cells to standard medium (known as the calcium switch). Very quickly (within minutes), cadherin receptors accumulate at sites of cell–cell contacts (*Figure 1d*; see also *Figure 2*). Following stable cadherin-mediated adhesion, a major reorganization of the cytoskeleton and formation of desmosomes and other adhesive structures occur. As a result, the morphology progressively changes to a cuboidal cell shape characteristic of epithelia. After 4–5 h, stratification begins, and the differentiating cells migrate upwards to form a suprabasal layer (*Figure 1e*; ref. 5). In this system, stratification is essentially complete within 24 h, with multiple layers being assembled. It has been shown that inhibition of cadherin function by specific antibodies can completely block the formation of intercellular contacts and stratification (5, 6).

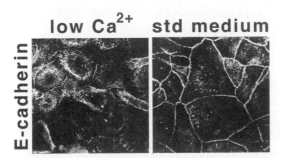

Figure 2 Immunostaining pattern of E-cadherin in normal human keratinocytes. Cells were grown in low calcium medium (low Ca^{2+}), and one set of coverslips was transferred to standard medium (std medium) for 1 h. Under this condition, E-cadherin receptors apparently accumulate at the boundaries between neighbouring cells, indicating functional cell–cell adhesion.

Thus, cadherin-mediated adhesion is essential for junction formation and the three-dimensional organization of keratinocytes.

With time after induction of cell–cell contacts and confluence, the stability of the cadherin receptors at junctions increases, in a process called maturation. It is characterized biochemically by an increased insolubility of the cadherin receptors in non-ionic detergents. Although the strength and functional properties of junctions change upon maturation, not much is known about the intracellular mechanisms that participate in this process in epithelial cells (see Section 4.1). Nevertheless, maturation is an important consideration when investigating cadherin-dependent adhesion, as it reflects a physiological property of cell–cell contacts of epithelial and endothelial cells.

2.3 Advantages of the model

The keratinocyte model has the following advantages:

(1) These are normal cells, and the physiological status of the junctions more closely resembles the conditions found *in vivo*.

(2) After primary cultures are isolated from human epidermis, keratinocytes can be frozen down and cultivated for up to 6–7 passages. Not much difference in cadherin staining at the junctions has been observed with increased passage number. However, the differentiation and proliferation properties of the cultures do change. Thus, when these aspects are investigated, it is best to use freshly isolated or early-passage keratinocytes.

(3) The keratinocyte model is unique since it allows the study of the regulation of cadherin function both in newly formed and in mature cell–cell contacts (see Section 3.5).

(5) Normal keratinocytes grow well for days in medium containing a low concentration of calcium ions, allowing the formation of a confluent layer of cells with negligible cell–cell contacts. Many other cell types cannot do this.

(6) The presence of a confluent monolayer means that during formation of new contacts there is no spreading or migration of cells towards each other. While junctions are established, there is only keratinocyte polarization as neighbour cells zipper up their membranes. This allows a tight control of the on/off status of cadherins during the calcium switch, and short-term time courses (minutes) can be performed.

(7) There are no changes in the surface expression levels of cadherins during the calcium switch and no requirement for new protein synthesis (5, 7).

2.4 Disadvantages of the model

The main disadvantages of the keratinocyte calcium switch system are:

(1) Because of the stratification process, a time course to investigate cadherin-dependent cell–cell adhesion can be only performed for up to 4–5 h. After that, the presence of stratifying keratinocytes can perturb the analysis by immunofluorescence.

(2) Normal keratinocytes do not express many different plasmids when introduced by microinjection. Under these conditions, the vector pcs2 is suitable
for expression in normal cultures. Another possibility is retroviral infections
(8). Alternatively, a keratinocyte cell line can also be used (HaCat cells), which
shows good transfection efficiency and expresses most of the plasmids tested
by microinjection. HaCat is a non-transformed, spontaneously immortalized
human keratinocyte cell line that shows good epithelial morphology and
cadherin staining at junctions. In addition, the calcium switch can also be
performed with HaCat cells. In this case, however, the formation of new contacts is not as efficient and the cells remain somewhat rounder when compared with normal keratinocytes.

2.5 Keratinocyte cultures

The method for growing normal keratinocytes is essentially as described by
Rheinwald and Green (standard medium, see 9). The isolation and culture of
primary keratinocytes have been described in detail elsewhere (9). The reader is
referred to the literature for details regarding the procedures, reagents, and
companies that provide them. This chapter will briefly review the basis of the
keratinocyte culture in standard medium and cover in more detail the differences with the calcium switch system.

Human epidermal keratinocytes can be isolated from skin biopsies from donors
of different ages, but we mostly use cells derived from neonatal foreskins. The
Rheinwald and Green technique seeds keratinocytes onto a feeder layer of fibroblasts to encourage attachment and provide growth factors. The feeder layer is
prepared by treating the fibroblasts with mitomycin C (Sigma) or irradiation to
prevent proliferation. Although the system requires some forward planning and
timing of the two cultures (keratinocytes and fibroblasts), in practice this is
easily achieved as feeder layers can be prepared up to 48 h in advance.

Keratinocytes are grown in a mixture of Dulbecco's modified Eagle's medium
(DMEM) and Ham's F12 (3:1, v/v, also known as FAD), supplemented with 5–10%
fetal calf serum (FCS), 10 ng ml^{-1} epidermal growth factor, 5 μg ml^{-1} insulin, 10^{-10} M
cholera toxin, and 0.4 μg ml^{-1} hydrocortisone. In this standard medium containing
approximately 1.8 mM calcium ions, keratinocytes grow as small colonies and
reach confluence within 7–10 days after plating. In colonies with sixteen or more
cells, stratifying, differentiating keratinocytes can be seen on top of the cells
attached to the dish. Thus, keratinocytes are cultured as a mixed population of cells
at distinct stages of differentiation and it is important to maintain the balance
between proliferating and differentiating cells. To avoid an increase in the
proportion of differentiated keratinocytes in culture, care should be taken to:

- trypsinize when 80–90% confluent, as upon confluence differentiation is
  stimulated; and

- seed the keratinocytes at low density, as plating at high density also favours
  differentiation. Seeding at 1–2 $\times$ 10^5 per 75 cm^2 flask is recommended.

2.5.1 Calcium switch system

2.5.1.1 Medium

The low calcium medium contains the same formulation and supplements as the standard medium described above, but without calcium ions. The powdered mixture of FAD without calcium can be bought from Biowhittaker. It is also necessary to remove calcium from the serum (see *Protocol 1*), but this procedure will also remove other divalent cations. Human keratinocytes do not grow in medium completely depleted of calcium, but survive well at concentrations of 0.1 mM. This must be done either by adding calcium back from a stock solution (1 M calcium chloride) or by adding a 1:1000 dilution of the standard medium.

Protocol 1

Chelation of divalent ions from serum

Equipment and reagents

- Chelex-100 resin (BioRad Laboratories)
- Concentrated hydrochloric acid (HCl)
- Whatman 3MM filter paper
- 0.2 μm tissue culture filter
- FCS, 200 ml

Method

1 Resuspend 80 g of Chelex-100 resin in 2 l of water in a beaker with agitation.
2 Add concentrated HCl (~0.7 ml). The pH will fall very rapidly and rise slowly. Adjust the pH to 7.4 (should take a couple of hours to stabilize).
3 Filter using Whatman 3MM .
4 Add 200 ml of FCS to the resin. Stir at room temperature for 3 h.
5 Filter with Whatman 3MM to remove the resin from the serum. These steps should be done on the same day, as the resin does not keep its binding affinity when hydrated for a long time.
6 Sterilize the serum using a 0.2 μm filter. Aliquot and freeze at −20°C until use.

2.5.1.2 Culture in low calcium medium

Keratinocytes initially are seeded onto a feeder layer in standard medium. For growth in low calcium medium, keratinocytes should be seeded at higher density (i.e. $0.3–1 \times 10^6$ per 75 cm^2). When small colonies are apparent (2–3 days later), cells are washed in 0.53 mM ethylenediamine tetraacetic acid (EDTA) in phosphate-buffered saline (PBS) and placed in low calcium medium. Keratinocytes should be confluent in 4 days and remain in good shape for up to 10 days under these conditions. When confluent, cultures can be induced to form new cell–cell contacts (*Figure 2*) by:

- transferring the keratinocytes to standard medium; or
- adding calcium ions to 1.8 mM from a stock solution (1 M calcium chloride), when changing medium is not appropriate (i.e. blocking experiments with antibodies).

3 Microinjection

3.1 Tools to modulate the activity of the small GTPases

Microinjection has been used extensively to study the function of the Rho family of small GTPases. A series of mutants has already been characterized that produce a constitutively active or a dominant-negative protein (see *Table 1*). The constitutively active mutations lock the protein in the GTP-bound mode, which is the form competent for signalling. Thus, upon microinjection, these mutants can immediately bind and activate their specific targets. How the dominant-negative molecules work is unclear. By analogy to Ras mutations, it is thought that the dominant-negative mutants may prevent the activation of endogenous small GTPases (reviewed in ref. 10).

A cautionary note should be sounded on the dominant-negative approach, as the inhibition of the endogenous activity may not be very efficient and the proteins may not be as stable as desired. For instance, the use of N17Ras is not able to inhibit the same spectrum of activities that are blocked by an inhibitory anti-Ras antibody (10). Thus, it is advisable to confirm the inhibition with an alternative approach, such as sequestration of endogenous small GTPases by the use of a specific target fragment, i.e. WASP fragment (containing the GTPase-binding domain).

Another approach is to use bacterial toxins that can either inhibit or activate small GTPases (toxin A and B, α-toxin, lethal toxin, cytotoxic necrotizing factor-1 (CNF-1), -2, haemorraghic toxin). The reader is referred to the literature for the

Table 1 Recombinant proteins used to manipulate the activity of the Rho subfamily of small GTPases[a]

Small GTPase		Mutants/proteins[b]	Effects
Rho	G14Rho	Wild-type	
	Activation	V14Rho	Constitutively active
		L63Rho	Constitutively active
	Inhibition	N19Rho	Dominant-negative
		C3 transferase	ADP-ribosylates and inactivates endogenous Rho
Rac	G12Rac	Wild-type	
	Activation	V12Rac	Constitutively active
		L61Rac	Constitutively active
	Inhibition	N17Rac	Dominant-negative
Cdc42	G12Cdc42	Wild-type	
	Activation	V12Cdc42	Constitutively active
		L61Cdc42	Constitutively active
	Inhibition	N17Cdc42	Dominant-negative
		WASP fragment	Binds and sequester endogenous Cdc42

[a] See text for details on the relative efficiency of the different mutants.

[b] Mutation notation is the amino acid single letter code followed by its position in the protein sequence, e.g. V14Rho is a Rho protein containing a mutation to valine at position 14.

different toxins and their mode of action (11). The major drawback of bacterial toxins is their specificity towards Rho proteins, as the same toxin can modify different members of the Rho subfamily. One exception is C3 transferase (see *Table 1*), which has much higher affinity for Rho than for Rac and Cdc42. The other problem is cell permeability and uptake. C3 does not show very good cell permeability and is usually microinjected into cells when it inactivates Rho efficiently and specifically. Other toxins such as toxin A, toxin B, CNF-1, etc. can be taken up by different cells. However, the uptake may vary considerably depending on the cell type and may produce a negative result. For instance, in normal keratinocytes, even though microinjection of N17Rac or C3 transferase readily inhibits cadherin-mediated adhesion, the latter is insensitive to toxin B 10463 treatment (blocks Rho, Rac, and Cdc42).

3.2 Methodology

Microinjection introduces the material of interest inside the cytoplasm or nucleus of individual cells by means of a fine needle. The contents flow from the needle at a constant, regulated pressure, and the material is injected by briefly touching the surface of the cell. The amount injected correlates with the pressure with which the liquid flows and the length of time the needle remains in contact with the cells.

Microinjection has several advantages:

(1) It can introduce proteins, DNA, or RNA inside a cell.

(2) It is ideal to study signal transduction pathways, as the activity of a given component can be manipulated very rapidly and an immediate response observed (within minutes).

(3) When compared with other methods of expression such as transfection or retroviral infection, microinjection is a powerful technique that provides results very quickly (hours as opposed to days).

The major drawback of microinjection is that although biochemical analysis can be performed, it is used mostly for analysis by immunofluorescence. Another issue is that the exact amount microinjected is somewhat variable from cell to cell, and the effects will also depend on the relative cell size. This factor is important, particularly in normal keratinocyte cultures, because as they differentiate their cellular volume and size also increase.

The microinjection technique has been described before in detail, and the reader is referred to additional literature (12–14). Only differences in the equipment and particular aspects that are applicable to the study of cell–cell adhesion will be discussed here.

3.3 Strategy

The strategy chosen for microinjection depends on the experimental protocol (incubation times, levels of expression necessary, etc.) and the characteristics of the protein of interest. Two possibilities are injection of DNA plasmids or injection of recombinant proteins, and the pros and cons of both approaches are considered below (13, 14).

Expression plasmids have the advantage that the protein produced is processed correctly, unlike bacterially produced proteins. Recombinant proteins can be produced in insect cells, which provide most of the post-translation modifications not found in bacteria. One advantage of recombinant proteins is a tighter control of the amount injected as the proteins are degraded with time according to their stability and intracellular half-life. When compared with recombinant protein microinjection, expression plasmids usually produce much higher expression levels, which may lead to undesired overexpression problems. In addition, the use of recombinant proteins allows short-term time courses to be performed, as the effect can be observed as early as a few minutes following microinjection. In contrast, injection of DNA needs at least a couple of hours to allow detectable expression levels.

In general, it is good to confirm results using microinjection of both DNA and recombinant protein, as this rules out effects due to minor contaminant proteins from bacteria or to overexpression of plasmid DNA. However, depending on the characteristics of the protein of interest, the production of recombinant proteins may be a difficult task. The most common problems are degradation (particularly for larger proteins), insolubility, and instability. Most of these problems can be solved by optimizing conditions such as temperature and length of induction of expression in bacteria, use of detergents, use of protease-deficient bacterial strains, changes to a different expression system, etc. Sometimes just using different constructs (deletion mutants, different portions of the protein) might solve the problem of protein purification from bacterial lysates. Another issue is when the protein is expressed well in bacteria, but is inactive (if activation requires a particular post-translation modification). In these circumstances, and if the purification problems are not solved, then microinjection of expression vectors must be done instead.

3.3.1 DNA

When injecting plasmids, it is important to perform a time course to determine the optimal expression time for every sample tested. This is necessary even if different DNA inserts are subcloned in the same expression vector, as the efficiency of transcription may vary with different inserts. In addition, an extra number of injected cells is needed because not every cell will express the exogenous DNA and also to account for variations in the expression levels. Staining for an epitope tag (such as Myc, Flag, or His) can identify which cells transcribed the injected DNA (13).

Optimal expression is achieved with high quality DNA. The best results are obtained with DNA purified using caesium chloride density gradients. Plasmids are normally injected at a concentration of 0.1 mg ml^{-1}, but this can vary from 0.05 to 0.5 mg ml^{-1}, depending on the efficiency of expression and the necessary incubation time. HaCat cells express plasmids well, such as pRK5 or pEF at 0.1 mg ml^{-1} within 2–3 h after microinjection.

9

3.3.2 Recombinant protein

Microinjection of recombinant proteins is the method of choice to manipulate the activity of small GTPases in normal keratinocytes, as these cells are not able to express many plasmids tested. The Rho family of small GTPases requires prenylation in order to be active. Although this lipid modification is not produced in bacteria, it is thought that, upon microinjection, the recombinant proteins are modified by intracellular components and become functional.

Small GTPases are easily purified as glutathione S-transferase (GST) fusion proteins, and the method used is described in detail elsewhere (15) . The yields are not as high as reported for other proteins and the limiting step is the digestion with thrombin to release the GST portion. Another problem is that only a proportion of the purified small GTPase has biological activity (14, 15). Thus, to normalize between different batches, the protein preparations are evaluated for their ability to bind GTP, and this is used as the active concentration (15). In addition, we also test different dilutions of the protein preparations in a biological system for their activity in the reorganization of the cytoskeleton (fibroblasts) or modulation of cadherin adhesion (keratinocytes).

3.4 Preparation of material to inject

It is essential that all preparations are clean and free of particles and precipitates that can clog the needles during microinjection. To achieve this, all solutions used for microinjection are filtered through spinex centrifuge tubes and kept free of dust/particles. To identify the injected patches of cells, a fluorescent dye is co-injected (conjugated to dextran, 10 000 Da; see *Protocol 2*). Different types of attached fluorophores are available, i.e. fluorescein isothiocyanate (FITC) or Texas red, and thus the best label can be chosen to fit with the immunostaining strategy afterwards (see Section 4.2.3).

Protocol 2

Preparation of samples for microinjection

Equipment and reagents

- Bench-top centrifuge
- Spinex centrifuge tubes (0.2 μM, Costar)
- Dextran–Texas red or dextran–FITC (mol. wt 10 000 Da, 25 mg, Molecular Probes)

A. Preparation of dextran label

1 Dissolve dextran dye in 500 μl of water to a final concentration of 20 mg ml^{-1}.

2 Pellet any insoluble particles by centrifugation (20 000 g for 5 min).

3 Filter through spinex tubes by centrifugation (20 000 g for 5 min).

4 Freeze small aliquots at $-20\,°C$; keep the working aliquot at $4\,°C$.

B. Preparation of samples (DNA or protein)

1   Freshly prepare appropriate dilutions of the material to be injected (in PBS or any other suitable buffer) and mix with the dextran label prepared above (1:10).

2   Immediately before microinjection, centrifuge the samples at 4°C (20 000 g for 5 min) to settle any particles.

3   Keep the samples on ice until and during the microinjection procedure.

3.4.1 Needles

Good needles and clean solutions are the key for a successful microinjection. The shape of the needle (sharp versus blunt) and the viscosity of the preparation determine the pressure necessary for a constant liquid flow during micro-injection. It is important to optimize the best needle for your cell type and the strategy chosen (DNA versus protein injection). Usually, nuclear microinjection is best achieved with thin, sharp needles; for protein injection into the cyto-plasm, this is not so crucial. On the contrary, because proteins in solution tend to aggregate and precipitate, sometimes a blunter needle is preferred to avoid blockage.

Needles are prepared by pulling borosilicate glass capillary tubes (100 mm length, 1.2 mm outside diameter $\times$ 0.69 mm internal diameter; Clark Electro-instruments) using a needle puller (Sutter Instruments, model P97). In order to optimize the shape of the needle obtained, parameters such as heat, velocity, pressure, and pull are tested empirically. The needles thus produced are then tested by microinjection. Once the best settings are found, this needle-puller model produces consistent and reproducible needles.

3.5 Microinjection procedure to study keratinocyte junctions

To investigate the regulation of cadherin-dependent adhesion, patches of adjacent keratinocytes plated on coverslips are microinjected with the appropriate dilu-tion of the reagents of interest. Following incubation for a period of time, cells are fixed and stained with anti-cadherin antibodies (*Figure 3*; see also Section 4.2). The non-injected, surrounding cells are used as controls. By comparing the levels of cadherin staining at junctions in the control cells and within the injected patch, the effect of the injected reagents can be evaluated. Initially, we test a given compound during the formation of cell–cell contacts (Section 3.5.1). If there is a perturbation of cadherin localization, the specificity is evaluated in keratinocytes containing mature junctions (see below; also Section 3.5.2). How-ever, it is possible that the regulation of cadherin function in newly formed or mature cell–cell contacts may be different.

To show a specific effect on cadherin receptors, staining for another adhesive component should be performed. Ideally, within the time frame studied, the localization of cadherin receptors at junctions should be perturbed, but not the

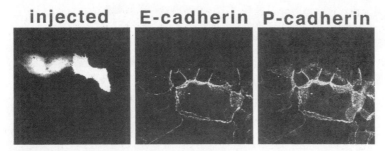

Figure 3 Microinjection of normal keratinocytes grown under standard conditions. Cells at the periphery of the colony were microinjected with C3 transferase to inhibit endogenous Rho activity. Following 2 h incubation, keratinocytes were double labelled for E-cadherin (ECCD-2 and anti-rat FITC conjugate) and P-cadherin (N-CAD-299 and anti-mouse Cy5 conjugate).

staining pattern of the control protein. This control preferably should be a transmembrane protein that localizes at intercellular contact and also interacts with actin filaments (16, 17). Good examples are integrin and CD44 molecules; markers for desmosomes are sometimes used, even though the latter associate with keratin filaments. With prolonged incubations following dismantling of cadherin contacts, all the other adhesive structures are also removed, illustrating the importance of performing a time course and choosing the appropriate incubation time point.

One should keep in mind that anything that disturbs the actin cytoskeleton would indirectly perturb cadherin-mediated adhesion. In addition to controls for cadherin receptor specificity, other ways to avoid non-specific, indirect effects due to cytoskeletal disturbance are:

- within the limitations of your experiment, to choose the earliest possible time point upon which an effect on cadherin adhesion can be observed; and

- to set up conditions so that, at the end of the experiment, cells must be touching each other. This is a good indication of how intact the cytoskeleton is: any signs of rounding up or retraction are a concern.

3.5.1 Formation of new contacts

The system to induce cell–cell contacts is the most sensitive of all, as the cytoskeletal reorganization and junction staining require prior cadherin-mediated adhesion. Thus, the parameters important for the formation of new cell–cell contacts are intrinsic to the regulation of the adhesive function of cadherin receptors.

Microinjection is performed in keratinocytes grown in low calcium medium. It is important to use post-confluent cells to obtain optimal staining of cadherins at junctions, and particularly if performing short time course incubations (up to 1 h). Patches of adjacent cells are microinjected (6–10 cells per patch). After microinjection, cells are transferred to standard medium to induce cell–cell adhesion for the necessary length of time.

Inhibition of endogenous Rho or Rac completely blocks newly formed cadherin-dependent contacts within 1 h (16). When using a dominant-negative approach (i.e. C3 transferase to inhibit endogenous Rho), it is advisable to incubate for a few minutes in low calcium medium before inducing cell–cell contacts (usually 5–15 min). This is important to ensure that a good proportion of the small GTPases is inhibited before cell–cell adhesion is initiated.

3.5.2 Mature contacts

In mature junctions, parameters relevant for the maintenance of stable cadherin-dependent adhesion are evaluated, among them the contribution of other adhesive components and the maturation process. This is also the best system to demonstrate the specificity of the regulation for cadherin receptors, as the other adhesive structures are already in place.

To investigate the regulation of mature contacts, we microinject medium sized colonies of keratinocytes grown in standard medium (between eight and 40 cells per colony). After microinjection, keratinocytes are incubated in the same medium. Usually, four to eight adjacent cells are microinjected at the periphery of the colonies (*Figure 3*). This is because stratification normally occurs in the middle of the colony, and the presence of cells in a suprabasal layer can perturb the confocal analysis. Inhibition of endogenous Rho or Rac in keratinocytes containing mature junctions requires at least 2 h to remove cadherin receptors completely, as opposed to only 1 h incubation in cells with newly formed contacts (17).

3.6 Analysis of microinjection experiments

Immunostaining is described in Section 4.2.3 and collection of confocal images from microinjected samples is detailed in Section 4.3.2.

In a single coverslip, up to 10–15 patches can be microinjected within 5 min, depending on the condition of the needles and quality of the injected material. It is also necessary to microinject a good number of patches in a single coverslip, as a number of patches may not be included in the analysis. Patches should be excluded when any of the following occur.

(1) Any retraction or gaps seen between injected cells.

(2) Poor cadherin staining in the control, non-injected surrounding cells, as this provides the levels for comparison with the injected patches.

(3) A high proportion of cell death. Usually, in a typical experiment, around 10% of injected cells die. Even after prolonged incubations (hours), sick and dying cells can still remain attached to the dish and their neighbours, and it is important to identify and exclude these cells from the analysis. With experience, dying cells can be identified by the cell morphology, shape of nucleus, and cytoplasmic texture.

Microinjection of buffer alone as a control is a good idea to check how sensitive the cells are during the injection procedure. Alternatively, a brief staining with propidium iodide before fixation can be performed.

In addition, only patches with a minimum of 3–4 injected cells are analysed. We usually focus our analyses on the localization of cadherin receptors at the borders between two injected neighbouring cells. Upon Rho or Rac inhibition, E-cadherin staining is first removed from the junctions between two injected cells and, subsequently, from junctions at the periphery of the injected patch (16–18). The reason for this difference is not clear. It is predicted that if there is cadherin staining at cell–cell contacts, there is a contribution from both cells to the homophilic binding. It seems that at the borders of the patch, because one neighbour cell does not have the small GTPase activity inhibited by microinjection, adhesion is more stable and the receptors are more resistant to extraction from junctions.

Usually, the results are straightforward if microinjection of an expression vector or protein gives a phenotype in approximately 80–90% of the injected patches. However, when the phenotype is observed in 70% or less of the patches, it is necessary to establish criteria to quantify objectively the amount of disruption observed. Baseline controls are used for comparison (i.e. buffer alone, bovine serum albumin (BSA) or immunoglobulin G (IgG) solution, an inactivating mutation, C3 transferase, etc.). Quantification will also help to discern between variations due to the amount of injected and expressed material.

4 Immunostaining and confocal analysis

4.1 Detergent extraction

The maturation of junctions is characterized biochemically by an increased insolubility of cadherin receptors in non-ionic detergents. This process occurs in different epithelial and endothelial cells, and the proportion of cadherin receptors that are detergent insoluble can vary depending on the cell type, confluence, and maturation of junctions. The mechanism that underlies the process of detergent insolubility is not clear. Although the interaction with the cytoskeleton is thought to play a role, some lipid interactions may also contribute to the insolubility of the receptors.

In some circumstances, it is informative to evaluate the detergent solubility of cadherin complexes as a readout of the effectiveness of cell–cell adhesion. In the absence of cell–cell contacts, the majority of cadherin receptors are solubilized by detergents. Upon formation of intercellular contacts, a pool of cadherin and associated proteins becomes insoluble in detergents. However, localization at the keratinocyte junctions does not correlate with detergent insolubility, as a percentage of receptors present at stable cell–cell contacts are also liable to detergent extraction (*Figure 4*; ref. 7). Thus, receptors engaged in a functional adhesive contact may also be solubilized, depending on the experimental conditions (see below). A good control is to compare the detergent-insoluble pool with the total amount of cadherin molecules present at cell–cell contacts (fixing and staining a sample in parallel without detergent extraction).

The detergent solubilization technique involves treatment of cells prior to fixation with a buffer to stabilize the cytoskeleton containing non-ionic detergent

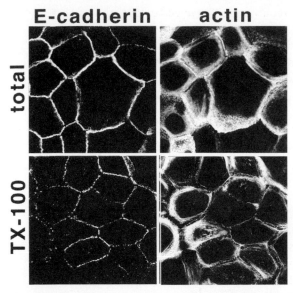

Figure 4 Detergent insolubility of E-cadherin receptors. Keratinocytes grown in low calcium medium were transferred to standard medium for 1 h. Cells were fixed, permeabilized, and double labelled for E-cadherin and actin filaments (total). Alternatively, cells were pre-extracted with non-ionic detergent (Triton X-100) as described in *Protocol 3* before fixation and immunolabelling.

(19). Parameters that influence the extent of extraction are: composition of the buffer (salt concentration, pH, etc.), concentration of detergent (usually Triton X-100 or Nonidet P-40 (NP-40), from 0.1 to 2%), temperature (on ice or at room temperature), and incubation time (2–15 min). *Protocol 3* is adapted from Shore and Nelson (20) and describes the method we use for keratinocytes.

Protocol 3
Detergent extraction prior to immunolabelling

Equipment and reagents

- Keratinocytes plated onto coverslips (13 mm) inside 4-well plates
- PBS
- Orbital shaker
- 3% Paraformaldehyde in PBS

- CSK buffer (10 mM piperazine-*N,N'*-bis-2-ethane sulphonic acid (PIPES) pH 6.8, 50 mM NaCl, 3 mM MgCl$_2$, 0.5% Triton X-100, 300 mM sucrose)

Method

1 Wash cells three times in PBS.

2 Drain all liquid excess carefully to avoid diluting the CSK buffer any further.

Protocol 3 continued

3 Add 500 μl of CSK buffer to the wells with the coverslips.

4 Incubate for 10 min at room temperature with mild agitation.

5 Carefully wash three times in PBS.[a]

6 Fix in 3% paraformaldehyde[b] for 10 min at room temperature.

7 Wash three times in PBS. Cells are ready for immunolabelling.

[a] These washes may also be done with CSK buffer (without Triton X-100)

[b] Paraformaldehyde is usually prepared in PBS; it may also be dissolved in CSK buffer (without Triton X-100).

4.2 Immunolabelling

4.2.1 Antibodies

There are many different monoclonal antibodies that recognize human E-cadherin. For single labelling, the mouse monoclonal anti-E-cadherin HECD-1 antibody is preferred due to its high affinity (21). ECCD-2, a rat monoclonal anti-E-cadherin, is also used, but gives weaker signal than HECD-1 after staining human keratinocytes (22). For P-cadherin, the monoclonal antibody N-CAD-299 produces good results (23). There is also a rabbit polyclonal anti-pan cadherin (7) that recognizes E-, P-, and N-cadherin because its epitope is on a highly conserved peptide sequence in the cadherin tail. This antiserum is not ideal for immuno-staining because of high background, but it is very useful for immunoprecipi-tation (see Section 6.3). Secondary antibodies can be bought from many different companies, but Jackson Laboratories (Stratech) offers a good range of secondary antibodies raised in different hosts and with a variety of attached fluorescent labels (important for double labelling, see Sections 4.2.3.1 and 4.2.3.2).

4.2.2 Single labelling

It is necessary to titrate the optimal dilution of both primary and secondary antibodies to avoid oversaturation of the signal, high background, and waste of material. Normally there is a range of dilutions suggested by the companies that provide antibodies. However, different factors influence the immunofluorescence output signal such as the expression levels in your cell type, pH buffer, length of incubation with the antibodies, and the detection system used (conventional versus confocal microscopy). Details about immunolabelling and trouble-shouting can be found elsewhere (24).

Prepare a humid chamber by wetting filter paper in water and placing a piece of Parafilm on top. Coverslips are placed in the humid chamber with the cell side upwards. Cells are permeabilized in 50 μl of 0.1% Triton X-100 dissolved in 10% FCS in PBS for 10 min at room temperature. After three washes in PBS, cells are incubated with an appropriate dilution of the primary antibody in 10% FCS for 1 h at room temperature. After three washes in PBS, cells are incubated with the secondary antibodies in the same manner. Cells are washed three times in PBS, and a last wash is performed in distilled water to eliminate any salts. After draining excess liquid, coverslips are mounted on slides using gelvatol (24). Samples are kept at 4 °C or −20 °C until analysis. Using this protocol, immunofluorescence signal can be detected after HECD-1 staining for months, even without the use of anti-fading agent in the mounting medium.

4.2.3 Multiple labelling

When performing multiple labelling for co-localization or other purposes, a false-positive result may arise due to cross-reactivity between the primary and secondary antibodies and bleed through between the different immunofluorescence channels. It is necessary to perform optimizations and controls to ensure that these two problems do not occur in your system. These issues are discussed below.

4.2.3.1 Selection of primary and secondary antibodies

For double labelling, it is important to choose the primary antibodies carefully. It is preferred to use antibodies raised in distantly related hosts to avoid cross-reaction. When the two different primary antibodies are raised in the same host, i.e. mouse monoclonals, then the best alternative is to attach the fluorophore directly to one of the primary antibodies (25). However, the signal obtained is usually weaker in this case, because the amplification given by a secondary conjugate is lost. Thus, this method works well only when the primary antibody has a high affinity and gives strong staining under normal indirect immunofluorescence conditions.

When choosing secondary antibodies, two considerations should be kept in mind: the cross-reaction between the different antibodies and the type of fluorescence label attached. There are now many types of secondary antibodies available commercially, that are pre-absorbed with epitopes from different species. In order to minimize cross-reaction, and if available, the different conjugates should be produced in the same host (i.e. donkey) so that the conjugates do not see each other.

4.2.3.2 Selection of attached fluorophore

Selecting the appropriate type of attached label on the conjugate is important to produce a well-resolved immunofluorescence signal and minimize bleed-through between the filters when collecting images during microscopy analysis. The characteristics of the most commonly used fluorophores are shown in *Table 2*.

For double labelling and co-localization experiments, we tend to use a Cy5-coupled conjugate together with a FITC conjugate as their emission peaks are the furthest apart. FITC is a widely used fluorophore but has the disadvantage of photobleaching. There are now alternative fluorophores, such as Cy2 (Jackson Laboratories) and Oregon green (Molecular Probes), that are good substitutes for FITC as they provide better photostability. Cy5 is brighter than most fluorophores and has the advantage of lower autofluorescence in biological samples. However,

Table 2 Absorption and emission peaks of most commonly used fluorophores

Fluorophore	Absorption peak (nm)	Emission peak (nm)
Cyanine, Cy2	350	450
Fluorescein, FITC	492	510
Indocarbocyanine, Cy3	550	570
Tetramethyl rhodamine, TRITC	550	570
Rhodamine red-X, RRX	570	590
Texas red	596	620
Indodicarbocyanine, Cy5	650	670

the major disadvantage is that, due to its long emission peak at 670 nm, Cy5 fluorecence can only be observed by confocal microscopy and not by eye in a conventional microscope.

Combinations of FITC- and Texas red-containing conjugates or, alternatively, Cy3 and Cy5 conjugates, can also be used. Under these conditions, good controls should be in place to test for bleed-through between the different channels (see Section 4.2.3.3). Cy5 immunofluorescence (680 nm filter) can also bleed through very brightly into the Texas red channel.

4.2.3.3 Optimization of multiple labelling experiments

Bleed-through can occur depending on the relative brightness of the two fluorescence signals and the type of filters present in the microscope/confocal set up (see also Section 4.3). The intensity of a fluorescence signal is a function of the primary antibody affinity for its epitope, the chosen fluorophore, and antibody dilutions.

Pilot experiments must be set up to test for different dilutions, bleed-through and the order of staining of the different antibodies. The optimization controls should be performed for each pair of primary antibodies tested, as the main problem is their relative immunofluorescence output. For example, consider the double staining for cadherin (rat monoclonal ECCD2) and integrin receptors (mouse monoclonal P5D2). Both conjugates chosen are raised in donkey, and dilutions for single labelling are already known. A typical pilot trial involves the sequential incubations in different coverslips:

(1) ECCD-2, anti-rat FITC

(2) P5D2, anti-mouse Texas red

(3) ECCD-2, anti-rat FITC, anti-mouse Texas red

(4) P5D2, anti-mouse Texas red, anti-rat FITC

(5) ECCD-2, anti-rat FITC, P5D2, anti-mouse Texas red

(6) P5D2, anti-mouse Texas red, ECCD-2, anti-rat FITC

(7) anti-rat FITC

(8) anti-mouse Texas red.

Coverslips 1 and 2 demonstrate any bleed-through between the two channels and whether there is a need for further dilution of the primary and secondary antibodies. Coverslips 3 and 4 shows any cross-reaction between the different antibodies. If there is any cross-reaction, this test may determine a preferred order of staining that does not present problems, i.e. ECCD2 incubation first (coverslip 5). If no cross-reaction is observed, then the staining can be done by incubation of both primary antibodies together (ECCD-2 and P5D2) followed by their secondary conjugates (anti-rat and anti-mouse IgG). Coverslips 7 and 8 demonstrate the background level of the conjugates by themselves.

Triple labelling is performed preferably as mentioned above for double labelling together with a third antibody directly conjugated to a particular fluorophore. If a direct conjugation is not possible, then controls for cross-reactivity between the three different conjugates must be performed as described above.

4.3 Confocal microscopy

It is beyond the scope of this chapter to provide specific details on confocal microscopy. This information can be found in the literature (26, 27). We discuss our basic confocal set up and how it is used to collect images after microinjection and co-localization experiments.

To determine co-localization between two different antigens in epithelial cells, it is imperative to obtain thin optical sections from a confocal microscope, as fluorescent microscopes do not have Z-axis resolution. Conventional microscopy may lead to a false-positive result, as co-localization may appear from epitopes localized on top of or below each other. The height of keratinocyte mature junctions can be as tall as 10–12 μm. In addition, the localization of different molecules can vary along the junctions. For example, cadherins and integrins have an overlapping but distinct distribution along the keratinocyte lateral domain (28).

4.3.1 Set-up

The confocal set-up we use is an MRC 1024 BioRad system, with argon/krypton laser attached to a Nikon Optiphot 2 microscope. Excitation is fixed at 488, 568, or 647 nm which are optimal for FITC, Cy3 (or equivalent dyes), and Cy5 fluorophores, respectively. Dichroic filters are set up with a narrow path of 32 nm at the emission wavelengths of 522, 605, and 680 nm (see *Table 2* for the emission peaks of the different fluorophores). Objectives used are 10×, 20×, 40× (numerical aperture (NA) 1.0), and 60× (NA 1.4).

4.3.2 Collection of images

Although the confocal set-up offers the possibility to collect immunofluorescence images with distinct fluorophores simultaneously, we prefer to collect them separately. This is more laborious time-wise, but offers the following advantages.

(1) When the fluorescence intensity in the two channels is not at similar levels, laser output can be optimized for each excitation wavelength individually. This eliminates a good amount of leakage between the different filters.

(2) The filter set-up for simultaneous collection of images might differ from that used for single labelling. We find this a problem with the above confocal set up. Controls should be performed to address any leakage during simultaneous imaging.

(3) Images can be collected at different levels that best suit the staining pattern for each fluorophore. This may be necessary for some types of experiment where co-localization is not addressed (see below).

Initially, the confocal parameters are set up so that the best image can be obtained for each fluorophore (laser power, aperture, background level, yield). Once these are optimized, the settings can be saved and they remain reasonably constant among coverslips from the same experiment. For co-localization experiments, it is important to set the aperture at similar sizes for both channels (27). The best level in the Z-axis is found and the images are collected for each fluorophore separately, without moving the stage in the Z-axis.

When analysing microinjection experiments, initially low-power objectives (10× or 20×) are used to localize the microinjected patches using the appropriate filter in the microscope (i.e. 605 nm if dextran–Texas red is co-injected). A good 40× objective is used for collecting confocal images so that the injected patch of cells appears in the microscopic field as well as the control neighbouring cells. Images are scanned in the Z-axis for the best level for aquisition of the confocal optical sections. The microinjected patch image is usually collected at the base of the cells to demonstrate that the cells are in contact with each other at the end of incubation. Cadherin staining is shown at the level containing the most intense staining in the injected patch. When double labelling for another antigen is performed, images are collected at the same level as the cadherin immunofluorescence.

5 Cadherin functional assays

To determine whether a specific effect results from cadherin-dependent cell–cell adhesion, two types of assays are performed: inhibition of cadherin function (Section 5.1) and artificially clustering cadherin receptors (Section 5.2). During formation of new contacts, inhibition of cadherin function is used to distinguish between cadherin-mediated effects and those due to increased extracellular calcium concentration. On the other hand, artificially clustering cadherins can identify effects resulting from the clustered receptors in the absence of calcium-dependent adhesion. Methods for both types of approaches are described below.

5.1 Inhibition by antibodies against E- and P-cadherin

Because keratinocytes express E- and P-cadherin, inhibition of cell–cell adhesion is best achieved by a mixture of blocking functional antibodies against both cadherins. It should be noted that binding of the antibodies may induce recycling of the receptors, reducing the amount of cadherins and antibodies available at the cell surface. Therefore, an excess of antibody should be used and a time course should be carried out to determine the minimum incubation necessary for appropriate inhibition.

When preventing formation of new contacts, keratinocytes grown in low calcium medium are pre-incubated with a cocktail of inhibitory antibodies (5 μg ml^{-1} HECD-1 and 2 μg ml^{-1} N-CAD-299) for 30–60 min at 37 °C. Cell–cell contacts are then induced by adding calcium ions to 1.8 mM. Inhibition by antibodies is efficient enough to disrupt even mature cell–cell contacts, although longer incubations are necessary in this case. Typically, incubation with the cocktail of blocking antibodies overnight is sufficient to disassemble mature junctions and remove stratified keratinocytes. Normally, purified IgG molecules are used, and it is advisable to reproduce the effects with F(ab)$'_2$ fragments to exclude steric hindrance effects.

5.2 Clustering cadherin receptors

Cadherin receptors can be clustered by two different methods, both using specific monoclonal antibodies. In addition, receptor clustering can be performed in the presence or absence of calcium-dependent adhesion. However, it is important to be cautious in the interpretation of the results obtained by clustering receptors, as they may not represent the full plethora of processes that follow a functional adhesive contact, since:

- the presence of calcium-dependent adhesion may induce a conformational change in the cadherin molecule, allowing its interaction with different subsets of proteins;

- the clustering of receptors outside junctions may induce only the initial reorganization of actin filaments, as there is no tension or contraction to form a polarized cellular structure; and

- other adhesive structures are also not formed, with the subsequent recruitment of cytoskeletal and transmembrane proteins.

5.2.1 Clustering with secondary antibodies

The first method involves incubation with 5 μg ml^{-1} HECD-1 for 15 min on ice. Then an excess of rabbit anti-mouse IgG (25 μg ml^{-1}) is added to cluster the bound antibodies for 30 min at 37 °C. The antibodies are washed extensively and the cells prepared for immunostaining (Section 4) or biochemical analysis (Section 6). In the subsequent steps, it should be noted that the bound immunoglobulins would remain associated with the receptors. Thus, immunostaining is limited and cannot be performed with an anti-mouse or anti-rabbit conjugate as it recognizes the HECD-1 and rabbit anti-mouse IgG, respectively. In this case, mostly phalloidin staining is used or staining with a directly conjugated primary antibody. If necessary, to stain by indirect immunofluorescence, i.e. with a rat monoclonal antibody, tests for cross-reaction between the anti-rat conjugate and the mouse and rabbit IgG should be performed (see Section 4.2.3.3). For biochemical analysis, the presence of the antibodies might interfere with the subsequent assays such as immunoprecipitation and/or Western blot (depending on the size of the protein of interest).

5.2.2 Clustering with beads

This method is adapted from Miyamoto *et al.* (29) and involves clustering cadherin receptors with beads coated with specific anti-cadherin antibodies (see *Protocol 4*). Because the antibodies are held in close proximity within the beads, the cadherin receptors are clustered without any additional treatment.

Beads can be obtained from Polysciences. Latex beads are found in a wide variety of diameters and with different attached fluorophores to facilitate localization. We selected beads with 12–15 μm diameters to avoid phagocytosis of smaller beads. Even non-professional phagocytic cells such as keratinocytes are able to engulf particles. Phagocytosis should be avoided because its early steps are similar to thoses triggered by cadherin clustering (i.e. actin recruitment, requirement for small GTPase activity). However, using the protocol described below (short-term incubation and 11.9 μm beads), we have never seen a fully internalized particle in keratinocytes (see *Protocol 4*). Additional controls are required to distinguish between the process of particle internalization and cadherin clustering (see below).

We optimized the concentration to 10^5 beads per coverslip, to achieve maximal binding to keratinocytes. The best concentration of beads per coverslip should be determined empirically and maximal binding is a function of how efficient the antibody coating is. The mechanism of protein interaction with the beads is not known. The optimal protein concentration for coating should be tested experimentally for each antibody, as the ability to coat latex beads can vary tremendously among different antibodies.

The following controls should be performed (16) to exclude non-specific effects and to determine:

- the negative basal line: beads are coated with a non-specific protein such as BSA (see *Protocol 4A* and *B*);
- the specificity to cadherin receptors: beads are coated with an antibody against another surface receptor (such as integrins or CD44).

Protocol 4

Clustering of cadherin receptors with latex beads

Equipment and reagents

- Keratinocytes grown onto coverslips under low calcium conditions
- 2 FAD (with and without calcium; Biowhittaker)
- PBS
- BSA

- Latex beads (Polysciences cat. no. 18328, diameter 15 μm, 10^5 beads μl^{-1} of suspension)
- HECD-1 purified IgG
- Water bath at 80°C

Protocol 4 continued

A. Heat-denaturing BSA

1 Prepare 10 ml of a 10 mg ml^{-1} BSA solution in water.

2 Heat the solution at 80°C in a water bath for 30 min.

3 Chill immediately on ice.

4 Aliquot into small aliquots in pre-chilled tubes. Freeze at −20°C.

B. Coating beads with proteins

1 Coat latex beads (10^5 beads) with HECD-1 (400 μg ml^{-1}) overnight at 4°C with a minimum volume of PBS. Coat controls with 10 mg ml^{-1} heat-denatured BSA overnight at 4°C.

2 Wash beads twice in PBS to eliminate non-associated antibody. Spin for 1 min at 20 000 g and take PBS out very carefully to avoid aspirating the beads.

3 Block with freshly thawed, heat-denatured BSA (10 mg ml^{-1}) for 1 h at room temperature (both HECD-1 and control beads).

4 Wash three times with PBS, spinning for 1 min at 20 000 g. Aspirate the PBS, leaving a 10 μl suspension.

5 Resuspend the beads by adding 50 μl of pre-warmed and gassed medium to each tube (with or without calcium), and keep them aside. The beads are ready for binding.

C. Clustering

1 Prepare a humid chamber with a layer of humid filter paper and a piece of Parafilm on top. Warm it up in the incubator.

2 Place coverslips containing keratinocytes grown in low calcium medium (cell side upwards) onto the Parafilm in the humid chamber.

3 Add the 50 μl beads suspension prepared in step B to each coverslip.

4 Incubate for 30–45 min at 37°C.

5 Wash three times in PBS carefully to avoid displacing attached beads.

6 Fix in 3% paraformaldehyde in PBS for 10 min at room temperature.

7 Wash carefully three times in PBS and stain as usual.

In addition, it is important to check whether the correct intracellular proteins are recruited by the clustered receptors, i.e. catenins around anti-cadherin beads but not focal adhesion proteins (i.e. talin, ref. 16). The latter should be recruited to anti-integrin beads (16, 29).

For analysing this type of experiment, it is best to obtain images of confocal optical sections at the base of the beads in contact with the keratinocytes (*Figure 5*). Conventional microscopy is not adequate to detect recruitment of proteins convincingly, as this process is variable. For example, recruitment of proteins can appear as discrete fine dots as well as a complete circle around the beads. It is also necessary to quantify the percentage of attached beads that triggers an

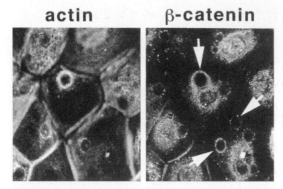

Figure 5 Clustering cadherin receptors with latex beads. Beads were coated with anti-E-cadherin monoclonal antibody and offered to keratinocytes (*Protocol 4*). After 45 min incubation in standard medium, cells were fixed and double labelled for actin filaments and β-catenin. Arrows point to different beads, showing the variability in the amount of intracellular protein recruitment. No uniform staining at cell–cell contacts is seen for β-catenin, because the confocal optical section was taken at the very top of the cells (where the beads are touching the cell surface).

effect, i.e. recruitment of a specific intracellular protein. For instance, BSA-coated beads attach to cells with very low efficiency but, even so, around 25% of attached BSA-coated beads can recruit actin. In contrast, the number of antibody-coated beads attached to keratinocytes is 10- to 50-fold more efficient and actin is recruited to around 80% of beads.

6 Biochemical analysis of cadherin complexes

6.1 Cadherin and catenins

Catenins are intracellular proteins that associate with the cytoplasmic tail of cadherin receptors (reviewed in ref. 1). Catenin is a group of proteins composed of α-catenin and the armadillo family of proteins (β-catenin, plakoglobin, p120, etc.). The armadillo family members β-catenin and plakoglobin bind directly to the C-terminus of cadherin cytoplasmic tail, while p120 interacts directly at a proximal membrane region. The other catenin, α-catenin, is an actin-binding and bundling protein. It does not associate directly with the cadherin tail, but rather with β-catenin or plakoglobin. In addition, other cytoskeletal and actin-binding proteins such as vinculin and α-actinin can also interact with cadherin complexes.

Since the cadherin–catenin complexes were discovered, there has been great interest in determining how the molecular composition and phosphorylation of the complex correlate with the functional status of cadherin receptors. However, following cell–cell contact formation, there is not a major qualitative or quantitative difference in the molecular composition of the cadherin–catenin complexes in keratinocytes or Madin–Darby canine kidney (MDCK) cells (28, 30). This is intriguing in light of the extent of the rapid cytoskeletal reorganization and the changes that occur in cadherin receptor solubility and localization. In contrast, in endothelial cells, there are clear changes in the localization of plakoglobin and phosphotyrosine-containing proteins at junctions upon confluence (see Chapter 2).

Thus, in epithelial cells, the correlation between the cadherin complex molecular composition and function appears to be more subtle than in endothelial cells. Two main issues are the intrinsic complexity of the interactions between components of cadherin complexes (*Figure 6*) and technical problems. The complexity is reflected in the ever increasing number of associated proteins, the

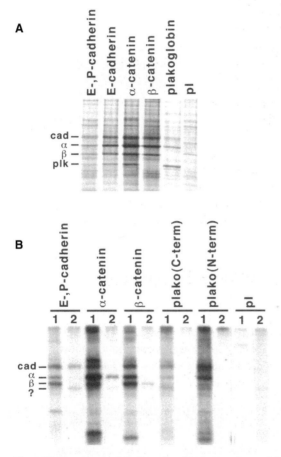

Figure 6 Immunoprecipitation of cadherin complexes. (**a**) Keratinocytes were labelled with [35S]methionine and cysteine and immunoprecipitation was performed with mild wash conditions as described in *Protocol 6*. (**b**) Keratinocytes were labelled with [32P]orthophosphate and immunoprecipitated as described in *Protocols 7* and *8*. Lysates were either precipitated directly with the different antisera (lanes 1) or boiled in the presence of 2% SDS before immunoprecipitation (lanes 2). Cadherin receptors were precipitated with anti-E,P-cadherin antibody (polyclonal anti-pan cadherin). Catenins were immunoprecipitated with the following polyclonal antisera: anti-α-catenin, anti-β-catenin, anti-plakoglobin (antisera raised against the C-terminus (C-term) or the N-terminus (N-term) of the molecule). Pre-immune antiserum (pl) was used as control. cad, cadherins; α, α-catenin; β, β-catenin; plk, plakogobin; ?, unknown phosphorylated band.

overlapping function, the distinct tissue-specific expression pattern of some catenins, and the differences in the proportion of associated catenins depending on the cell type and culture conditions. In addition, while some catenins are necessary for the cadherin adhesive function, other catenins appear to regulate cell–cell adhesiveness negatively (such as p120 and δ-catenin).

Technically, there are three main problems. First, detergent-soluble extracts contain a mixture of cadherin receptors originating from distinct cellular localizations (i.e. cell surface versus intracellular pools; junctional versus non-junctional receptors, etc.). Thus, minor differences in the molecular composition of cadherin complexes at cell–cell contacts might be diluted out or masked. Second, a problem is posed by the insolubility of a proportion of cadherin receptors engaged in functional adhesive contacts (*Figure 4*). This may be the more interesting pool of cadherin receptors, but this pool can only be extracted by boiling in sodium dodecyl sulphate (SDS; which dissociates any interaction between proteins). This technical difficulty has been a major hurdle in understanding the function of both classical and desmosomal cadherins. Third, most of the associated proteins are found in a narrow molecular weight range (at least six different proteins within 80–120 kDa). This, together with a number of different isoforms for p120, makes the cadherin immunoprecipitation pattern difficult to analyse without controls to identify the associated proteins (i.e. Western blot or re-precipitation with a different antibody; see *Protocol 8* and *Figure 6b*).

The reader is referred to the literature for alternative techniques to overcome the above-mentioned problems. For instance, diverse techniques have been attempted such as a cross-linking procedure to analyse the insoluble pool of complexes, sucrose density fractionation, cell fractionation, and pulse–chase labelling (19, 30).

6.2 Protein extracts

The protocol below is used to obtain soluble and insoluble extracts from keratinocytes. Under these conditions, the soluble pool contains around 85–90% of the total amount of keratinocyte proteins; the insoluble pool corresponds to only 10–15%.

Protocol 5

Protein extraction of keratinocytes

Equipment and reagents

- Keratinocytes plated onto 75 cm^2 flasks
- Triton X-100 buffer (15 mM Tris–HCl pH 7.5, 0.5% Triton X-100, 120 mM NaCl, 25 mM KCl, 2 mM CaCl$_2$, 1 mM phenylmethylsulphonyl fluoride (PMSF), 10 μM leupeptin, 1 μM pepstatin, 100 μM NaVO$_3$)
- PBS
- SDS buffer (2% SDS, 15 mM Tris–HCl pH 7.5, 2 mM CaCl$_2$, 100 μM NaVO$_3$)
- Orbital shaker
- Bench-top ultracentrifuge
- Refrigerated low-speed centrifuge

Protocol 5 continued

Method

1 Wash keratinocytes three times in 10 ml of cold PBS + 100 μM NaVO$_3$. Drain excess PBS well to avoid diluting the lysis buffer any further.

2 Extract in 1 ml of Triton X-100 extraction buffer (75 cm^2 flasks) for 10 min at 4 °C on an orbital shaker.

3 Scrape cells from the flask. Pellet the cell residue at 48 000 g for 10 min.

4 Transfer the supernatant to a clean tube labelled 'Triton-soluble fraction'.

5 Resuspend the pellet in 0.2 ml of SDS buffer.

6 Boil for 5 min to extract proteins. Chill on ice.

7 Centrifuge 20 000 g for 5 min.

8 Transfer the supernatant to a separate tube labelled 'Triton-insoluble fraction'.

9 Take an aliquot for determination of protein concentration for both soluble and insoluble fractions.

10 Keep all protein extracts at −70 °C until use.

6.3 Immunoprecipitation and Western blotting

Cadherin immunoprecipitation can be performed from radiolabelled keratinocyte extracts ([^{35}S]methionine and cysteine) followed by fluorography and autoradiography (5, 28). Alternatively, unlabelled keratinocyte extracts can be immunoprecipitated followed by Western blotting to identify specific interacting proteins. The latter is described below.

Cadherins are very sticky molecules, and we had problems in the past with batches of protein A–Sepharose that showed a very high background without any antibody added. This is important, particularly if a co-precipitation experiment is performed. Before use, we routinely test each new batch of protein A–Sepharose for non-specific interaction with cadherin molecules. The antibody used for immunoprecipitation may be recognized by the conjugate in the final steps of the Western blot. Depending on the background obtained and/or the size of your protein of interest, it may be necessary to cross-link the immunoprecipitating antibody to the beads to avoid this (24). Under reducing conditions, the immunoprecipitating IgG runs at around 60 kDa, and does not usually interfere with detection of catenins during the Western blotting.

For immunoprecipitation of cadherin complexes, we use HECD-1 (mouse monoclonal) or anti-pan-cadherin antiserum (rabbit polyclonal, *Figure 6*). The latter is preferred for immunoprecipitating cadherin receptors from the insoluble pool. Catenins are immunoprecipitated with anti-peptide antisera raised against their C-terminal amino acid sequence (ref. 7, *Figure 6*). The co-precipitation of different proteins can be regulated depending on the stringency of the washes, and this should be optimized (i.e. salt and detergent concentrations, ref. 19). The interaction of cadherin tail with catenins can survive a high salt concentration such as 0.5 M NaCl. However, the association of α-catenin with keratinocyte cadherins is more labile and can be reduced by the presence of 1% sodium deoxycholate in the wash buffer (28).

For comparison of the composition of cadherin complexes, normalization of the amount immunoprecipitated should be carried out (equal amount of protein). This is important because low calcium cultures differ from standard keratinocyte cultures in the number of cells per confluent dish (depending on the presence/absence of stratifying cells). Quantification of the bands obtained by Western blot can be performed by densitometry.

Protocol 6

E-cadherin immunoprecipitation and Western blot from soluble and insoluble protein extracts

Equipment and reagents

- Keratinocyte protein lysates (soluble and insoluble fractions prepared as in *Protocol 5*)
- Protein A–Sepharose slurry (1:1 in PBS)
- Relevant antibodies (i.e. anti-pan cadherin antiserum, pre-immune serum, horseradish peroxidase conjugates, etc.)
- Triton X-100 buffer (15 mM Tris–HCl pH 7.5, 0.5% Triton X-100, 120 mM NaCl, 25 mM KCl, 2 mM $CaCl_2$, 1 mM PMSF, 10 μM leupeptin, 1 μM pepstatin, 100 μM $NaVO_3$)
- PBS
- Semi-skimmed milk
- Blocking solution (5% semi-skimmed milk, 0.1% ovalbumin, 5% FCS in PBS)

- Enhanced chemiluminescence (ECL) reagents (Pierce)
- Transfer buffer (48 mM Tris–HCl, 39 mM glycine, 0.037% SDS)
- India ink (1:500 in PBS)
- Polyvinylidene difluoride (PVDF) membranes (Millipore)
- Rotating wheel
- Refrigerated bench-top centrifuge
- Orbital shaker
- SDS–polyacrylamide gel electrophoresis (PAGE) apparatus and solutions
- Semi-dry transfer apparatus (Hoeffer)

A. Immunoprecipitation

1 Pre-clear protein extracts by adding 100 μl of protein A–Sepharose slurry/tube.

2 Incubate on a rotating wheel at 4°C for 30 min minimum.

3 Spin down to separate the supernatant from protein A–Sepharose beads.

4 Place equal amounts of protein from the soluble and insoluble fractions in different tubes. For immunoprecipitation followed by Western blotting, 200 μg of protein per tube is sufficient to obtain a good signal, for cadherin catenins.

5 Dilute the insoluble fraction 10-fold with Triton X-100 buffer so that the final SDS concentration is around 0.1%.

6 Add 20 μl of protein A–Sepharose slurry and the necessary amount of anti-cadherin antiserum. Add pre-immune serum to control tubes.

Protocol 6 continued

7 Incubate on a rotating wheel at 4 °C for 60 min minimum.

8 Spin down in bench-top centrifuge (20 000 g for 30 s).

9 Aspirate off the supernatant and keep the beads.

10 Wash the protein A–Sepharose beads four times with 1 ml of Triton X-100 buffer per tube, repeating steps 8 and 9.

11 Add 20 μl of 2× sample buffer under reducing conditions to the beads.

12 Boil at 100 °C for 5 min in a heating block.

13 Separate the proteins in an SDS–polyacrylamide gel using a pre-stained molecular weight marker to check for transfer efficiency during the Western blot. Load an aliquot of keratinocyte extract as control for the Western blot as well.

B. Western blotting

1 After electrophoresis, transfer the immunoprecipitated proteins to PVDF membranes for 25–30 min using the semi-dry apparatus and the transfer buffer described above.

2 Stain the membranes with India ink solution for 5–10 min.[a]

3 Wash in PBS until transferred proteins are visible.[b] Mark and cut the membrane as required.

4 Block with blocking solution for 30 min.[c]

5 Wash briefly in PBS.

6 Incubate with an appropriate dilution of anti-catenin antibodies in 5% milk for 1 h at room temperature in an orbital shaker.

7 Wash three times for 5 min in PBS.

8 Incubate with the secondary antibody (horseradish peroxidase conjugate) diluted in 5% milk for 1 h at room temperature in an orbital shaker.

9 Wash three times for 5 min in PBS.

10 Wash briefly in 0.1 M Tris–HCl pH 8.[d]

11 Incubate with ECL substrate as described by the manufacturer.

12 Obtain chemiluminescence exposures as necessary.

[a] This step is necessary to check the evenness of the transfer across the membrane. India ink staining is more sensitive than Ponceau S and is not erased as easily from the membrane.

[b] Immunoprecipitates may not be visible but rather the IgG used for immunoprecipitation and the lysate control loaded.

[c] A richer blocking solution is necessary following immunoprecipitation to reduce the background. For standard blots, blocking in 5% milk alone is usually enough.

[d] This wash equilibrates the membrane with a more alkaline pH buffer, which is optimal for the enzymatic activity of the peroxidase.

7 Phosphorylation of cadherin complexes

There is controversy in the literature as to how phosphorylation may regulate the function of cadherin receptors. In spite of much circumstancial evidence, so far key residues in either cadherin tail or catenins whose phosphorylation is essential for cell–cell adhesion have not been identified. The reader is referred to a recent review that covers this issue (31). Potential pitfalls are:

- difficulty in dissecting the effects of phosphorylation on cadherin complexes as opposed to their perturbation of overall cytoskeletal structures;

- most reports studied only the tyrosine phosphorylation of components of the cadherin complex; however, cadherins and catenins contain many serine and threonine residues that potentially can participate in the regulation of adhesion;

- prolonged pre-incubation with phosphatase inhibitors which can produce artefacts;

- a critical comparison between distinct experimental systems is very difficult due to differences in cell type, cadherin complex composition, and methodology used (see also Section 6.1).

We used orthophosphate labelling of keratinocytes to identify phosphorylated peptides and the type of phosphorylated amino acids in the cadherin complexes. The principles and general methodology are described in the literature and we refer the reader to key important papers. Below, I detail the optimization performed for the orthophosphate labelling of keratinocytes, preparation of extracts, and isolation and characterization of cadherin complexes.

7.1 Orthophosphate labelling

Comprehensive reviews of orthophosphate labelling and separation of phosphorylated proteins can be found elsewhere (32, 33). The intracellular pool of phosphate is labelled very slowly, and a few hours is needed to achieve equilibrium. We label keratinocytes overnight, but at least 4 h may be necessary to achieve optimal labelling. The drawback is that DNA and RNA are also labelled and can increase the background following immunoprecipitation and SDS–PAGE. If this poses a problem, try reducing the amount of label or labelling time. Alternatively, samples can be treated with protease-free DNase and RNase before loading onto the gel (test for protein degradation first before adding it to precious samples!).

Protocol 7

Orthophosphate labelling and cell lysis

Equipment and reagents

- Keratinocytes grown onto 75 cm² flasks under low calcium conditions
- 1 M CaCl₂
- PBS
- FAD without calcium and phosphate (Biowhittaker)
- [³²P]Orthophosphate

- Lysis buffer (15 mM Tris–HCl pH 7.5, 50 mM NaCl, 20 mM β-glycerol phosphate pH 7.5, 5 mM EDTA, 2.5 mM EGTA, 1% Triton X-100, 20 mM sodium pyrophosphate, 50 mM NaF, 100 μM NaVO₃, 1 μM microcysteine, 1 mM PMSF, 10 μM leupeptin, 1 μM pepstatin)
- Refrigerated bench-top centrifuge

Method

1 Wash cells three times with warm medium without phosphate. Add 3 ml of fresh medium without phosphate per 75 cm² flask.

2 Add [³²P]orthophosphate (0.5–1 mCi ml⁻¹) and incubate overnight.

3 Next day, induce cell–cell contacts by adding calcium ions to 1.8 mM from a stock of 1 M CaCl₂.

4 Incubate at 37 °C for the necessary amount of time.

5 Stop labelling by chilling the cells on ice.

6 Wash three times with cold PBS containing 20 mM sodium pyrophosphate, 50 mM NaF, 100 μM NaVO₃.

7 Drain and remove the PBS completely to avoid diluting the lysis buffer.

8 Add 1 ml of cold lysis buffer and scrape the cells.

9 Transfer to a screw-top tube and centrifuge for 5 min 20 000 g at 4 °C.

10 Transfer the supernatant to a clean tube and discard the pellet.

11 Immunoprecipitate (100–200 μl lysate per tube) with the relevant antibodies as described in *Protocol 6A*.

7.2 Isolation of phosphorylated proteins

Orthophosphate-labelled keratinocyte lysates are immunoprecipitated with anti-cadherin and anti-catenin antibodies as described in *Protocols 6A* and *8*. In order to avoid the action of phosphatases during the immunoprecipitation, the procedure is carried out as quickly as possible, samples and solutions kept cold at all times, and phosphatase inhibitors added to the immunoprecipitation wash buffer as well.

Because cadherin and catenin immunoprecipitation pattern is very complex, controls are performed to be certain of the identity of the phosphorylated band (*Figure 6b*). This is necessary because:

- minor protein bands are not detected by [^{35}S]methionine labelling and can be detected readily by orthophosphate labelling; and

- a phosphorylated band may have a molecular weight similar to that of the catenins yet it may be a distinct associated protein (28).

These controls can be performed by reprecipitation (see *Protocol 8*) or by transferring the immunoprecipitated proteins to nitrocellulose and immunoblotting with the relevant antibody (*Protocol 6B*). Re-precipitation has the disadvantage that the antibody used must be able to recognize denatured proteins (*Figure 6b*). Polyclonal antisera usually work well, but this may be a problem with some monoclonal antibodies. The main disadvantage with Western blots is that high levels of radioactivity are used during orthophosphate labelling and may still be present in the immunoprecipitates. By using the protocol below, we found that cadherins and α- and β-catenin, but not plakoglobin, are phosphorylated in keratinocytes (28).

Protocol 8

Identification of phosphorylated proteins

Equipment and reagents

- Relevant antibody[a]
- Orthophosphate-labelled keratinocyte protein extracts
- 20% SDS
- Rotating wheel
- Refrigerated bench-top centrifuge
- 100°C water bath or heat block
- Apparatus and solutions for SDS–PAGE

- Lysis buffer (15 mM Tris–HCl pH 7.5, 50 mM NaCl, 20 mM β-glycerol phosphate pH 7.5, 5 mM EDTA, 2.5 mM EGTA, 1% Triton X-100, 20 mM sodium pyrophosphate, 50 mM NaF, 100 μM NaVO$_3$, 1 μM microcysteine, 1 mM PMSF, 10 μM leupeptin, 1 μM pepstatin)
- Cassettes and X-ray films

Method

1 Pre-clear the orthophosphate-labelled protein lysates by adding 100 μl of protein A–Sepharose slurry and incubating for 30 min at 4°C in a rotating wheel.

2 Separate the beads by centrifugation and place 100 μl of pre-cleared lysates in different tubes.

3 Add 10 μl of 20% SDS to one set of tubes and boil at 100°C for 5 min to denature and dissociate the components of cadherin complexes.

4 Chill on ice quickly and dilute with 900 μl of lysis buffer.

5 Add 20 μl of protein A–Sepharose slurry and the relevant antibody to the boiled and non-boiled tubes.

6   Incubate at 4 °C for 1 h in a rotating wheel.

7   Wash immunoprecipitates four times with lysis buffer containing phosphatase inhibitors.

8   Add 20 µl of 2× SDS buffer and boil the samples.

9   Separate precipitated proteins in an SDS–polyacrylamide gel and perform auto-radiography.

[a] Antibody or antiserum must be able to precipitate denatured proteins.

7.3 Phosphopeptide mapping

The method used is essentially as described by Millar *et al.* (34) and I describe below the optimization performed for the separation of cadherin and catenin phosphopeptides.

Briefly, phosphopeptide mapping involves the digestion of the purified protein with a protease (usually trypsin) and separation of the phosphorylated peptides via high-voltage thin-layer electrophoresis and subsequent chromatography. The protein is immunoprecipitated and separated by SDS–PAGE. Following auto-radiography, the band of interest is cut out of the gel (mimimum of 500–2000 c.p.m. per band necessary to start with). The dried gel is then rehydrated and the sample solubilized out of the gel by digestion with trypsin. Because cadherin and catenins have methionine and cysteine residues in their amino acid sequence, we perform an oxidation step following the tryptic digest (see *Protocol 9*). This is the main difference from the protocol described (34).

Protocol 9

Methionine and cysteine oxidation

Equipment and reagents

- Hydrogen peroxide (33%)
- Formic acid
- Speed-Vac concentrator/drier
- Scintillation counter

Method

1   Prepare performic acid (PA) fresh for each experiment by adding 100 µl of hydrogen peroxide (33%) to 900 µl of 100% formic acid and leave at room temperature for 45 min. Store on ice.

2   Put some water to cool down to be used to stop the reaction later on.

3   Add 100 µl of ice-cold PA to trypsin-digested and dried samples. Vortex. Keep samples ice-cold so that peptide bond cleavage cannot occur.

4   Incubate for 60 min on ice.

Protocol 9 continued

5 Dilute samples with 300 μl of cold water. Freeze and dry in the Speed-Vac.

6 Wash once with water. Dry in the Speed-Vac.

7 Cherenkov count the amount of radioactivity in the tube.

8 Resuspend dried samples in 10 μl of water and vortex.

9 Heat at 65 °C for 1 min and vortex.

10 Centrifuge at 20 000 g for 10 min.

11 Carefully transfer the supernatant to a fresh tube, avoiding aspirating any pellet. Supernatant is ready for separation of the phosphopeptides.

12 Count the radioactivity in the pellet again to check the efficiency of the solubilization.

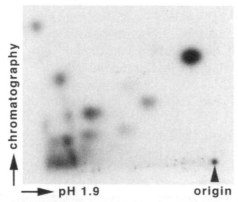

Figure 7 Phosphopeptide map of E-cadherin immunoprecipitate. E-cadherin was immunoprecipitated from orthophosphate-labelled keratinocyte extracts (*Figure 6b*). The band corresponding to 120 kDa was isolated, digested, and the phosphopeptides separated as described in Section 7.3. Arrows point to the direction of migration during high-voltage electrophoresis (pH 1.9) and chromatography; the arrowhead points to the origin where the sample was applied.

Samples are spotted onto thin-layer chromatography (TLC) plates and high-voltage electrophoresis is performed at 500 V. The best buffer for separation of the keratinocyte cadherin phosphopeptides is pH 1.9 buffer (50 ml of formic acid (88%), 156 ml of acetic acid (glacial), and 1794 ml of water). Plates are dried and ascending chromatography is performed (500 ml of pyridine, 750 ml of butanol, 150 ml of acetic acid, 600 ml of water). A typical phosphopeptide map for keratinocyte E-cadherin is shown in *Figure 7*.

7.4 Determination of the type of phosphorylated amino acid

The type of phosphorylated amino acid residue (serine, threonine, or tyrosine) can be determined from the bands isolated from the SDS–polyacrylamide gels or from the spots separated during phosphopeptide mapping. The method is described in detail by Millar *et al.* (34). Samples are spotted onto TLC plates along with phosphoamino acid standards. Phosphorylated residues are separated by sequentially running electrophoresis in pH 1.9 buffer (see above) and pH 3.5 buffer (100 ml of acetic acid, 10 ml of pyridine, 1890 ml of water; ref. 34). Under the conditions described in *Protocols 7* and *8*, keratinocyte cadherins have mostly serine-phosphorylated residues, while α- and β-catenin contain serine-, threonine-, and tyrosine-phosphorylated residues.

References

1. Yap, A. S., Brieher, W. M., and Gumbiner, B. M. (1997). *Annu. Rev. Cell Dev. Biol.*, **13**, 119.
2. Braga, V. M. M. (1999). *Mol. Pathol.*, **52**, 197.
3. Van Aelst, L. and D'Souza-Schorey, C. (1997). *Genes Dev.*, **11**, 2295.
4. Watt, F. M. (1989). *Curr. Opin. Cell Biol.*, **1**, 1107.
5. Hodivala, K. J. and Watt, F. M. (1994). *J. Cell Biol.*, **124**, 589.
6. Lewis, J. E., Jensen, P. J., and Wheelock, M. J. (1994). *J. Invest. Dermatol.*, **102**, 870.
7. Braga, V. M. M., Hodivala, K. J., and Watt, F. M. (1995). *Cell Adhes. Commun.*, **3**, 201.
8. Zhu, A. J. and Watt, F. M. (1996). *J. Cell Sci.*, **109**, 3013.
9. Watt, F. M. (1994). In *Cell biology: a laboratory handbook* (ed. J. E. Celis). p. 83. Academic Press, London.
10. Feig, L. A. (1999). *Nature Cell Biol.*, **1**, E25.
11. Machesky, L. M. and Hall, A. (1996). *Trends Cell Biol.*, **6**, 304–310.
12. Graessmann, M. and Graessmann, A. (1983). In *Methods in enzymology* (ed. R. Wu, L. Grossman, and K. Moldave). Vol. 101, p. 482. Academic Press, London.
13. Paterson, H., Adamson, P., and Robertson, D. (1995). In *Methods in enzymology* (ed. W. E. Balch, C. J. Der, and A. Hall). Vol. 256, p. 162. Academic Press, London.
14. Ridley, A. J. (1995). In *Methods in enzymology* (ed. W. E. Balch, C. J. Der, and A. Hall). Vol. 256, p. 313. Academic Press, London.
15. Self, A. J. and Hall, A. (1995). In *Methods in enzymology* (ed. W. E. Balch, C. J. Der, and A. Hall). Vol. 256, p. 3. Academic Press, London.
16. Braga, V. M. M., Machesky, L. M., Hall, A., and Hotchin, N. A. (1997). *J. Cell Biol.*, **137**, 1421.
17. Braga, V. M. M., Del Maschio, A., Machesky, L. M., and Dejana, E. (1999). *Mol. Biol. Cell*, **10**, 9.
18. Takaishi, K., Sasaki, T., Kotani, H., Nishioka, H., and Takai, Y. (1997). *J. Cell Biol.*, **139**, 1047.
19. Nelson, W. J., Wilson, R., and Mays, R. W. (1992). In *Cell–cell interactions: a practical approach* (ed. B. R. Stevenson, W. J. Gallin, and D. L. Paul). p. 227. IRL Press, Oxford.
20. Shore, E. and Nelson, W. J. (1991). *J. Biol. Chem.*, **266**, 19672.
21. Shimoyama, Y., Hirohashi, S., Hirano, S., Noguchi, M., Shimosato, Y., Takeichi, M., *et al.* (1989). *Cancer Res.*, **49**, 2128.
22. Hirai, Y., Nose, A., Kobayashi, S., and Takeichi, M. (1989). *Development*, **105**, 271.
23. Shimoyama, Y., Yoshida, T., Terada, M., Shimosato, Y., Abe, O., and Hirohashi, S. (1989). *J. Cell Biol.*, **109**, 1787.
24. Harlow, E. and Lane, D. (1988). *Antibodies: a laboratory manual.* p. 1. Cold Spring Harbor Laboratory Press, Cold Spring Harbor, NY.
25. Geiger, B. and Volberg, T. (1994). In *Cell biology: a laboratory handbook* (ed. J. E. Celis). p. 297. Academic Press, London.

26. Brejle, T. C., Wessendorf, M. W., and Sorenson, R. L. (1993). In *Methods in cell biology: cell biological applications in confocal microscopy* (ed. B. Matsumoto). Vol. 38, p 98. Academic Press, London.

27. Pawley, J. B. and Centouze, V. E. (1994). In *Cell biology: a laboratory handbook* (ed. J. E. Celis). p. 44. Academic Press, London.

28. Braga, V. M. M., Najabagheri, N., and Watt, F. M. (1998). *Cell Adhes. Commun.*, **5**, 137.

29. Miyamoto, S., Akiyama, S., and Yamada, K. M. (1995). *Science*, **267**, 883.

30. Hinck, L., Nathke, I. S., Papkoff, J., and Nelson, W. J. (1994). *J. Cell Biol.*, **125**, 1327.

31. Daniel, J. M. and Reynolds, A. B. (1997). *BioEssays*, **19**, 883.

32. Garrison, J. C. (1993). In *Protein phosphorylation: a practical approach* (ed. D. G. Hardie). p. 1. Oxford University Press, Oxford.

33. van der Geer, P., Luo, K., Sefton, B. M., and Hunter, T. (1993). In *Protein phosphorylation: a practical approach* (D. G. Hardie). p. 31. Oxford University Press, Oxford.

34. Millar, N. S., Moss, S. J., and Green, W. N. (1995). In *Ion channels: a practical approach* (ed. R. H. Ashley). p. 191. Oxford University Press, Oxford.

Chapter 2
Analysis of intercellular adhesion molecules in endothelial cells

Gianfranco Bazzoni, Maria G. Lampugnani, Ofelia M. Martinez-Estrada
Laboratory of Vascular Biology, Istituto di Ricerche Farmacologiche Mario Negri, via Eritrea 62, 20157 Milano, Italy

Elisabetta Dejana
Università degli Studi dell' Insubria, Dipartimento di Scienze Cliniche e Biologiche, Facolta di Medicina e Chirurgia, Varese, Italy

1 Introduction

Intercellular adhesion between endothelial cells (ECs) is mediated by members of the cadherin, immunoglobulin, integrin, and proteoglycan families of cell adhesion molecules. Some of these molecules are concentrated at intercellular structures known as adherens junctions and tight junctions. Other molecules, such as platelet–endothelial cell adhesion molecule-1 (PECAM-1), are localized at cell–cell contacts, but outside of the junctional complexes. Although the morphological and molecular organization of these complexes are similar in ECs and in other cell types, some components are more specific for the endothelium. For instance, vascular endothelial-cadherin (VE-cadherin) is expressed exclusively in ECs at the adherens junctions, while claudin-5 and the alpha minus isoform of zonula occludens-1 (ZO-1), two tight junction components, are expressed preferentially in the endothelium (1, 2). By interacting (either homophilically or heterophilically) with their receptors, junctional adhesive molecules provide physical links that hold ECs together and regulate the paracellular passage of ions, solutes, and transmigrating leukocytes. Also, by interacting with intracellular partners, the adhesive receptors may strengthen their own linkage with the cytoskeleton. Finally, besides providing attachment for neighbouring cells and the cytoskeleton, junctional molecules may also impinge upon cell signalling and trigger responses that are translated into changes in cell morphology and eventually in the organization of three-dimensional networks of patent tubes (3).

Here, we will focus on methodological issues relevant to the analysis of endothelial junctional molecules, such as isolation and culture of ECs, measurement of paracellular permeability, and immunofluorescence microscopy of ECs. Emphasis will be given to the expression of endothelial markers and to those experimental conditions that may affect the expression and distribution of intercellular molecules.

2 Endothelial cell culture

2.1 Human umbilical vein endothelial cells (HUVECs)

Among ECs of different origin, HUVECs are commonly used for studies of endothelial adhesion molecules and intercellular junctional complexes. The method for isolating HUVECs is based on the separation of ECs from the vessel wall using collagenase to digest the subendothelial basement membrane (4–6). ECs can be obtained from other tissues and organs of different animal species, but description of their isolation is beyond the scope of this chapter. Also, generation of mouse EC lines using the polyoma middle T oncogene has been reported previously (7).

Protocol 1

Isolation of human umbilical vein endothelial cells (HUVECs)

Equipment and reagents

- Umbilical cords (at least 20 cm in length)
- Collagenase solution (0.1% in HBSS containing Ca^{2+} and Mg^{2+}; collagenase A, Boehringer Mannheim)
- Medium M199 supplemented with 20% newborn calf serum (Gibco-BRL/Life Technologies)
- Trypsin solution (1.5 U ml^{-1}) containing ethylenediamine tetraacetic acid (EDTA) 0.02%

- Ca^{2+}- and Mg^{2+}-free Hank's balanced salt solution (HBSS)
- Endothelial cell growth supplement,bovine, culture grade from bovine brain (Roche)
- Heparin from porcine intestinal mucosa (Sigma)
- Culture flasks coated with 1.5% gelatin (Sigma)
- Bench centrifuge

Method

1 Collect umbilical cords from normal deliveries or cesarean sections in sterile plastic bags and excise any crushed area. Cords can be stored at 4°C for 1 week before processing.

2 In a tissue culture hood, excise both ends of each cord (~1 cm long) with a sterile scalpel to expose a sterile surface. Perfuse the umbilical vein with Ca^{2+}- and Mg^{2+}-free HBSS.

3 Perfuse the umbilical vein with the collagenase solution until the vein is distended, clamp the two ends of the cord with sterile clamps, and incubate the cords for 20 min at 37°C. Then, gently massage the cords to facilitate EC detachment from the vessel wall.

4 Flush out the solution and wash the lumen of the vein with HBSS. Collect cells.

Protocol 1 continued

5 Centrifuge the cell suspension at 1200 r.p.m. for 10 min.

6 Resuspend the pellet in medium M199 containing 20% serum.

7 Seed the cells in culture flasks at a concentration of $20–40 \times 10^3$ per cm². The following day, wash the cell layer to remove red blood cells and cell debris that may have detached with ECs. Usually primary culture cells will reach confluence in 4–6 days.

8 Passage the cells using trypsin-EDTA. After the primary culture, HUVECs can only be maintained for further passages in the presence of endothelial cell growth supplement (5–50 μg ml^{-1}) and heparin (100 μg ml^{-1}) and should always be grown on culture flasks previously coated with 1.5 % gelatin.

Before reaching confluence, HUVEC may resemble fibroblasts or smooth muscle cells. Once confluent, however, a trained observer can easily recognize the typical 'cobblestone' morphology. 'Healthy' cultures are composed of uniform cells of relatively small dimensions. The appearance of giant cells is a signal of senescence. To retard cell senescence and to use cells until the fourth *in vitro* passage, it is critical to change the medium every second day.

Cell contamination usually is caused by fibroblasts. Gross fibroblast contamination is easily detectable by the presence of spindle-shaped cells piled in a criss-cross fashion. Single fibroblasts may be identified as they often lie on the top of EC monolayers. Contamination with smooth muscle cells is unusual, as these cells grow rather slowly, and their growth is inhibited by the presence of ECs. In the authors' experience, however, contamination of primary HUVEC cultures is not a frequent event.

If not purchased, the endothelial cell growth factor can also be prepared according to the method described by Maciag *et al.* (8).

2.1.1 Characterization of HUVECs and ECs from specific lineages

Antibodies directed against endothelial markers allow immunofluorescence or fluorescence-activated cell sorting (FACS) analysis to confirm that the cells obtained are indeed ECs and, more in general, to identify ECs both *in vitro* and *in vivo*. Endothelial-specific markers can be either constitutive or inducible. It is noteworthy that some markers are shared by ECs and haematopoietic precursors, thus indirectly supporting the hypothesis that ECs and blood cells derive from a common precursor. Endothelial markers comprise adhesion molecules (VE-cadherin, PECAM-1, intercellular cell adhesion molecule-2 (ICAM-2), the $\alpha v \beta_3$-integrin, CD36, S-Endo 1/Muc 18), a miscellaneous class of molecules (factor VIII-related antigen, angiotensin-converting enzyme, type-I scavenger receptor for acetylated low-density liporotein (LDL), agglutinins, thrombomodulin), receptors for vascular endothelial growth factor (VEGF) (KDR/Flk-1 and Flt-1) and angiopoietin (Tie-1 and Tie-2), the Weibel–Palade bodies, and adhesion molecules expressed upon activation with cytokines or growth factors (ICAM-1, VCAM-1, and E- and P-selectin).

To date, relatively few markers for ECs of specific origin have been identified, probably reflecting the difficulty in isolating microvascular ECs from most organs. Surface molecules (e.g. GlyCAM-1, MAdCAM-1, CD34, MECA-79, VAP-1 and -2) are expressed preferentially on the high endothelium of post-capillary venules in lymph nodes and Peyer's patches. Molecules specific for brain ECs (one of the components of the blood–brain barrier) comprise the multidrug resistance protein Mdr 1a and transporters for glucose (Glut-1) and amino acids (see ref. 9 for a detailed description). It is noteworthy that mRNA for the VEGF receptor Flt-4 becomes restricted during later stages of development to ECs of lymphatic vessels and to some high endothelial venules (10).

2.1.2 Modulation of the expression and distribution of endothelial junctional molecules

Besides providing intercellular cohesion, endothelial junctions may vary in composition and adhesive strength in response to several agents that increase microvascular permeability. It has been proposed that these agents might activate the contractile cytoskeleton and open up small gaps between adjacent ECs (11). Only recently, however, direct evidence has shown that intercellular adhesive molecules themselves are targets of signals initiated by permeability-inducing agents. As *in vivo* changes in junctional organization and vascular permeability can be reproduced *in vitro*, knowledge of experimental conditions that modify the expression and distribution of junctional molecules may help to define the dynamic regulation of vascular function. In *Table 1* we have summarized the effect of vasoactive mediators on some endothelial junctional molecules. This brief survey is not exhaustive, but allows us to draw some general conclusions (for a detailed analysis see refs 12, 13).

First, not all mediators always cause the formation of intercellular gaps, but rather they induce subtle changes in the distribution of some junctional molecules (from a continuous to a more fragmentary or even diffuse staining), that are nonetheless associated with increases in permeability, indicating that formation of frank discontinuities is not absolutely required for the permeabilizing action. Secondly, tyrosine phosphorylation is emerging as a common mechanism, and it has been suggested that phosphorylation of junctional molecules may cause their dissociation from the cytoskeleton. Thirdly, other mediators may reorganize junctions by either inducing internalization of junctional molecules or inhibiting their synthesis. Finally, heterogeneity exists between ECs from different tissues, animal species, and states of maturation. For instance, adherens junction components show a higher degree of phosphorylation (and a looser junctional organization) in early than in late stages of EC confluency.

3 Vascular permeability and morphological changes associated with increased permeability

Hereafter, methods to measure paracellular permeability and to study morphological changes associated with increased permeability are described.

Table 1 Modulation of junctional molecule expression and/or distribution in ECs

Molecule	Agent	Endothelial cell type	Action	Reference
PECAM-1	TNF-α + IFN-γ	HUVEC	↓ Expression	(18)
		Bovine aorta/glomeruli	Disappearance from intercellular contacts and internalization	(19)
	TNF-α	HUVEC	↑ Phosphorylation, redistribution	(20)
JAM	TNF-α + IFN-γ	HUVEC	Disappearance from intercellular contacts	(21)
VE-cadherin	TNF-α + IFN-γ	Rat mesenteric venules	Focal loss at intercellular junctions	(22)
		HUVEC	Junction disassembly	(23)
	Thrombin		Junction disassembly	(16)
	Histamine		↑ Phosphorylation, dissociation from junctions/cytoskeleton	(14)
	VEGF		↑ Phosphorylation, intercellular contacts loosening	(15)
Occludin	PUFAs	Human ECV304 line	↑ Expression	(24)
	HGF/SF		↓ Expression and discontinous distribution	(25)
	VEGF	HUVEC	Disappearance from the junctions	(26)
		Bovine retina	Phosphorylation (short term); ↓ expression (long-term effect)	(27)
ZO-1	TNF-α + IFN-γ	Human microvessels	Discontinous distribution	(28)
	Histamine	Bovine retina	↓ Expression	(29)
	HGF	Human ECV304 line	Discontinous distribution	(25)
	Dexamethasone	Schlemm's channel	↑ Expression	(30)

HGF, hepatocyte growth factor; IFN-γ, interferon-γ; JAM, junction adhesion molecule; PUFA, polyunsaturated fatty acid; SF, scatter factor; TNF-α, tumour necrosis factor-α; VEGF, vascular endothelial growth factor.

3.1 Vascular permeability

An *in vitro* model in which ECs are grown on porous filters has proven to be a valuable tool for evaluating the role of ECs in the regulation of paracellular permeability in response to vasoactive compounds, such as histamine, thrombin, and cytokines (14–16).

Protocol 2

Method for the analysis of vascular permeability

Equipment and reagents

- Human fibronectin (Sigma, code F2006, store in 200 µg ml^{-1} aliquots at −20 °C)
- Fluorescein isothiocyanate (FITC)-conjugated dextrans of different molecular weights (Sigma)
- Medium M199
- Trypsin solution (1.5 U ml^{-1}) containing EDTA 0.02%
- Transwell system: Transwell polycarbonate filters (pore size 0.4 µm, filter area 0.33 cm^2) in multiwell plates (Costar)
- CO$_2$ incubator

Method

1 Coat the Transwell filters with 50 µl of fibronectin (at a concentration of 5–10 µg ml^{-1}). Incubate for 1 h at room temperature.

2 Using an automatic pipette, aspirate off the fibronectin solution immediately before seeding the cells. Avoid needles and be careful not to damage the filter. Rinse the filter once with serum-free medium M199.

3 Detach confluent cells with trypsin-EDTA.

4 Resuspend the detached cells with culture medium. Seed HUVECs in a volume of 100 µl per filter at a concentration of 2.5–3.0 × 10^5 cells ml^{-1}, which corresponds to the cell density of a confluent cell layer. Add 600 µl of medium in the lower compartment, according to the manufacturer's instructions.

5 Incubate the cells in an incubator at 37 °C in the presence of 5% CO$_2$/95% air.

6 Every other day, change the culture medium, adding 100 and 600 µl in the upper and lower compartment, respectively.

7 Use the cells for experiments 4–7 days after seeding.

8 Add dextran–FITC to the upper compartment (5 µl of a 20 mg ml^{-1} solution in phosphate-buffered saline (PBS)). Immediately transfer the Transwell system to the incubator.

9 At regular time points, remove 50 µl of medium from the lower compartment and replace it with 50 µl of fresh culture medium.

10 Dilute the 50 µl aliquot with 950 ml of PBS and determine fluorescence with a fluorimeter using a wavelength of 492 and 520 nm for excitation and emission, respectively.

Measurement of permeability can be performed using different techniques and tracers, such as horseradish peroxidase (HRP). If HRP is used, add 2 μl of a 5 μM solution of HRP (type VI-A, 44 kDa, specific activity 250 U mg^{-1}, Sigma) and determine its enzymatic activity using o-phenylenediamine dihydrochloride (Sigma) as a substrate, according to the manufacturer's instructions.

Ethylene glycol-bis(2-aminoethyl) tetraacetic acid (EGTA; at a concentration of 2 mM in serum-free medium M199) can be used as a positive control in the upper chamber to quantify the maximal level of permeability of the HUVEC monolayer.

3.2 Immunofluorescence microscopy analysis of ECs

Protocol 3 describes our methodology for immunolabelling ECs for the localization of junctional proteins.

Protocol 3

Staining of HUVECs for analysis of junctional molecules

Equipment and reagents

- Tissue culture plates, glass coverslips, microscope slides
- Human fibronectin (Sigma; see *Protocol 2*)
- Permeabilization buffer (HEPES-Triton buffer): 20 mM N-(2-hydroxyethyl) piperazine-N'-(2-ethanesulphonic acid) (HEPES), 300 mM sucrose, 50 mM NaCl, 3 mM MgCl$_2$, 0.5% Triton X-100 (add just before use), pH 7.4. Sterile filtered and stored at 4°C
- PBS and PBS supplemented with bovine serum albumin (BSA) (PBS–1% BSA as washing and blocking buffer)
- Thimerosal (Sigma)
- Fresh formaldehyde solution in PBS prepared from paraformaldehyde (pH 7.6)
- Methanol
- Primary and fluorochrome-tagged secondary antibodies
- Mounting media (polyvinyl alcohol (Mowiol 4-88, Hoechst) 20% in PBS containing 0.02% sodium azide)
- Epifluorescence microscope

Method

1   Place glass coverslips (13 mm diameter) in 24-well tissue culture plates. Pre-coat coverslips with fibronectin[a]. Seed cells and grow them to the desired confluence. Then remove culture medium and gently add 0.8–1.0 ml of formaldehyde solution[b]. Incubate with the fixative for 15 min at room temperature.

2   Aspirate off fixative and wash twice with PBS. If not used immediately for staining, store the plate in PBS containing 0.01% Thimerosal.

3   To permeabilize the cells (when visualization of internal antigens is required), just before staining, put the plate on ice and add 0.8–1.0 ml of chilled HEPES-Triton buffer. Incubate for 3 min. Then, remove the permeabilization buffer and wash the cells twice with PBS.

Protocol 3 continued

4   Alternatively (as required to maintain optimal reactivity of individual antibodies), fix–permeabilize cells with methanol, as follows. Remove culture medium and immediately add methanol (previously chilled at −20°C). Incubate for 3 min on ice. Wash twice with ice-cold PBS. Store in PBS containing 0.01% Thimerosal (if not used immediately).

5   Transfer the coverslips to a 12-well plate. To block non-specific binding, add PBS–1% BSA and incubate for 30 min to 1 h at room temperature.

6   Add the solution of the primary antibody (usually at a concentration of or below 10 μg ml^{-1} in PBS–1% BSA^c) in a volume of 20 μl on the top of the coverslip and homogenously smear it with a dispensing tip to facilitate diffusion. Incubate at 37°C for 60 min in a humidified chamber^d). Remove the primary antibody solution by washing two or three times with PBS–1% BSA (5 min for each wash).

7   Add the fluorochrome-tagged secondary antibody in a volume of 20 μl (in PBS–1% BSA), smear it, and incubate for 45 min at 37°C as above. Wash thoroughly three times (5 min each) with PBS–1% BSA to remove non-specifically bound secondary antibody.

8   To mount coverslips, add 7 μl of Mowiol 4-88 on a microscope slide. Then, remove each coverslip from the wells with forceps, and invert onto the Mowiol drop on the microscope slide. Gently press the slide and leave to dry overnight.

^{a} To pre-coat coverslips with fibronectin, dilute the stock solution in PBS with Ca^{2+} and Mg^{2+} to 7 μg ml^{-1}. Then, add 70 μl drops onto a piece of Parafilm set in a Petri dish. Set a coverslip on each drop and incubate either overnight at 4°C or for 2 h at 37°C. Before cell seeding, set the coverslips in a 24-well plate with the coated surface upwards and wash once with PBS. Gently press the coverslip along the border to remove air bubbles that may get entrapped under the coverslip, in order to avoid coverslip floating and streaming of the cell suspension under the coverslip.

^{b} Once culture medium is aspirated off and before fixative is added, avoid washing to preserve cell morphology. Different fixation methods can be used depending on the specific antigen to be stained.

^{c} When using anti-VE-cadherin antibodies, we include Ca^{2+} and Mg^{2+} in the buffers used at all the steps (including washing after fixation, storage, and staining). VE-cadherin and PECAM-1 can be observed using either paraformaldehyde or methanol fixation. The former method, however, offers the advantage of preserving F-actin, a useful marker of the general cell morphology (of both cell layers and individual cells). We use rhodaminated antibodies (DAKO or Jackson Laboratories) in combination with fluoresceinated phalloidin (Sigma code P-5282, at a final concentration of 2 μg ml^{-1}). Phalloidin is aliquoted (400 μg ml^{-1} in PBS with 5% dimethylsulphoxide (DMSO)) and stored at −20°C.

^{d} Incubation in the humidified chamber is meant to avoid evaporation of the small volume of antibody. A tight box containing a wet paper can be used for this purpose. For a detailed description, the reader is referred to ref. 17.

Acknowledgements

This study was supported by grants from Human Frontiers Science Program (RG 0006/1997-M), EEC (BIO4 CT 980337, BMH4 CT 983380, QLG1-CT-1999-01036, and QLK3-CT-1999-00020), CNR (97.01299.PF49), MURST (9906317157-003), and AIRC. O.M.M.E. is a recipient of a Fellowship from ICGEB, Trieste, Italy.

References

1. Dejana, E., Bazzoni, G., and Lampugnani, M. G. (1999). *Exp. Cell Res.*, **252**, 13.
2. Lampugnani, M. G. and Dejana, E. (1997). *Curr. Opin. Cell Biol.*, **9**, 674.
3. Bazzoni, G., Dejana, E., and Lampugnani, M. G. (1999). *Curr. Opin. Cell Biol.*, **11**, 573.
4. Balconi, G. and Dejana, E. (1986). *Med. Biol.*, **64**, 231.
5. Gimbrone, M. A. J., Cotran, R. S., and Folkman, J. (1974). *J. Cell Biol.*, **60**, 673.
6. Ryan, U. S., Ryan, J. W., Whitaker, C., and Chiu, A. (1976). *Tissue Cell,* **8**, 125.
7. Allavena, P., Dejana, E., Bussolino, F., Vecchi, A., and Mantovani, A. (1995). In *Cytokines* (ed. F. R. Balkwill). p. 225. Oxford University Press, New York.
8. Maciag, T., Cerundolo, J., Ilsley, S., Kelley, P. R., and Forand, R. (1979). *Proc. Natl Acad. Sci. USA*, **76**, 5674.
9. Garlanda, C. and Dejana, E. (1997). *Arterioscler. Thromb. Vasc. Biol.*, **17**, 1193.
10. Kaipainen, A., Korhonen, J., Mustonen, T., van Hinsbergh, V. W., Fang, G. H., Dumont, D., *et al.* (1995). *Proc. Natl Acad. Sci. USA*, **92**, 3566.
11. Lum, H. and Malik, A. B. (1994). *Am. J. Physiol.*, **267**, L223.
12. Dejana, E., Valiron, O., Navarro, P., and Lampugnani, M. G. (1997). *Ann. NY Acad. Sci.*, **811**, 36.
13. van Hinsbergh, W. M. (1997). *Arterioscler. Thromb. Vasc. Biol.*, **17**, 1018.
14. Andriopoulou, P., Navarro, P., Zanetti, A., Lampugnani, M. G., and Dejana, E. (1999). *Arterioscler. Thromb. Vasc. Biol.*, **19**, 2286.
15. Esser, S., Lampugnani, M. G., Corada, M., Dejana, E., and Risau, W. (1998). *J. Cell Sci.*, **111**, 1853.
16. Rabiet, M. J., Plantier, J. L., Rival, Y., Genoux, Y., Lampugnani, M. G., and Dejana, E. (1996). *Arterioscler. Thromb. Vasc. Biol.*, **16**, 488.
17. Marchisio, C. and Trusolino, L. (1999). In *Adhesion protein protocols* (ed. E. Dejana and M. Corada). p. 85. Humana Press, Totowa, NJ, USA.
18. Stewart, R. J., Kashxur, T. S., and Marsden, P. A. (1996). *J. Immunol.*, **156**, 1221.
19. Rival, Y., Del Maschio, A., Rabiet, M. J., Dejana, E., and Duperray, A. (1996). *J. Immunol.*, **157**, 1233.
20. Ferrero, E., Villa, A., Ferrero, M. E., Toninelli, E., Bender, J. R., Pardi, R., *et al.* (1996). *Cancer Res.*, **56**, 3211.
21. Ozaki, H., Ishii, K., Horiuchi, H., Arai, H., Kawamoto, T., Okawa, K., *et al.* (1999). *J. Immunol.*, **163**, 553.
22. Lampugnani, M. G., Resnati, M., Raiteri, M., Pigott, R., Pisacane, A., Houen, G., *et al.* (1992). *J. Cell Biol.*, **118**, 1511.
23. Wong, R. K., Baldwin, A. L., and Heimark, R. L. (1999). *Am. J. Physiol.*, **276**, H736.
24. Jiang, W. G., Bryce, R. P., Horrobin, D. F., and Mansel, R. E.(1998). *Biochem. Biophys. Res. Commun.*, **244**, 414.
25. Jiang, W. G., Martin, T. A., Matsumoto, K., Nakamura, T., and Mansel, R. E. (1999). *J. Cell. Physiol.*, **181**, 319.
26. Kevil, C. G., Payne, D. K., Mire, E., and Alexander, J. S. (1998). *J. Biol. Chem.*, **273**, 15099.
27. Antonetti, D. A., Barber, A. J., Hollinger, L. A., Wolpert, E. B., and Gardner, T. W. (1999). *J. Biol. Chem.*, **274**, 23463.
28. Blum, M. S., Toninelli, E., Anderson, J. M., Balda, M. S., Zhou, J., O'Donnell, L., *et al.* (1997). *Am. J. Physiol.*, **273**, H286.
29. Gardner, T. W., Lesher, T., Khin, S., Vu, C., Barber, A. J., and Brennan, W. A. (1996). *Biochem. J.*, **320**, 717.
30. Underwood, J. L., Murphy, C. G., Chen, J., Franse-Carman, L., Wood, I., Epstein, D. L., *et al.* (1999). *Am. J. Physiol.*, **277**, C330.

Chapter 3

Gap junction-mediated interactions between cells

David Becker

Department of Anatomy and Developmental Biology, University College London, Gower Street, London WC1E 6BT, UK

Colin Green

Department of Anatomy with Radiology, University of Auckland School of Medicine, Private Bag 92019, Auckland, New Zealand

1 Introduction

Gap junction channels are expressed in almost all types of cell. Their roles are diverse and are crucial for normal embryonic development and the functioning of adult organs. Point mutations, in a growing number of members of the gap junction family, appear to underlie several human diseases including profound non-syndromic deafness, heart disease, peripheral nerve myopathy, skin disorders, and cataract promotion. This has led to a rapid increase in the number of researchers investigating gap junctional communication in these disease conditions.

In order to study gap junctional interactions between cells, there are three basic approaches which, when integrated, provide a rounded picture of communication in any particular system.

(1) Establish the pattern and extent of communication between cells.

(2) Identify the connexins that mediate the coupling.

(3) Manipulate the communication in order to determine connexin roles.

There are many ways and levels of sophistication in which to carry out these investigations. In this chapter, we will cover a variety of techniques, both old and new, which currently are used for such studies. We would also like to refer readers to 'Methods in molecular biology: connexin channels' (1), which covers some of these topics in more detail.

2 Establishing the pattern and extent of communication between cells

There are two basic ways in which to monitor communication between cells: electrophysiologically, measuring the passage of ions; or by the transfer of low molecular weight marker molecules between cells. Two electrophysiological approaches were covered by Veenstra (2) and Dahl (3) in great detail in the first edition of this text and therefore do not need to be covered here. It is also possible, in some systems, to monitor the progress of waves of calcium between cells, though it must be noted that whilst in some cell types these waves propagate through gap junctions, in others the propagation involves extracellular routes.

2.1 Dye transfer

2.1.1 Types of dyes and technical considerations

Molecules up to about 1 kDa in size can pass through vertebrate gap junctions. However, not all communicating gap junction channels will pass such large molecules, and some, such as those between the photoreceptors of the retina, have only been shown to be electrically, and not tracer, coupled. The molecular weight of a tracer is only a guide to how a molecule will pass through gap junctions; its structure, charge, and hydration state will also come into play.

Low molecular weight fluorescent dyes such as Lucifer yellow (mol. wt 453 Da) or Alexa 488 (mol. wt 570 Da) are very popular for dye transfer studies as they can be visualized immediately on a fluorescence microscope and their passage between cells can be monitored directly. More recently, the biocytin family of molecules (mol. wt range 286–372 Da) have been used extensively in tracer studies because they will often go through gap junction channels that dyes such as Lucifer yellow will not pass through. The disadvantage of the biocytins is that they cannot be visualized immediately and the tissue has to be processed as for immunohistochemistry with avidin conjugated to either a fluorophore or horseradish peroxidase (HRP) in order to determine its distribution. However, biocytins are of particular use in studies of coupling in the nervous system where tracer coupling is often restricted to the smaller molecules.

When making intracellular dye injections into complex preparations such as the nervous system, it is possible to impale more than one cell with the electrode. It is therefore often advisable to have two dyes in the electrode, a fluorophore conjugated to a lysinated dextran (3–10 kDa, which will not pass through gap junctions) and a smaller tracer such as neurobiotin (that can pass through gap junctions). The neurobiotin can be visualized later using avidin conjugated to a fluorophore of a different colour. This approach has several advantages. First, it enables you to see immediately that the injection has been successful. Secondly, you can later identify the injected cell as well as those to which it is coupled. This enables identification of the artefactual appearance of coupling that can occur if two cells are injected at the same time. A more detailed account can be found in ref. 4.

2.1.2 Sharp electrodes

Most microelectrode glass now has an internal filament which increases the capillary attraction of the electrode, allowing the electrode tip to be filled by inserting the back of the electrode into the solution of dye (typically 2–5%). The fluid remaining along the filament is often sufficient to make an electrical connection with a wire inserted into the back of the electrode. Using a sharp microelectrode has the advantage that it is relatively easy to penetrate 100–150 μm into a tissue in order to inject a cell. Once inside the cell, its membrane potential, and 'well being', can be monitored electrically. Suitable electrophysiological amplifiers are relatively cheap and simple to use. If the contents of the electrode are charged, they can be introduced to the cell iontophoretically by applying a small positive or negative current to the electrode (max $\pm$ 100 mV to avoid damage); positive pulses for positive molecules. Continuous application of current can result in electrode blockage, so small pulses of reversed polarity are recommended. Once an electrode has started to block, it should be replaced as it rarely unblocks. The higher the resistance of the electrode, the less damage it will do to the cell but the harder it will be to eject the tracer. Uncharged or large molecules, such as antibodies or constructs, can often be ejected from the electrode by vibrating the end of the electrode with capacity compensation (zapping or buzzing). Alternatively, applying air pressure to the back of the electrode can force some of the contents of the electrode out of its tip and into the cell. Such devices should be calibrated carefully so that not too much pressure is applied or the cell will explode.

Maintaining the tissue or cells in optimal conditions on the microscope stage should not be overlooked. Many gap junctions are temperature sensitive and will close down if they become too cold. The media perfused over the preparation should be well oxygenated and buffered appropriately so that the pH does not drop (which will cause pH-sensitive junctions to close). These considerations are particularly important for delicate neuronal tissues, but cultured cells can often be handled successfully at room temperatures without the need for perfusing oxygenated media over them constantly.

Protocol 1

Injection of neurobiotin and Lucifer yellow into avian retinal ganglion cells using sharp electrodes

Equipment and reagents

- 2–4% solution of neurobiotin (Vector labs) with a 5% solution of Lucifer yellow potassium salt (Molecular probes) in 3 M potassium acetate (for the lithium salt of Lucifer, use 1 M LiCl)
- Avian Ringer (136.9 mM NaCl, 5.3 mM KCl, 3.0 mM $CaCl_2$, 0.41 mM $MgCl_2$, 3.0 mM N-(2-hydroxyethyl) piperazine-N'-(2-ethanesulphonic acid) (HEPES), 5.6 mM glucose, pH 7.5, bubbled with O_2)
- Glycerol-based antifade mountant

- Sharp electrodes, pulled from borosilicate glass (GC150TF, Clark Electromedical) with a resistance of 60–90 MΩ
- Fresh solution of 4% paraformaldehyde in 0.1 M phosphate-buffered saline (PBS), pH 7.4
- PBS for washes and PBS with 0.1% Triton X-100 (or Tween-20)
- 1:400 solution of streptavidin-Cy3 (or Cy5) (Amersham)

Protocol 1 continued

Method

1 Keep the retina flat-mounted, ganglion cell side uppermost, on the injection microscope stage, and superfuse continually with oxygenated avian Ringer.

2 Fill the tip of a microelectrode with a solution of 2–4% neurobiotin mixed with 5% Lucifer yellow by inserting the back of the electrode into the solution and allowing capillary attraction to draw up the fluid.

3 Identify ganglion cells using Nomarski interference contrast optics by virtue of their size and the possession of an axon.

4 Inject the cells with 1–2 nA of alternating positive and negative 5-s pulses for 10 min. Allow a period of 20–30 min for the neurobiotin to diffuse through any gap junctions between the recorded cell and those coupled to it.

5 Fix the retina in 4% paraformaldehyde ($\sim$12 h at 4°C).

6 Wash the retina three times in PBS (10 min per wash) and then permeabilize in 0.1% Triton X-100 in PBS (1 h at room temperature).

7 Incubate the retina in 1:400 streptavidin-Cy3 or -Cy5 overnight at 4°C.

8 Wash the retina in PBS (three changes, 30 min) and mount the retina in a glycerol-based mountant containing antifade agents.

9 Examine on a confocal laser scanning microscope (CLSM) using the 488 line to excite the Lucifer yellow and the 514 or 655 line to excite Cy3 or Cy5.

2.1.3 Patch electrodes as a means of introducing tracers

Using a patch electrode to introduce tracers into cells has advantages and disadvantages. One advantage is that in the whole-cell mode, the contents of the electrode rapidly move into the cell and through the cell's processes. More dye can therefore get into the cell more rapidly than with a sharp microelectrode, and dyes can be used at 10-fold lower concentrations. Combined with electrical recording, the cells 'well being' can be recorded and data from any cells that are dying can be discarded. The disadvantage of the technique is that it can only be used on exposed cells in tissue culture or organ slice preparations as cells deep within a tissue are inaccessible to the patch electrode. The electrophysiological equipment can be quite expensive and requires more expertise to use. A sample protocol for injecting retinal ganglion cells in avian retina is given below.

The approach in *Protocol 2* can be adapted easily to other preparations ranging from brain slices to tissue culture. Buffers, internal solutions, and temperatures should be altered for each particular application. Care should be taken to maintain the appropriate temperature and pH in the preparation as both of these factors can affect gap junction coupling. If desired, immunohistochemistry for connexins (*Protocol 6*) can be carried out after step 8.

Protocol 2

Patch injection of neurobiotin and fluorescein isothiocyanate (FITC)–dextran into avian retinal ganglion cells[a]

Equipment and reagents

- 0.5% neurobiotin (Vector labs) with 5% FITC–dextran (mol. wt 3000–10 000 Da; Molecular Probes) w/v in a solution of 103 mM Csgluconate, 0.1 mM $CaCl_2$, 2.0 mM $MgCl_2$, 40.0 mM HEPES, 5.0 mM N-methyl-D-glucamine-ethylene glycol-bis(2-aminoethyl) tetraacetic acid (EGTA), 5.6 mM glucose, pH 7.0

- Fresh solution of 4% paraformaldehyde in 0.1 M PBS, pH 7.4

- Avian Ringer (see *Protocol 1*)

- Patch pipettes pulled from borosilicate glass to have a resistance of between 10 and 20 MΩ

- PBS for washes and PBS with 0.1% Triton X-100 (or Tween-20)

- 1:400 solution of streptavidin-Cy3 (or Cy5) (Amersham)

- Glycerol-based antifade mountant

Method

1 Keep the retina flat-mounted, ganglion cell side uppermost, on the injection microscope stage, and superfuse continually with oxygenated avian Ringer.

2 Load the neurobiotin/FITC–dextran solution into the patch electrode.

3 Identify ganglion cells using Nomarski interference contrast optics by virtue of their size and the possession of an axon.

4 Hold the ganglion cells in whole-cell mode for approximately 2 min before pulling away the patch pipette. Done carefully, the membrane seals over and the dyes are left trapped within the cell. Allow a period of 20–30 min for the neurobiotin to diffuse through any gap junctions between the recorded cell and those coupled to it.

5 Fix the retina in 4% paraformaldehyde ($\sim$12 h at 4°C)

6 Wash the retina three times in PBS (10 min per wash) and then permeabilize in 0.1% Triton X-100 in PBS (1 h at room temperature).

7 Incubate the retina in 1:400 streptavidin-Cy3 or -Cy5 (overnight at 4°C).[b]

8 Wash the retina in PBS (3 × 30 min) and mount the retina in a glycerol-based mountant containing antifade agents.

9 Examine on a CLSM using the 488 line to excite the FITC and the 514 or 655 line to excite Cy3 or Cy5.

[a] Adapted from the methods in refs 5 and 6. Buffers and internal solutions should be altered for each specific application.

[b] Using FITC and Cy5 will give maximal signal separation, reducing signal overlap or cross-talk.

2.1.4 Scrape loading

This is a quick, if somewhat dirty means of monitoring the extent of gap junctional communication. It is particularly useful in tissue culture experiments when dealing with confluent cells and has the advantage that it is very cheap and easy to do and does not require expensive equipment or training in order to carry it out. The principle is that by scraping a sharp instrument such as a microelectrode or sharpened tungsten needle across the cultured cells in the presence of tracers, slightly damaged cells will take up the tracers through their disrupted membranes before they reseal and the cell recovers. After a fixed time period, say 10 min, the number of cells that the dye has transferred to, away from scrape, is determined. Addition of a high molecular weight marker such as 3 kDa FITC–dextran to the medium along with the low molecular weight tracer molecule will allow differentiation between cells which were damaged and took up both dyes, and those cells to which they were coupled, and only contain the low molecular weight tracer. Analysis is quicker if fluorescent tracers are used.

Protocol 3

Scrape loading

Equipment and reagents

- Standard culture medium and a loading culture medium containing a 5% solution (w/v) of the fluorophore to be transferred (Lucifer yellow, Cascade blue, fluorescein).[a] Optionally add a 5% high molecular weight fluorophore or dextran conjugate (>3 kDda), with different spectral properties

- Scraping tool such as a microelectrode, a sharpened tungsten wire, scalpel, etc.
- PBS
- Fresh 4% paraformaldehyde in PBS pH 7.4
- Glycerol-based antifade mountant (see *Protocol 1*)

Method

1   Change the medium on the cultured cells to that of the loading medium.

2   Make a clean scrape across the culture plate with a sharp fine instrument and leave the cells to load and transfer for 10 min.

3   Remove the fluorescent medium and rinse twice in PBS before fixing with 2–4% fresh paraformaldehyde for 20 min.

4   Wash in PBS (2×5 min), mount the preparation in a suitable glycerol-based antifade mountant, and examine on a fluorescence microscope or CLSM in order to determine the degree of tracer spread.[b,c]

[a] Neurobiotin may also be used but this will require an additional step of permeabilization and then incubation with a streptavidin-bound fluorophore before washing and mounting (see *Protocols 1* and *2*).

[b] Only cells that contain just the low molecular weight tracer should be counted as coupled. The degree of coupling can be gauged from the number of cells back from the scrape wound edge, that contain the tracer.

[c] The act of wounding the cells may cause sensitive gap junction channels to close rapidly and therefore not transfer tracers that they normally would. However, this technique has been shown to work well in many publications and should be considered for its speed, ease of use, and its lack of requirement for expensive equipment.

2.1.5 Cell settlement

Cell settlement represents a rapid and reliable method for monitoring cell–cell communication in tissue culture. The advantages are that it does not require expensive equipment and can give reasonably large N numbers quite quickly. The concept is to load one population of cells with two dyes, a larger, non-transferable one to identify the population of labelled cells, and a second smaller dye that will pass through gap junctions and reveal coupling. A variety of dyes and tracers can be used, for instance, the carbocyanine dye DiI, 4′,6-diamidino-2-phenylindole dihydrochloride (DAPI), or Hoescht can be used to label the cells, whilst amine modified (AM) forms of tracer dyes are readily taken into cells where they are cleaved into a smaller form that can pass through any gap junctions. The dye-loaded cells are then plated onto other cultured cells (or vice versa) at a ratio of about 1:100 and left for a period of 20–60 min to form gap junctions. Passage of dyes from the loaded to unloaded cells reflects the formation of functional gap junctions and the extent of coupling.

Protocol 4

Assessing coupling via cell settlement[a]

Equipment and reagents

- Culture medium containing 2 μM calcein-AM tracer and 10 μM DiI (Molecular Probes)
- 4% fresh paraformaldehyde in PBS
- Trypsin-ethylenediamine tetraacetic acid (EDTA) (1× liquid) (Gibco-BRL)
- Glycerol-based antifade mountant

Method

1. Incubate cells cultured on coverslips in medium containing tracers for 5 min at 37°C.[b]

2. Wash cultures three times for 5 min in normal medium.

3. Trypsin-EDTA treat cells for 5 min then wash cells three times for 5 min in normal medium and re-suspend the cells.

4. Take 50 ml of cell suspension containing about 100 cells, add to an untreated cell culture grown on a coverslip at a ratio of about 1:100 and allow to settle. Return to the incubator.

5. One hour later, quickly wash the preparation in PBS and then fix in 4% paraformaldehyde for 20 min.

6. Mount in glycerol-based mountant and analyse the extent of coupling using fluorescence microscopy or confocal microscopy. Loaded cells will appear red and green; any cells they are coupled to will be just green.

[a] Adapted from ref. 7.

[b] AM and hydrolysable dyes potentially reduce pH in their action, this has the potential to reduce communication but probably does not as in most cases dye transfer can be seen.

3 Gap junction density and connexin isoform expression

In order to determine which connexins are expressed where and when in tissues or cells, there are several alternative approaches available: polymerase chain reaction (PCR), *in situ* hybridization, Western blotting, and immunohistochemistry. PCR and *in situ* hybridization will indicate which mRNAs are being produced, whereas Western blotting and immunohistochemistry will demonstrate that the proteins are being produced. Western blotting techniques were covered in detail in the previous edition of this text and so will not be repeated here. Of these techniques, perhaps immunohistochemistry is the most informative as it has the ability to show the precise location of the protein within the cell.

3.1 Immunohistochemistry: connexin-specific antibodies

New members of the connexin family of proteins are being discovered rapidly and it is therefore not surprising that many of the first antibodies to be raised against gap junctions or specific connexins recognize more than one member of the family due to conservation of some protein sequences. Antibodies now tend to be raised to short peptide sequences rather than the whole protein, and these can be screened easily on the computer to check for possible cross-reactivity with other family members. However, cross-reactivity with an as yet unsequenced member of the family will always remain a possibility and must be borne in mind. Since the N-terminal, transmembrane domains and extracellular loops tend to have the highest degree of conservation, antibodies raised to the more variable regions of the cytoplasmic loop and C-terminal tail are the most connexin specific. Ultimately, the best test for antibody specificity is to screen them on knockout mice deficient for that particular connexin to check that they do not label anything artefactually.

Immunohistochemistry for light microscopy can be carried out on a number of preparations: cultured cells, frozen sections, wax sections, brain slices, and even small whole-mount tissue or embryos. The protocols are essentially variations on a theme making changes in fixation, permeab-lization, and permeation times to accommodate for the characteristics of the antibody and the preparation. The first step of fixing the tissue is often the most crucial as many antibodies fail to work when strong fixatives such as glutaraldehyde have been used. Light fixation with fresh paraformaldehyde or cooled methanol is therefore more common. Fixation guidelines for the antibody should be obtained and adhered to. Blocking of non-specific binding sites with either fetal sera or L-lysine is generally carried out in the same step as the tissue permeabilization, generally with 0.05% Triton X-100 or Tween-20. The length of time for permeabilization will vary with the thickness of the tissue and the fixation, as will the time for incubation in primary antibody, and can be in the range of hours to days. Excess primary antibody then needs to be washed away before incubation in an appropriate secondary antibody. Secondary antibodies target the species in which the primary antibody was raised. These are usually conjugated to a label which allows their visualization. Fluorescent labels are particularly useful, especially in combination with confocal microscopy. It is possible to carry out antibody labelling for multiple connexin proteins on the same tissue provided that the primary antibodies are raised in different species and care is taken to ensure that the secondary antibodies do not cross-react with one another.

Finally, it is advisable to use confocal microscopy to examine connexin staining as plaques can be so small that they are hard to detect by eye. In addition, surface dirt on sections can appear punctate and it is useful to be able to focus through thick preparations to ensure that staining is consistent throughout the tissue.

Protocol 5

Immunohistochemistry of frozen tissue sections[a]

Equipment and reagents

- Glass slides subbed with either poly-L-lysine, gelatin, or Tespa
- Blocking and permeabilizing solution of PBS with 0.1 M L-lysine (Sigma) and 0.05% Triton X-100 (Sigma)
- Wax (PAP) pen or Vaseline and a cotton bud
- PBS, primary antibodies, fluoresceinated secondary antibodies
- Glycerol-based antifade mountant

Method

1 Cut cryostat sections 8–15 μm thick and mount on gelatin-subbed slides. Sections can be from fresh or fixed tissue.[b]

2 Circle each frozen section using either a wax pen (PAP) or a cotton bud with Vaseline[c] and drip PBS onto the section to rehydrate. Leave for 5 min to equilibrate.

3 Place the slides on moistened tissue paper in a plastic box with a lid. Carefully replace the PBS with the blocking and permeabilizing solution and incubate for 45 min at room temperature.

4 Place the box on a gently rotating/shaking table and rinse sections in PBS for 5 min.

5 Add drops of primary antibody at the appropriate dilution and incubate at 37°C for 1 h.[d]

6 Wash away excess primary antibody with PBS, 3 × 15 min on the rotating table.

7 Incubate in appropriate secondary antibody, conjugated to a fluorophore, for 1 h in the dark (or longer for thicker specimens).

8 Wash away excess secondary antibody with PBS, 3 × 15 min on the rotating table, whilst minimizing exposure to light.[e]

9 Mount and coverslip slides using a glycerol-based mountant containing an antifade agent. Seal around the coverslip with nail varnish.

Protocol 5 continued

10 View immediately on a CLSM or standard fluorescence microscope. Fixed tissue will last some time if stored at 4 °C in the fridge, but unfixed tissue should be examined immediately.

[a] Adapted from refs 8 and 9.

[b] Fixed tissue can be cryo-protected, to some extent, by immersing the tissue in a 20% sucrose solution until it sinks.

[c] The wax or Vaseline ring should hold the liquid in a small drop over the section, 50–100 µl, preventing the sections from drying out and minimizing the use of reagents.

[d] Most common antibodies label well within 1–2 h at room temperature. Cleaner backgrounds, better penetration through the tissue, and greater economy can be achieved by labelling at 2–10× lower concentrations overnight at 4 °C.

[e] At this stage, 0.001% propidium iodide (Sigma) in PBS can be added to the first of the washes. This will stain the nuclei red and can be useful in determining the tissue structure providing a different colour: green (FITC, Alexa 488, Cy2), blue (Cascade blue), or far red (Cy5), is used for the secondary antibody.

Protocol 6

Immunohistochemistry of brain slices or whole-mount retina

Equipment and reagents

- Glass slides subbed with either poly-L-lysine, gelatin, or Tespa
- Blocking and permeabilizing solution of PBS with 0.1 M L-lysine (Sigma) and 0.05% Triton X-100 (Sigma)

- Wax (PAP) pen or Vaseline and a cotton bud
- PBS, primary antibodies, fluoresceinated secondary antibodies
- Glycerol-based antifade mountant

Methods

1 Fix neuronal tissue, such as whole-mount retina or 2–300 mm thick brain slices, free floating, for 30–180 min (depending on thickness) in either −20°C methanol or 4% fresh paraformaldehyde.[a]

2 Wash the tissue in PBS 2 × 10 min and then incubate the tissue, free floating in the blocking and permeabilizing solution overnight.

3 Wash the tissue in PBS, 3 × 30 min on a gently rotating/shaking table.

4 Incubate in primary antibody in the fridge for 24 h.[b]

5 Wash the tissue in PBS, 3 × 30 min on a gently rotating/shaking table

6 Incubate in secondary antibody in the fridge for 24 h.[c]

7 Wash the tissue in PBS, 3 × 30 min on a gently rotating/shaking table.[d]

Protocol 6 continued

8 Mount the tissue in a glycerol-based antifade mountant such as Citifluor. Seal with nail varnish and view on a confocal microscope.[e]

[a] Processing should be carried out on free-floating tissue in order to maximize access of reagents and minimize damage to the tissue through handling.

[b] Antibody should be at a 2–10× lower concentration than for normal section staining.

[c] A fluoresceinated secondary antibody works quite well for most purposes. However, if this is not sensitive enough, a biotin secondary antibody followed by streptavidin–fluorophore may help.

[d] At this stage, 0.001% propidum iodide (Sigma) in PBS can be added to the first of the washes (see *Protocol 5*).

[e] When mounting whole-mount tissue such as retina or a brain slice, it is often useful to create a spacer between the slide and coverslip so that the specimen does not get squashed. A thick layer of nail varnish in two bands along each side of the slide and left to dry overnight is an easy way of making a spacer.

Protocol 7

Zamboni's method for immunohistochemistry on wax sections[a]

Equipment and reagents

- Fixative (0.2% picric acid, 2% paraformaldehyde, 0.1 M phosphate buffer, pH 7.4)
- Series of alcohols: 50, 70, 90, and 100%, xylene, and chloroform

- Wax for embedding
- Trypsin solution (20 mM Tris, 0.1% trypsin (Sigma T-8728 type II crude Ex Porcine pancreas), 0.1% $CaCl_2$, pH 7.2 with HCl

Methods

1 Fix tissue for 2–4 h at room temperature.

2 Wash in tap water overnight.

3 Dehydrate tissue: 2 × 1 h in 70% ethanol, 2 × 1 h in 90% ethanol, 4 × 1 h in absolute ethanol, and chloroform overnight.

4 Wax embed (using standard protocol and times for the tissue type).

5 Section wax blocks and mount on slides.

6 De-wax sections: 2 × xylene, 5 min, then 100, 90, 70, and 50% ethanol, 5 min each, and water, 5 min.

7 Trypsin treat the tissue on the slides for 10 min.

8 Wash with tap water, 5 min.

9 Continue to process as for frozen sections in *Protocol 5*.

[a] Adapted from ref. 10.

Protocol 8

Methanol freeze substitution[a]

Equipment and reagents

- Liquid nitrogen, 2-methyl-butane (isopentane), and liquid nitrogen-chilled methanol
- Series of alcohols: 50, 70, 90, and 100%, xylene, and chloroform
- Wax for embedding

Methods

1 Freeze tissue rapidly by immersion in isopentane slush.

2 After about 20–30 s, transfer tissue to methanol at the same temperature and store at −80 °C for at least 3 days.[b]

3 Transfer tissue, in the same methanol, to −20 °C for at least 1 day, then to 4 °C for at least 1 day, and then to 20 °C for 1 day.

4 Transfer tissue to chloroform for 1 h, change the chloroform, and leave overnight.

5 Wax embed using standard protocol and times for the tissue type.

6 Section wax blocks and mount on slides.

7 De-wax sections: 2 × xylene, 5 min, then 100, 90, 70, and 50% ethanol, 5 min each, and water, 5 min.

8 Continue to process as for frozen sections in *Protocol 5*.

[a] This technique can help preserve delicate epitopes that a wax sectioning protocol often destroys: connexins are often very sensitive to wax embedding protocols.

[b] The volume of methanol should be large relative to the tissue.

3.2 Quantification of connexin plaque size and number

A fairly accurate guide to the extent of connexin protein expression, in terms of size and number of plaques, can quickly be gained by combining immuno-histochemistry with confocal microscopy. This technique is particularly useful for examining the effects of different treatments on connexin expression. As a caveat, this only provides a guide to the size and number of plaques. Whilst this technique can produce measurements very close to those from the electron microscope, over- or under-immunostaining or inappropriate settings on the confocal microscope when collecting data can make plaque sizes appear larger or smaller.

Protocol 9

Quantification of connexin plaque size and number

Equipment and reagents

- Confocal laser scanning microscope, computer, and NIH image software (http://www.nih.gov)

Method

1 Following standard fluorescence immunohistochemistry, specimens should be examined blind on the confocal microscope and the brightest group selected for setting up of data collection parameters under a high-power high-numerical aperture (NA) objective (e.g. ×63, 1.4 NA).

2 Laser power should be sufficient to excite the fluorophore without causing rapid fading. Set the pinhole aperture at the optimal size for the objective.

3 Optimize the settings of the offset and gain by adjusting to provide close to maximal grey scale range so that for an 8-bit image (256 levels of grey) there would be some pixel values close to 0 and some close to, but not quite at, 255 so that there is no burnout in the image. If the image needs to be averaged, the number of scans should be determined and kept constant. Once determined, all of these factors should not be adjusted.

4 Capture a series of images from at least five different areas for each specimen.[a] These should ideally be identical in nature, so if a uniform tissue culture is being examined then areas can be selected at random. Non-uniform specimens should have areas selected for quantification that are similar.

5 Store images digitally in 'tiff' format and analyse using NIH image. Each section should be thresholded at a constant level and a binary (black and white) image created. NIH can then be used to count the number of plaques and their size. These data can then be exported to Excel or Mini Tab for graphical presentation and statistical analysis.[b]

[a] More may be required for statistical analysis if variation is great.

[b] In some cases, projections of small z-series can be analysed. However, as the number of sections increases, this will introduce errors as separate plaques become fused in the reconstructions, resulting in the appearance of fewer, but larger plaques than there actually are (for further details, see ref. 11).

3.3 Expression analysis of mRNA by quantitative RT–PCR

Quantitative reverse transcriptase–PCR (RT–PCR) can be divided into two steps. The first step involves the production of complementary DNA (cDNA) from total RNA extracted from the tissue of interest using avian myeloblastosis virus (AMV) reverse transcriptase. Step two involves the amplification of two specific mRNA sequences, the first corresponding to a sequence of the transcript of interest and the second representing an endogenous transcript that is uniformly expressed in most tissue types. The latter acts as an internal standard when simultaneously amplified in the same tube, and the level of production can be used to equalize for variability in individual reactions. Routinely used internal standards include glyceraldehyde-3-phosphate dehydrogenase (GAPDH) (12) or β-actin (13). The size of both products should be fairly close, between 200 and 400 bp, and both products must be produced during exponential amplification for true quantitative deductions to be made.

Protocol 10

Quantitative RT–PCR for connexins[a]

Equipment and reagents

- Thermocycler
- Spectrophotometer
- Phosphoimager
- TRIzol reagent (Gibco-BRL)
- Random hexamers (Amersham Pharmacia Biotech)
- RNase inhibitor (Promega)
- Deoxynucleotides (dNTPs) (Promega)
- AMV reverse transcriptase (RT) (Promega)

- [α-^{32}P]dCTP (Amersham Pharmacia Biotech)
- *Thermus aquaticus* (*Taq*) DNA polymerase (Promega), primer sets (Genosys)
- RT solution (10 mM Tris–HCl, pH 8.3, 50 mM KCl, 5 mM MgCl$_2$, 2.5 mM random hexamers, 1 mM of each dNTP, 50 U of RNase inhibitor, 12.5 U of AMV RT).[b]
- PCR buffer (10 mM Tris–HCl, pH 8.3, 50 mM KCl, 2.5 mM MgCl$_2$)

Method

A. Extraction of total RNA

1 Use the TRIzol reagent (based on guanidium thiocynate) method appropriate to the kit to extract total RNA.

2 Determine the RNA concentration spectrophotometrically by UV absorbance at 260 nm.

B. First-strand cDNA synthesis

1 Adjust volume so that 15 µl of RNase-free water contains 1.25 µg of total RNA.

2 Denature RNA by heating to 65 °C for 10 min and then cool on ice.

3 Add 10 mM Tris–HCl (pH 8.3), 50 mM KCl, 5 mM MgCl$_2$, 2.5 mM random hexamers, 1 mM of each dNTP, 50 U of RNase inhibitor, 12.5 U of AMV RT.[b]

4 Incubate tubes at 30 °C for 10 min and 42 °C for 30 min, then terminate reactions by heating to 94 °C for 10 min.

5 Store cDNA at −20 °C or proceed immediately to PCRs.

C. Amplification of target mRNA and internal control mRNA

1 Set up a PCR containing 10 mM Tris–HCl (pH 8.3), 50 mM KCl, 2.5 mM MgCl$_2$, 2 µl of cDNA, 0.2 µM of each primer set (one for amplification of the test mRNA and the other for the internal control), 0.2 mM dNTPs, 5 µCi of [α-^{32}P]dCTP, 0.25 U of *Taq* DNA polymerase.

2 Denature cDNA at 94 °C for 1 min and then carry out PCR amplification cycles with a denaturing step at 94 °C for 30 s, annealing at 56 °C for 30 s, and extension at 72 °C for 90 s.[c]

3   Run the PCR products on a 6% polyacrylamide gel, dry, and expose to photographic
     film. Quantify the amount of radioactivity incorporated using a phosphoimager.

4   Normalize the incorporation of radioactivity in the product by the incorporation of
     radioactivity in the internal control product.

[a] Adapted from refs 12–14.

[b] As a control, the same reaction can be set up omitting RNA.

[c] Initial tests should be carried out to determine the optimal number of cycles resulting in
exponential amplification of both products.

3.4 Whole-mount *in situ* hybridization for connexins using digoxigenin

In situ hybridization is based on the complementary pairing of labelled DNA or
RNA probes with nucleic acid sequences in intact chromosomes, cells, tissue
sections, or even whole embryos (15). Probes may be radiolabelled or, more commonly, they may incorporate non-isotopic markers such as biotin or digoxigenin
(DIG). Whole-mount techniques have been used widely to localize transcripts in
whole chick or mouse embryos and map their developmental regulation (16–18).
This technique is excellent for the identification of connexin mRNA molecules.
The variable cytoplasmic loop or tail regions are the best targets for probe
sequences.

Protocol 11

Whole-mount *in situ* hybridization for connexins in embryos using digoxigenin[a]

Equipment and reagents

- Spectrophotometer
- Dithiothreitol (DTT; 100 mM; Promega)
- RNase inhibitor (40 U μl^{-1}; Promega)
- DIG RNA labelling mix ($\times$10; Boehringer Mannheim)
- RNase-free DNase (1 U μl^{-1}; Promega)
- T3, T7, and SP6 RNA (5 U μl^{-1}) polymerase enzymes are provided with ($\times$5) transcription buffer (Promega)
- 4 M LiCl
- PBS tablets (Oxoid)
- Paraformaldehyde (TAAB)

- Glutaraldehyde, 25% solution (Agar Scientific)
- Proteinase K (1 mg ml^{-1} stock solution in RNase-free H_2O; Promega)
- Formamide (Fluka)
- Blocking powder (Boehringer Mannheim)
- CHAPS (3-[(3-cholamidopropyl)-dimethylammonio]-1-propanesulphonate; Sigma)
- Tween-20 (polyoxyethylenesorbitan monolaurate; Sigma)
- Heparin, sodium salt (Sigma)

Protocol 11 continued

- Yeast tRNA (Sigma)
- Heat-inactivated sheep serum (Gibco-BRL)
- Pre-hybridization mix (for 1 l: 500 ml of formamide, 20 g of blocking powder, 5 g of CHAPS, 2 ml of Tween-20, 250 ml of 20× standard saline citrate (SSC), 10 ml of 0.5 M EDTA, 50 mg of heparin, 1 mg/ml yeast tRNA, 240 ml of RNase-free H_2O)
- Anti-digoxigenin-AP-Fab fragments (0.75 U μl^{-1}; Boehringer Mannheim)
- NBT (4-nitro blue tetrazolium chloride; 100 mg/ml; Boehringer Mannheim)
- BCIP (5-bromo-4-chloro-3-indoyl-phosphate; 100 mg/ml; Boehringer Mannheim)
- Formalin (Sigma)
- Wash buffer (100 mM Tris–HCl, 50 mM $MgCl_2$, 100 mM NaCl, 1% Tween-20)

A. Probe synthesis[b]

1 Mix the following reagents in a microfuge tube at room temperature: 4 μl (×5) of transcription buffer, 2 μl of 100 mM DTT, 1 μl of RNase inhibitor, 2 μl of DIG-RNA labelling mix, 1 μg of linearized DNA template, and RNase-free H_2O to a final volume of 19 μl.

2 Add 10–20 U of RNA polymerase (T3, T7, or SP6) and incubate reactions at 37 °C for 60 min.

3 Run 1 μl of the reaction on a 1% agarose gel. The RNA transcript band should be at least 10 times more intense than the template DNA band.

4 Digest template DNA with 1–5 U of RNase-free DNase at 37 °C for 15 min.

5 Extract with phenol/chloroform and then precipitate the RNA probe with 1/10 vol. of 4 M LiCl and 2.5 vols of ethanol at −20 °C for at least 30 min.

6 Collect the probe by centrifugation at 4 °C for 10 min, wash the pellet with 70% ethanol, and allow to air dry before resuspending in 20–30 μl of RNase-free H_2O.

7 Determine the RNA concentration spectrophotometrically by UV absorbance at 260 nm. Probes can be stored at −70 °C for several months.

B. Pre-treatment of embryos

1 Rinse embryos in PBS. Fix in 4% paraformaldehyde for 2 h to overnight at 4 °C.

2 Wash twice for 10 min in PBS containing 0.1% Tween-20 (PBT).

3 Dehydrate through methanol series, 25, 50, and 75% in PBT, and finally 100%, and store overnight at −20 °C.[c]

4 Before use, rehydrate through a reverse methanol series, 75, 50, and 25%, and wash twice with PBT for 10 min.

5 Add 5–10 μg of proteinase K in PBT and incubate at room temperature for 5–20 min depending on the size of the embryo. Care must be taken at this stage due to the mechanical fragility of the specimen.

6 Wash twice for 5 min in PBT and fix for 20 min at room temperature in 4% para-formaldehyde in PBT with 0.25% glutaraldehyde.

7 Wash twice for 5 min in PBT.

C. Hybridization

1 Transfer embryos into 2 ml microfuge tubes and add 1 ml of pre-hybridization mix, incubate for 3 h to overnight at 55–70 °C, depending on the probe.

2 Prepare hybridization mix consisting of pre-hybridization mix plus DIG-labelled probe added at 0.1–1 μg ml^{-1}. Remove the pre-hybridization mix from the samples and replace with 1 ml of warmed hybridization mix. Incubate at the temperature determined by the probe overnight.

D. Post-hybridization

1 Wash embryos in a series of pre-warmed SSC/CHAPS solutions for 20 min each, twice in 2× SSC, twice in 0.1% CHAPS in 2× SSC, and twice in 0.1% CHAPS in 0.2% SSC, all at 65 °C.

2 Wash twice in PBT for 5 min then block by gently rocking in 15% heat-inactivated sheep serum in PBT for 2–3 h at room temperature.

3 Rock overnight with a 1:1000 dilution of anti-DIG antibody conjugated to AP in fresh blocking solution at 4 °C.

4 Wash over the day in PBT at room temperature changing every hour. Wash overnight in fresh PBT at 4 °C.

5 Wash three times for 10 min each in wash buffer then transfer embryos to solid watch glass bottles and incubate in the dark with gentle rocking in 1 ml of fresh wash buffer containing 3.5 μl of NBT and 3.5 μl of BCIP. Monitor rate of colour development.

6 Wash by rocking in wash buffer for 30 min with several changes. Refix in 10% formalin in PBS for 1 h.

[a] This method is adapted from ref. 19.

[b] All solutions up to the post-hybridization stage should be RNase free.

[c] Embryos can remain in methanol for several weeks.

4 Manipulating gap junctional communication

Manipulation or perturbation of gap junction communication can be used to investigate the role of this junction type in particular systems. Perturbation may be brought about in several ways. Channels can be opened or closed with the use of a variety of pharmacological agents. They can be blocked with connexin-specific antibodies and peptides. Alternatively, the connexin gene can be completely knocked out in transgenic animals or spatially and temporally knocked down in its expression using antisense. Viral strategies also provide a means of up- or down-regulating connexin proteins in infected cells by the insertion of normal or dominant-negative constructs. Knockout animals and viral transfection approaches are beyond the scope of this chapter.

4.1 Pharmacological agents

There are several pharmacological agents which will block gap junction communication (*Table 1*). For most of these, the mode of action of the agents is uncertain but it is likely to be through disruption of membrane organization. Because of this, they are not specific in their action and the possibilities that the drugs have effects other than those on the gap junctions must be borne in mind. For instance, halothane blocks not only gap junctions but also calcium-induced calcium release, so its effects on the spread of calcium waves through gap junction channels may be ambiguous.

Table 1 Pharmacological agents for blocking gap junctions

Agent	Concentration range	References
18 α-Glycyrrhetinic acid	10–100 μM	(34–36)
18 β-Glycyrrhetinic acid	10–100 μM	(37, 38)
Carbenoxalone	75–150 μM	(39, 40)
Anandamide	5–50 μM	(35)
Oleamide	50–100 μM	(41)
Volatile anaesthetics including halothane, octanol, and heptanol	0.2–2 mM	(42, 43)

This short list reflects some common pharmacological agents used for blocking gap junctional communication.

- Most of these blocking agents are effective within 10–20 min but reports on the completeness and longevity of block varies with the preparation, as does the rate of recovery after wash off.
- Some of the agents appear to be harmful to cells in long-term applications (24 h or more).
- Anandamide and oleamide show strong blocking effects on some cell types but others are less effective; recovery of communication is rapid after their wash off.
- Problems have been reported with some agents precipitating out in oxygenated/bubbled medium.
- Common controls including glycyrrhizic acid and oleanolic acid are non-blocking analogues of 18 α- and β-glycyrrhetinic acid.

4.2 Antibody blockers

There have been several reports in which antibodies to various epitopes of connexins have been utilized in order to block or reduce intercellular communication (9, 20). The drawback to this approach is in the antibodies gaining access to their epitopes, inside or outside the cells. Antibodies with intracellular epitopes can either be injected into individual cells or bulk loaded into cells or tissues using, for example, osmotic shock, dimethylsulphoxide (DMSO), or electroporation. This approach has been very successful in simple systems and in early embryonic development.

Antibodies to the extracellular loops of connexins still have problems of access to their epitopes as communicating cells are often in very close contact. This approach is therefore much more effective in preventing communication being established than in reducing it in closely packed communicating tissues.

A similar approach to that of extracellular loop antibodies is to introduce short peptide motifs which correspond to the adhesive regions of the extracellular loops (21). These short peptides appear to penetrate even intact tissues and, presumably through interacting with the extracellular loops, interfere with connexon–connexon interactions and reduce intercellular communication.

Antibodies to cytoplasmic domains can be introduced into cells using sharp electrodes or patch electrodes by adding the antibodies to the dyes as in *Protocols 1* and *2*. If using patch electrodes, the antibody will diffuse into the cell rapidly once whole-cell mode is reached. Sharp electrodes can eject antibodies into a cell either by pressure injection or by applying capacity compensation (zapping or buzzing) to the electrode. For more details, see refs 4, 9, and 20.

It is important to include controls with antibody block experiments. These might include the following.

Injection of non-related connexin antibodies which should not block dye transfer or affect cell division.

Titrating out the blocking response with different concentrations of antibody.

Adding the parent peptide back to the antibody which should compete out its binding sites and prevent its blocking

Checking that surrounding cells are communicating well with each other but not with the injected cell.

Protocol 12

Bulk loading of blocking antibodies using DMSO[a]

Equipment and reagents

- 5% DMSO, in culture medium at 4°C
- Normal medium at 4°C
- Blocking antibody in an aliquot that will give 0.4 mg ml^{-1} in volume of medium.

Method

1 Rinse the embryo or tissue in 5% DMSO in culture medium at 4°C.

2 Add the appropriate volume of blocking antibody and incubate at 4°C for 30 min.[b]

3 Transfer to normal culture medium at 4°C for 15 min and then allow the tissue to return to room temperature.

[a] Adapted from ref. 22.

[b] Specimens should be processed on ice in small multiwell dishes so that the volume used is about 10 μl.

4.3 Extracellular loop peptides

Addition to the medium of peptide sequences corresponding to the extracellular loops of connexins have been shown to interfere with the formation of gap junction channels and also gap junctional communication in intact tissues such as blood vessels. However, the precise mode of action of these peptides is not clear (see ref. 21 for more details).

4.4 Manipulation of gap junction gene expression

There are several ways in which to alter gene expression through transfection by a variety of means, transgenic knockouts or antisense knockdown. Retroviral strategies (23) and electroporation (24) have been used successfully to insert and express connexins in chick embryo tissue. Transgenic knockouts are now commonly used for gap junction analysis (25), and antisense approaches have been developed (26). There are advantages and drawbacks for each of these approaches, as outlined in *Table 2* for knockout and knockdown techniques. It would be inappropriate to attempt to describe all of these approaches in the limited space of this chapter, and only the latter is covered here.

4.4.1 Antisense strategies

Antisense deoxyoligonucleotides (ODNs) are particularly useful for the manipulation of specific connexin gene expression, especially in embryonic development. Blocking translation, rather than transcription, appears to eliminate compensation by other homologous connexin proteins. There are still some difficulties with the approach such as the short half-life of ODNs and how to deliver the ODNs consistently and reliably to target tissues. Unmodified phosphodiester oligomers (POs) typically have an intracellular half-life of only 20 min due to intracellular nuclease degradation. Phosphorothioate and methylphosphonate

Table 2 Comparison of the advantages and disadvantages of antisense knockdown and gene knockout approaches

Antisense knockdown	Gene knockout
Advantages of antisense knockdown compared with gene knockout	
Easy to use	Technically difficult
Applicable to multiple species	Applicable to limited species
Extent of knockdown controllable available.	Complete knockout so no internal controls
Inexpensive	Expensive to set up and maintain
Temporal control	Early lethalities may obscure later roles
Spatial control	Compensation by related genes likely
Compensation by related genes unlikely	
Natural	
Potentially clinically relevant	
Disadvantages of antisense knockdown compared with gene knockout.	
Rapid breakdown	Permanent
May be non-specific	Very specific
Several modes of action	Single mode of action
Transient action may be of little use	Total knockout of target protein
if protein turnover is slow	Stable and comparable population can be obtained
Incomplete knockdown may be difficult to interpret	
Variation between treatments	
Access or delivery may be difficult	

ODNs are often used, being more nuclease resistant, although they have a weaker affinity than POs, ODNs can be less efficient at entering the cells, and can cause non-specific inhibition by binding to essential proteins (27). Because the probes are rapidly broken down within the cells, they need to be replenished constantly. This is linked to a second problem, that the probes need to be delivered accurately and consistently to the target tissues over a period of time if the treatment is going to be effective. One novel method is to use Pluronic gel to deliver antisense ODNs to the tissue (26, 28, 29). Pluronic F-127 gel (BASF Corp.) has the advantage that it is liquid between 0 and 4 °C but sets at higher temperatures and is a mild surfactant aiding ODN entry into the cell. This provides a delivery system which is applicable to a number of different systems including the developing chick embryo (*in ovo*), tissue culture, and organ culture, or topical application to adult organs or tissues *in vivo*. This approach overcomes the effect of intracellular nuclease activity by providing, for a period, a constant source of unmodified ODNs from the gel at relatively low concentrations. This decreases artefacts which can occur at higher dose levels including non-specific, imperfectly matched, mRNA binding (30, 31) and accumulation of nucleotide and nucleoside breakdown products which can affect cell proliferation and differentiation (27). In addition, the gel can be placed accurately onto the target tissue, thereby providing precise spatio-temporal control of gene expression. The Pluronic gel–antisense oligonucleotide approach may have a wide variety of applications provided that sufficient attention is paid to the antisense ODN design and appropriate controls are carried out (detailed in ref. 27). When targeted towards tissue or organ culture preparations, a serum-free medium such as Opti-MEM (Gibco) should be used to reduce the rate of ODN breakdown (detailed in ref. 28).

Protocol 13

Pluronic gel and antisense application to chick embryos *in ovo*

Equipment and reagents

- 25–30% solution of Pluronic F-127 gel (BASF Corp.) in PBS (molecular grade water)[a]
- Fertilized white leghorn eggs, incubated at 38 °C in a humid environment and staged according to Hamburger and Hamilton stages (32)
- Sterile forceps

- Pluronic gel containing ODNs, antisense, sense, or random controls, thoroughly vortexed at the desired concentration (0.5–1.0 μM is most effective)[b], aliquoted into 1.5 ml molecular grade microfuge tubes and stored at −80 °C until required[c]

Method

1   Clean eggs with 70% alcohol and then make a small hole with the forceps into the blunt end of the egg where the air sac is found. Insert a hypodermic needle vertically through the hole to the bottom of the egg, withdrawing 1 ml of albumin.

Protocol 13 continued

2 Carefully window the egg (33) and open the vitelline and amniotic membranes over the area to be treated using fine forceps and taking care not to damage the embryo or associated blood vessels.[d]

3 Apply 5-10 µl of chilled Pluronic gel to the target site using a cooled 0.5-10 µl pipette tip.[e]

4 After Pluronic gel application, seal the eggs with Sellotape and replace in the incubator until ready for analysis. In most cases, embryos are analysed after 48 h.

[a] The concentration of gel determines its softness and persistence. Higher concentrations of gel set more quickly and remain in place longer before dissolving. Different concentrations should be tested for each system, but for most purposes a 25-30% solution is best. The gel is best made up on a shaker in a cold room and can take 24–48 h to dissolve because it is only liquid between 0 and 4°C.

[b] At concentrations above 10 µM, non-specific and toxic effects are seen which vary with the ODN sequence . Screening of ODN efficacy is essential and should be carried out at concentrations between 0.05 and 50 µM.

[c] Whilst Pluronic gel is a mild surfactant, for some ODNs addition of 1% DMSO to the gel increases the efficacy of the antisense, presumably by elevating entry to the tissue.

[d] 30% India ink/PBS solution can be injected under the amniotic sac to improve contrast of the translucent embryo.

[e] The gel sets rapidly at temperatures above 4°C so it is essential to work swiftly. Drops of gel, between 5 and 10 µl, can be placed with high accuracy on parts of an embryo and will mould around the site of application and set in place. The volume of the drops can be varied as required for the experiment and applications can be repeated at later time points if required. A fresh tip should be used for each application to reduce contamination of the ODNs.

4.4.1.1 Tips and controls

Several control experiments should always be carried out in parallel to the knock-down experiment.

(1) The use of Pluronic gel with the matching sense, with random ODNs of a similar base composition, or Pluronic gel alone.

(2) In some cases, matching sense sequences can also block protein formation by forming stable DNA triplets. Some researchers recommend the reverse sequence as a control.

(3) Application and analysis of the experiment should be carried out blind.

(4) Use immunohistochemistry to determine where and when your target protein is expressed, if possible quantitate the levels at which the protein is normally expressed. Determine both the specificity and time course of knockdown and recovery of protein expression directly using immunohistochemistry and confocal microscopy. This is rapid and has distinct advantages over Western blotting as it provides information on the spatial distribution of the protein

knockdown in relation to the target site and it can be quantitated rapidly from the confocal microscope image (see Section 3.2, *Protocol 9*).

(5) Check that the expression of other related proteins is not affected by the antisense or is undergoing compensatory up-regulation for the knocked down protein.

(6) Northern blots and RNase protection assays do not provide a direct measure of antisense effects and are not recommended.

(7) FITC-tagged ODNs can be used to demonstrate the entry of the ODNs into the target tissue, the time course of entry, and the depth of penetration of the tagged probes.

(8) Addition of ink to the gel allows one to determine how long the gel remains in place. Gel placed on a tissue undergoing a lot of movement will dissipate more quickly.

(9) Antisense sequences between 18 and 30 bases long work well with connexins. Sequences should be analysed for their specificity, formation of stem–loop structures or stable secondary structures, and homodimers.

In this chapter, we have outlined protocols for the three key steps in studying gap junctions: establishing the pattern and extent of communication; identifying the connexins involved; and manipulating their function to determine their roles. Gap junction communication is vital in a number of areas, and the approaches outlined will, we hope, help many researchers achieve their goals.

References

1. Bruzzone, R. and Giaume C. (ed.) (2000). *Methods in molecular biology: connexin methods and protocols*. Humana Press Inc., NJ, USA.

2. Veenstra, R. D. and Brink, P. R. (1992). In *Cell–cell interactions: a practical approach* (ed. B. R. Stevenson, W. J. Gallin, and D. L. Paul). p. 167. IRL Press, Oxford.

3. Dahl, D. (1992). In *Cell–cell interactions: a practical approach* (ed. B. R. Stevenson, W. J. Gallin, and D. L. Paul). p. 143. IRL Press, Oxford.

4. Mobbs, P., Becker, D. L., Williamson, R., Bate, M., and Warner, A. (1994). In *Microelectrode Techniques. The Plymouth Workshop Handbook* (2nd edn), (ed. D. Ogden). p. 361. Company of Biologists, Cambridge, UK.

5. Becker, D. L., Bonnes, V, Catsicas, M., and Mobbs, P. (1996). *J. Physiol (Lond)*, **494**, 11P.

6. Becker, D. L., Bonnes, V., and Mobbs, P. (1998). *Cell Biol. Int.*, **22**, 781.

7. Goldberg, G. S., Bechberger, J. F., and Naus C. C. G. (1995). *Biotechnology*, **18**, 490.

8. Green, C. R., Bowles, L., Crawley, A., and Tickle, C. (1994). *Dev. Biol.*, **161**, 12.

9. Becker, D. L., Evans, W. H., Green C. R., and Warner, A. E. (1995). *J. Cell Sci.*, **108**, 1455.

10. Toshimori, H., Toshimori, K., Oura, C., and Matsuo, H. (1987). *Histochemistry*, **86**, 595.

11. Green, C. R. G., Peters, N. S., Gourdie, R. G. G., Rothery, S., and Severs, N. J. (1993). *J. Histochem. Cytochem.*, **41**, 1339.

12. Dukas, K., Sarfati, P., Vaysse, L., and Pradayrol, L. (1993). *Anal. Biochem.*, **215**, 66.

13. Kinoshita, T., Imamura, J., Nagai, H., and Shimotohno, K. (1992). *Anal. Biochem.*, **206**, 231.

14. Tokuyama, W., Hashimoto, T., Li, Y. X., Okuno, H., and Miyashita, Y. (1999). *Brain Res. Protocols*, **4**, 407.

15. Herrington, C. S. (1998). *Mol. Pathol.*, **51**, 8.

16. Bell, S. M., Schreiner, C. M., and Scott, J. S. (1998). *Mech. Dev.*, **74**, 41.

17. Christen, B. and Slack, J. M. W. (1997). *Dev. Biol.*, **192**, 455.

18. Yokouchi, Y., Ohsugi, K., Sasaki, H., and Kuroiwa, A. (1991). *Development*, **113**, 431.

19. Wilkinson, D. G. and Nieto, M. A. (1993). In *Methods in enzymology* (ed. P. M. Wassaman and M. L. DePamphilis). Vol. 225, p. 361. Academic Press, London.

20. Warner, A. E., Guthrie, S., and Gilula, N. B. (1984). *Nature*, **311**, 127.

21. Warner, A. E., Clements, D. K., Parikh, S., Evans, W. H., and DeHaan, R. L. (1995). *J. Physiol. (Lond.)*, **488**, 721.

22. Fraser, S. E., Green, C. R., Bode, H. R., and Gilula, N. B. (1987). *Science*, **237**, 49.

23. Jiang, J. X. and Goodenough, D. A. (1998). *Invest. Opthalmol. Vis. Sci.*, **39**, 537.

24. Momose, T., Tonegawa, A., Takeuchi, J., Ogawa, H., Umesono, K., and Yasuda, K. (1999). *Dev. Growth Differ.*, **41**, 335.

25. Willecke, K. Kirchhoff, S., Plum, A., Temme, A., Thonnissen, E., and Ott, T. (1999). *Novartis Found. Symp.*, **219**, 76.

26. Becker, D. L., McGonnel, I., Makarenkova, H., Patel, K., Tickle, C., Lorimer, J., *et al.* (1999). *Dev. Genet.*, **24**, 33.

27. Wagner, R. W. (1994) *Nature*, **372**, 333.

28. Becker, D. L., Lin, J. S., and Green C.R. (1999). In *Antisense techniques in the CNS: a practical approach.* (ed. R. Lindsey). p. 149. Oxford University Press, Oxford.

29. Green C. R, Law, L. Y., Lin, J. S., and Becker, D. L. (2000). In *Methods in molecular biology: connexin methods and protocols* (ed. R. Bruzzone and C. Giaume), Humana Press Inc., NJ, USA.

30. Boiziau, C., Moreau, S., and Toulme, J. J. (1994). *FEBS Lett.*, **340**, 236.

31. Woolf, T, M., Melton, D. A., and Jennings, C. G. B. (1992). *Proc. Natl Acad. Sci. USA*, **89**, 7305.

32. Hamburger, V. and Hamilton, H. (1951). *J. Morphol.*, **88**, 49.

33. Tickle, C. (1993). In *Essential developmental biology* (ed. C. Stern). p. 119. IRL Press, Oxford.

34. Woolsin, J. M. (1991). *Am. J. Phys.*, **261**, C857.

35. Venance, L., Piomelli, D., Glowinski, J., and Giaume, C. (1995). *Nature*, **376**, 590.

36. de Curtis, M., Manfridi, A., and Biella, G. (1998). *J. Neurosci.*, **18**, 7543.

37. Yamamoto, Y., Fukuta, H., Nakahira, Y., and Suzuki, H. (1998). *J. Physiol. (Lond.)*, **511**, 501.

38. Bani-Yaghoub, M., Bechberger, J. F., Underhill, T. M., and Naus, C. (1999). *Exp. Neurol.*, **156**, 16.

39. Davidson, J. S. and Baumgarton, I. M. (1988). *J. Pharmacol. Exp. Ther.*, **246**, 1104.

40. Travagali, A., Dunwiddie, T. V., and Williams, J. T. (1995). *J. Neurophysiol.*, **74**, 519.

41. Guan, X., Cravatt, B. F., Ehring, G. R., Hall, J. E., Boger, D. L., Lerner, R. A., *et al.* (1997). *J. Cell Biol.* **18**, 7543.

42. Burt, J. and Spray, D. (1989). *Circ. Res.* **65**, 829.

43. Mantz J., Cordier, J., and Giaume, C. (1993). *Anaesthesiology*, **78**, 892–901.

Chapter 4

Functional analysis of the tight junction

Marcelino Cereijido, Liora Shoshani and Ruben G. Contreras

Centro de Investigaciones y Estudios Avanzados, Av. Instituto Politécnico Nacional 2508, México, DF 07300, Mexico

1 Introduction

The very names used to describe the tight junction (TJ) since it was discovered a century and a half ago reveal that, actually, it was not expected to function: 'Schlussleisten', 'terminal bars', 'bandelettes de fermeture', 'hoops', 'occluding junctions', 'tight junctions', 'gaskets', 'attachment belts', etc. (1). Our understanding of TJs has drastically changed subsequently due to a series of, now, well-known facts (2).

(1) Transepithelial electrical resistance (TER) ranges from 10 (proximal kidney tubule) to more than $10\,000\ \Omega\ cm^{-2}$ (urinary bladder), indicating that in the so-called 'leaky epithelia', such as the mucosa of the small intestine and the gall bladder, or the proximal tubule of the kidney, most permeating molecules do not cross through the cytoplasm of epithelial cells but circumvent them, and their flux is therefore controlled by the TJ (3, 4).

(2) While the plasma membrane has pores with a radius of around 0.4 nm, some epithelia have pore radii of some 3–4 nm used by water and small solutes to cross the epithelium, supporting the idea that transepithelial fluxes proceed through paracellular routes.

(3) A series of protein species (zonula occludens 1 (ZO-1), ZO-2, ZO-3, cingulin, occludin, claudin, etc.) have been found to integrate at or be closely and specifically associated with the TJ, forming a molecular network stretching from the lips of the TJ to the cytoskeleton (5, 6).

(4) Epithelial cells incubated without Ca^{2+} do not establish TJs but, when this ion is added, it acts on the extracellular segment of E-cadherin, stimulates phospholipase C, protein kinase C, and calmodulin, and, through this cascade, triggers the synthesis and assembly of TJs (7, 8), revealing that the TJ integrates a complex cellular machinery.

(5) In keeping with this view, some TJ-associated molecules change their degree of phosphorylation (9), and the permeability of the TJ varies in response to physiological conditions and pharmacological challenge (10–12).

(6) Some TJ proteins contain nuclear addressing signals and have been observed to switch from the nucleus to the vicinity of the TJ, depending on the concentration of Ca^{2+} (13, 14).

(7) At a given moment, TJs may relax their tightness to allow for the passage of macrophages toward an infected spot (15), or of spermatozoa in their journey from the Sertoli cell to the lumen of the seminiferous tube (16).

(8) Some TJ-associated molecules contain consensual segments with tumour suppressor proteins (17).

Obviously, the lips of the TJ are the tips of a molecular iceberg. Today, as different molecular components of the TJ are being identified, isolated, and sequenced, we are starting to learn which molecular species is associated with another one, and their phosphorylation state is found to vary in response to phorbol esters, hormones, and stages of the cell cycle. However, in spite of this wealth of information, little is known of the functional role of molecules, consensus segments, and diversity of phosphorylation states. Furthermore, the increasingly complex picture of TJ organization has begun to be associated with diseases due to the malfunctioning of a given component (18). Therefore, this is a time when functional analysis of the TJ is urgently needed. This chapter reviews the functional techniques available, and describes a few of them in detail, focusing mainly in those employing cultured cell lines as model systems for transporting epithelia.

2 Model systems

The permeability of the TJ has been studied in:

- whole organs, e.g. the mucosae of the gall bladder (19–21), intestine (22), and testis (23);
- early stages of development (24, 25);
- cell suspensions (26, 27);
- primary cultures (28); and
- culture of epithelial cell lines (29).

Culturing conditions include cells suspended in spinners without Ca^{2+} (30), monolayers on glass coverslips that can be coated with attaching–enhancing (e.g. polyornithine, see ref. 31) or attaching–weakening molecules (e.g. collagen, see ref. 32), permeable supports of various kinds (29, 33), totally defined culture media (34), and transfilter induction (35).

2.1 Monolayers of epithelial cells

For a while, it was expected that cells would exactly reproduce *in vitro* the epithelium where they came from, so one would have several square centimetres of proximal or distal tube, ciliated epithelium, or any other epithelial cell type whose anatomy and size pose formidable difficulties to study in terms of permeability properties. Yet, it was soon realized that for reasons that have to do with the

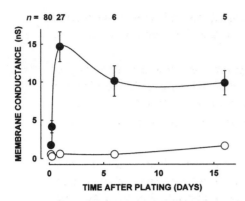

Figure 1 Specific (filled circles) and non-specific K$^+$ conductance (open circles) as a function of time after plating MDCK cells at confluence. Cells were harvested with trypsin-EDTA, plated on coverslips, and membrane currents were measured in the whole-cell configuration. Standard errors for the initial specific conductance and for all non-specific conductances are smaller than symbols. Notice that harvesting has almost completely abolished K$^+$ conductance, presumably by clipping most K-channels, but cells quickly synthesize new ones and deliver them to the membrane.

composition of culture media, disassembly of specific cell–cell contacts, stage of the cell cycle, etc., cells seldom, if ever, reproduce the epithelium of origin. Nevertheless, it is taken for granted that cells treated with trypsin, ethylenediamine tetraacetic acid (EDTA), and other artificial procedures would make TJs and polarize by switching on the same genes and cellular mechanisms used in the natural situation. In this respect, the damage itself may be used to obtain information. Thus, when studying the synthesis and polarized delivery of ion channels, Bolívar *et al.* (36) faced the alternative of either using as little trypsin-EDTA as possible so as to avoid artefacts or, on the contrary, resorting to a drastic treatment so as to force the cell to make entirely new channels and afford an opportunity to study the mechanisms involved (*Figure 1*).

Protocol 1
Cell culture and calcium switching

Equipment and reagents

- Plastic bottles, 175 cm^2 (Corning Costar)
- CO$_2$ incubator (Forma Scientific CO$_2$ incubator, Steri-Cult 200)
- Transwell inserts (Corning Costar)
- Dulbecco's modified Eagle's medium (DMEM; Gibco-BRL)
- Ca^{2+}-free minimal essential medium (Ca-free MEM; Gibco-BRL)
- Phosphate-buffered saline (PBS; Gibco-BRL)
- Ca^{2+}-free phosphate-buffered saline (Ca-free PBS; Gibco-BRL)
- 0.5% Trypsin-0.5 mM EDTA in Ca- and Mg-free Hank's balanced salt solution (HBSS) (Gibco-BRL)
- 5 mg/ml trypsin inhibitor in PBS (Sigma)
- CDMEM: DMEM supplemented with 100 U ml^{-1} penicillin, 100 µg ml^{-1} streptomycin, 10% fetal calf serum (Gibco-BRL) and 0.8 U ml^{-1} insulin (Eli Lilly)[a]

Protocol 1 continued

Method

1 Wash cells growing in plastic bottles three times with Ca-free PBS. Cells with a strong attachment to the substrate can be detached easily if the last wash is prolonged for 30–60 min.

2 Remove Ca-free PBS.[b]

3 Add 1–2 ml per 175 cm^2 bottle of trypsin-EDTA solution. Remove the excess to avoid cell damage.

4 Incubate at 36.5 °C in a 5% CO$_2$ atmosphere for 5–15 min, depending on the degree of attachment of the cell type. Follow the trypsinization process by frequent observation at the microscope.

5 Once cells detach and become rounded, add CDMEM to the bottle to stop trypsinization.[c]

6 Suspend the cells by pipetting several times, avoiding foam formation and then quantify their number.

7 To plate the cells, add the desired volume of suspension to the vessel (Petri dishes, plastic bottles, multiwell plates, glass coverslips, filters, or inserts). Incubate for 45–60 min as in step 4.

8 Wash the monolayers with fresh medium to discard unattached cells and incubate for 12–24 h to obtain a fully polarized and attached monolayer of cells.

9 To perform a calcium, switch, wash the monolayers five times, 5 min each, with Ca-free PBS and then add Ca-free MEM.[d] Incubate for 10–16 h.

10 Switch Ca-free cells to normal Ca^{2+} medium by replacing the Ca-free MEM with CDMEM.[e]

[a] Use the recommended medium, i.e. for Madin–Darby canine kidney (MDCK) cells. Some cells also require specific substances, such as β-mercaptoethanol, glutamine, or essential amino acids.

[b] Only about 0.5 ml must remain in the bottle to keep the cells wet.

[c] Alternatively, trypsin can be stopped with 20 ml of PBS, DMEM, or a medium of specific composition, supplemented with 50 μl of trypsin inhibitor (5 mg/ml).

[d] MEM without Ca^{2+} can be supplemented with serum extensively dialysed against Ca-free PBS.

[e] TJs will be fully formed in 4 h as can be gauged through the measurement of the TER (see *Protocol 7* below).

2.2 Co-cultures

Co-cultures consist of a mixture of suspended cells from different origins, that are mixed and seeded together to form a patchwork monolayer. This was used by González-Mariscal *et al.* (37) to demonstrate that TJs can be established between neighbouring cells from different organs and even from different animal species. Prior to mixing, one of the cell types can be stained with an intra-vital dye that remains in the cytoplasm for several days, thus allowing easy recognition. In keeping with the observation of González-Mariscal *et al.* (37), the expression of the TJ protein occludin is independent of whether a given cellular cleft is a homologous or heterologous one (Plate 1), but the expression of E-cadherin is instead detected in homologous borders only (2). As mentioned below, the establishment of TJs depends on the Ca^{2+} activation of E-cadherin, and a subsequent cascade of intracellular events. Therefore, the fact that a TJ can be established in homologous and heterologous borders, but E-cadherin is only evident in homogeneous ones suggests that the cascade leading to TJ formation includes some diffusible messenger, and does not depend on mechanically bringing together the borders of the two neighbouring cells. Our methodologies for co-culture and for immunofluorescent examination of cell cultures are described in *Protocols 2* and *3* below.

Protocol 2
Co-culture of epithelial cells

Equipment and reagents

- 1.3 mM Cell Tracker orange (CMTMR, Molecular Probes)[a]
- PBS with Ca^{2+} (Gibco-BRL)

- CDMEM (see *Protocol 1*)
- Transwell inserts (see *Protocol 1*)

Method

1   Plate each different cell type at confluence on small Petri dishes or T25 plastic bottles as in *Protocol 1*.

2   Incubate one cell type with CMTMR for 1 h at 37 °C in a CO_2 incubator with gentle shaking.[b]

3   Wash the cells with PBS three times rapidly and three more times, 5 min each.

4   Incubate for 1 h in CDMEM to allow enzymatic processing of CMTMR.

5   Trypsinize both the labelled and unlabelled cells, as described in *Protocol 1*.

6   Quantify the cell number.

7   Mix both cell types thoroughly at the desired proportion.

8   Plate the mixture on the desired support (glass coverslips or inserts).

9   Cells can now be used for immunofluorescence or lipid diffusion experiments.

[a] The Cell Tracker CMTMR is available in a number of variants with different fluorophores to fit any experimental design. Determine the optimal concentration experimentally using a physiological indicator such as the TER measurement.

[b] Protect from light.

Protocol 3

Immunofluorescence of epithelial cell cultures

Equipment and reagents

- TBST (10 mM Tris pH 7.4 at room temperature, 100 mM NaCl, and 0.1% Tween-20)
- PBS (see *Protocol 1*)
- Fixative: methanol (−20 °C) or 4% paraformaldehyde in PBS

- 0.2% Triton X-100 in TBST
- 0.25% saponin in TBS
- 10 mM propidium iodide in TBST
- Fluorguard (Bio-Rad)
- Glass slides and coverslips

Method

1 Rinse the glass coverslips or the inserts containing cells twice with PBS.

2 Fix and permeabilize with methanol at −20 °C for 45 s in TBST. Alternatively, fix cells with paraformaldehyde for 10–20 min at room temperature and subsequently permeabilize with Triton X-100 or saponin solutions for 5 min with gentle agitation.[a]

3 Rinse cells once with TBST.

4 Block[b] non-specific sites by incubating for at least 30 min at 37 °C with either (i) 3% fetal bovine serum in TBST, (ii) 3% immunoglobulin-free serum bovine albumin (BSA) in TBST, or (iii) 5% fat-free dry milk in TBST.[c]

5 Incubate for 1 h at room temperature with the specific first antibody diluted in TBST.[d] For 25 mm^2 coverslips, place 10 μl of the diluted antibody on a Parafilm sheet. Pick up the coverslip with thin tweezers and turn it over to float on the antibody. Make sure cells are in direct contact with the antibody solution.[e]

6 Rinse monolayers three times with TBST for 5 min each.

7 Incubate with a flurophore-labelled secondary antibody for 30 min at room temperature.[e,f]

8 Rinse as in step 6.

9 To stain nuclei, soak the cells in propidium iodide solution and then rinse as in step 6.

10 Mount cells grown on glass coverslips on glass slides with a small drop of the Fluorguard mounting medium.[g] For cells attached to filters, mount between a slide and a coverslip with the cells orientated toward the coverslip.

11 Examine with a fluorescence or confocal microscope.

[a] Fixation and permeabilization methods depend on the type of antigen. Some antigens are very well fixed and cells are permeabilized by an incubation with methanol, while other antigens are better fixed with paraformaldehyde.

[b] Blocking depends on the type of antigen, antibodies, and cell. Try several protocols to find out the best for a given cell type and a specific antibody.

[c] Some antibodies require overnight incubation at 4 °C.

[d] The suitable antibody dilution must be explored.

[e] Two antibodies obtained from different species can be added at once.

[f] New flourophores with higher quantum efficiency, such as Cy5 (Researches Organics) or Alexa (Molecular Probes), strongly improve detection.

[g] There are several excellent media similar to Fluorguard, such as Vectashield (Vector Labs Inc.) and ProLong (Molecular Probes).

3 Electron microscopy

In a plane perpendicular to the surface of the epithelium, the TJ appears in transmission electron microscopy (TEM) as a series of punctate contacts. In preparations stained with OsO_4, the transverse section of the plasma membrane of each cell has a typical dark–light–dark image, corresponding to the hydrophilic–hydrophobic–hydrophilic structure of the membrane. At the TJ itself, the outermost (dark/hydrophilic) domain of the two neighbouring membranes fuse ('kiss'), so the point of contact shows five instead of six bands. However, it is often difficult to ensure that the image is not caused by a small degree of obliquity, so most researchers prefer to identify the TJ as the point where the diffusion of electron-dense tracers (horseradish peroxidase, ruthenium red (RR), etc.) is clearly

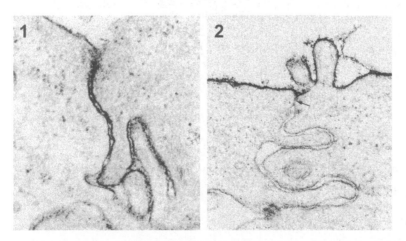

Figure 2 The TJ blocks diffusion of extracellular markers, in this case ruthenium red (RR) (black), added to the solution bathing the apical domain (above). Monolayers of MDCK cells were kept without (**A**) or with 2.0 mM calcium (**B**). In the first case, the two neighbouring cells have established no TJs, as revealed by the fact that RR penetrates and freely stains lateral membranes. The two neighbouring MDCK cells on the right were incubated in Ca^{2+}-containing medium, have established TJs, and blocked the passage of RR, that only stains apical domains.

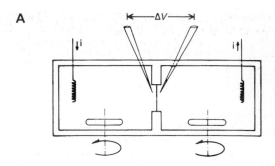

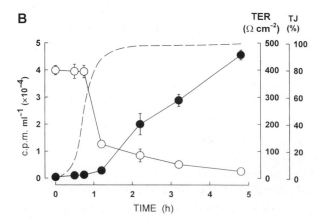

Figure 3 (**A**) A Millipore filter on which a monolayer of MDCK cells pre-incubated in calcium-free medium and switched to Ca^{2+}-containing medium at $t = 0$ was mounted between two Lucite chambers. Two silver wire electrodes (i) serve to pass current, and another set of electrodes measure the voltage deflection (ΔV). Each chamber is stirred by a small magnetic bar.
(**B**) Apical-to-basolateral [^{3}H]mannitol flux (left ordinate, open circles), TER (right ordinate, filled circles), and degree of sealing of the TJs (f_c, dashed line) as calculated with Equation 1 during the calcium switch started at $t = 0$. Labelled mannitol was added to the apical side and periodic samples were taken from the opposite side, and counted. TER was recorded in the same set of monolayers at the end of each sampling period. f_c was calculated with Equation 3.

stopped (38; see *Figure 2*). This technique has been used by González-Mariscal *et al.* (39) to score the number of junctions sealed as a function of time after adding Ca^{2+}, and relates the degree of sealing to the development of a TER (40; *Figure 3*, filled circles).

3.1 Electron-dense tracers

TJs enable transporting epithelia to separate fluid compartments of different composition. This sealing capacity is demonstrated in electron microscopy (EM) using electron-dense dyes such as RR, or that contain a heavy element (Fe), such as ferritin and haemoglobin.

Protocol 4

Use of electron-dense tracers to evaluate tight junction sealing

Equipment and reagents[a]

- Transwell® polycarbonate membrane inserts (Corning Costar®)
- 25% glutaraldehyde in distilled water
- 0.1 and 0.2 M sodium cacodylate buffers, pH 7.3
- 4% osmium tetroxide in distilled water
- Fixative: add 9 ml of 0.1 M sodium cacodylate buffer and 1 ml of 25% glutaraldehyde to 0.06 g of RR (final concentrations: 2.5% glutaraldehyde, 0.6% RR in 0.1 M sodium cacodylate buffer, pH 7.3)
- 0.5% uranyl acetate in 0.1 M Na$^+$ acetate buffer

- Osmium tetroxide incubation solution: add 0.06 g of RR to 2 ml of 0.2 M sodium cacodylate buffer and 2 ml of 4% osmium tetroxide
- 0.1 M Na$^+$ acetate buffer, pH 5.0
- Ethanol/water mixtures (50, 75, 95, and 100%)
- Spurr resin
- Spurr resin/propylene oxide
- Carbon/Formvar®-coated grid (Ted Pella)
- 7.5% uranyl acetate
- Lead citrate

Method

1 Grow cells to confluence on polycarbonate Transwell inserts.

2 Wash the monolayer (once for 2 min) with 0.1 M sodium cacodylate buffer, pH 7.3.

3 Fix the apical surface with the fixative solution. Incubate for 30 min at room temperature.

4 Wash in 0.1 M sodium cacodylate buffer, pH 7.3 (once for 2 min).

5 Add 1 ml of osmium tetroxide incubation solution at the apical surface and incubate for another 30 min at room temperature.

6 Repeat step 4.

7 Cut out the filter from the insert frame and process it for EM.

8 Wash in 0.1 M Na$^+$ acetate buffer, pH 5.0, and leave overnight in uranyl acetate solution.

9 Dehydrate the filters by soaking them sequentially in ascending concentrations of ethanol/water mixtures (50, 75, 95, and 100%).

10 Infiltrate the filters in a mixture of Spurr resin/propylene oxide overnight and embed in Spurr resin overnight.

11 Collect thin sections on carbon/Formvar-coated grids, double stain with uranyl acetate and lead solutions.

12 Photograph thin sections in a transmission electron microscope [b].

[a] All reagents for EM are from Electron Microscopy Sciences.

[b] TJ permeability to RR is determined by the presence of electron-dense deposits in the intercellular space between epithelial cells.

3.2 Freeze-fracture

Since the TJ is primarily a membrane structure, it has been studied extensively with freeze-fracture techniques (41). The TJ appears as a belt of anastomozing strands that completely surrounds the cell, marking the limit between the apical and the lateral domain of the cell membrane. Each strand in freeze-fracture corresponds to a 'kiss' in TEM. The plane of fracture always crosses through the hydrophobic matrix of one of the two neighbouring membranes. Therefore, depending on whether the replica corresponds to the moiety that remains attached to the cell (and seen from outside) or to the one that is peeled off (and seen from the cell side), the replica offers a P- or an E-face. Strands appear as strings in P-faces and as furrows in E-face (*Figure 4*).

Since 'kisses' are the points where the diffusion of extracellular tracers stop, and the value of TER increases with the number of strands, an effort was made to identify strands with resistors (42). The fact that the observed relationship is not a linear one, as expected from resistors arranged in series, led Claude (43) to assume that strands are traversed by channels that can be in an 'open' or 'closed' state. In fact, Simon *et al.* (44) found that paracellin-1, a member of the claudin family of TJ membrane proteins, is responsible for Mg^{2+} resorption. The diffusion of macromolecules through the paracellular route is inversely proportional to the electrical conductance (39). Yet transfection of deletants of occludin demonstrated that these parameters can vary independently, and even in opposite directions (45). This prompted further elaboration to take into account the role that anastomozing strands may play (46).

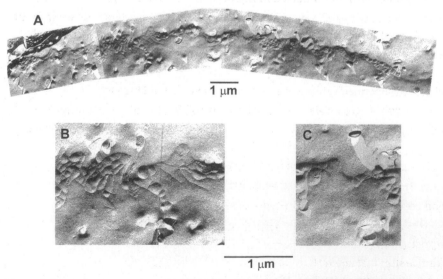

Figure 4 Freeze-fracture replicas of MDCK cells. (**A**) A fracture spanning a significant fragment of the perimeter of an MDCK cell (16 out of 60 mm). The TJ appears as filaments that vary in number and intermeshing. (**B**) A segment of the same TJ showing several strands, and (**C**) a segment reduced to only one strand.

Protocol 5

Freeze-fracture imaging of the tight junction

Equipment and reagents[a]

- Balzers BAF 400 (Balzers)
- 300-mesh size copper EM grids
- Gold specimen support
- Plastic culture flasks (Becton Dickinson)
- PBS (see *Protocol 1*) filtered through 0.22 mm
- 25% glutaraldehyde in PBS (see *Protocol 4*)
- 10, 20, and 30% glycerol in PBS
- Rubber policeman (Corning Costar®)
- Liquid nitrogen
- Platinum
- Carbon
- Chromic mixture

Method

1 Grow cells in plastic flasks.

2 Fix with 2.5% glutaraldehyde in PBS for 30 min at 37°C.

3 Wash three times with PBS.

4 Cryoprotect by successive incubations in 10, 20, and 30% glycerol for 30, 30, and 60 min, respectively.

5 Detach from the substrate as a sheet by gently scraping with a rubber policeman.

6 Place on gold specimen holders. Absorb the excess solution carefully with filter paper.[b]

7 Immerse the sample rapidly in liquid nitrogen.[c]

8 Place gold specimen supports with the samples on the Balzers specimen table. Take care that samples remain immersed in liquid N_2 all the time.

9 Rapidly introduce the specimen table with the counterflow loading device into the Balzers chamber, previously stabilized at -150°C and 5×10^{-6} bar of pressure.

10 Increase the temperature of the chamber to reach -100°C by limiting liquid N_2 supply.

11 Cut with the Balzers knife at 0.1–0.2 mm over the gold holder base.[d]

12 Shadow the fractured samples with platinum at 45° for 20 s at 1900 V and 70 mA.

13 Immediately shadow the fractured samples with carbon at 90° for 20 s at 2400 V and 110 mA.

14 Take the fractured samples out of the Balzers chamber.[e]

15 Clean replicas with chromic mixture and rinse with distilled water.

16 Place on copper grids and examine with an electron microscope.

[a] Reagents are from Electron Microscopy Sciences (unless otherwise indicated).

[b] Do not touch the sample.

[c] Samples can be stored in liquid N_2.

[d] This will give a larger area of study.

[e] To process another sample immediately, decrease the temperature of the Balzers chamber to -150°C and proceed as indicated in step 9.

When the epithelium is fractured as if it were an open book, the TJ can be observed in its P- and counterpart E-faces. These double replicas confirmed that the TJ consists of an anastomozing network of strands, because when these do not appear continuous in a P-face, but segmented, the missing part of the segment shows up in the complementary part of the E-face (47). Freeze-fracture techniques can be also combined with immuno-techniques to identify the specific molecules composing the strands (48).

4 Information derived from the physical properties of permeating molecules

The size of pores in a TJ, the radius of solvent and solutes, their electric charges, and the nature of the driving forces give rise to a multitude of natural and experimental phenomena, such as simple diffusion, single file diffusion, dilution potentials, streaming potentials, electro-osmosis, solvent drag, etc., that are used to investigate the properties of the TJ (3, 20, 21, 49). The degree of discrimination between molecules permeating through a TJ is not as delicate as that through a plasma membrane. Thus, while the membrane of a muscle can exhibit a K^+-to-Na^+ permeability (P) ratio of 80, in a monolayer of MDCK cells the permeability ratio is only 2 or 3. *Figure 5* shows that these monolayers may have a P_{Na} higher than P_{Cl}, and that this difference may be decreased reversibly by an acidic pH, suggesting that the discrimination between Na^+ and Cl^- is due to acidic groups that can be protonated.

4.1 Solute size and charge assay for paracellular transport

Paracellular flux is measured using tracers of known molecular weight. The tracer can be a radiolabelled molecule such as D-[^{14}C]mannitol (mol. wt 182.2 Da)

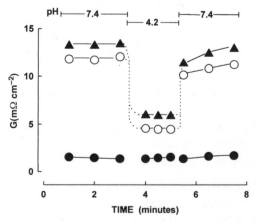

Figure 5 The TJ discriminates between different ion species. These curves show the conductances (G) across a monolayer of MDCK cells as measured with dilution potentials. Triangles represent total conductance. Most of this conductance is accounted for by the conductance of Na^+ (open circles). When the pH of the solution bathing the apical side is lowered to 4.2, Na^+ conductance is reduced, but Cl^- conductance (filled circles) is not changed.

or [^{3}H]inulin (mol. wt 5200 Da), or fluorescein isothiocyanate (FITC)-labelled dextran which is available in a range of molecular weights. *Figure 3* shows an example in which the flux of mannitol decreases and the value of TER increases as the TJ is progressively sealed, as calculated using Equation 1.

Protocol 6

Paracellular flux of FITC–dextran (J_{DEX})

Equipment and reagents

- Spectrofluorometer (Perkin Elmer)
- Transwell inserts of 4.5 cm^2 (see *Protocol 1*)
- P buffer (145 mM NaCl, 10 mM HEPES, pH 7.4, 1.0 mM Na-pyruvate, 10 mM glucose, 3.0 mM CaCl$_2$)

- 12-well plastic cell culture plates (Corning Costar®)
- Tracer solution: 10 µg/ml FITC–dextran 4000 Da (Sigma)^a in P buffer

Method

1 Grow cells to confluence on filter inserts (as in *Protocol 1*). The area of monolayer exposed is 4.5 cm^2.

2 One hour prior to the start of the experiment, replace apical and basal solutions with P buffer.

3 To start, transfer the inserts to 12-well plastic cell culture plates containing P buffer (2.0 ml in the bottom chamber).

4 Add 500 ml of tracer solution to the apical side (upper chamber).

5 Incubate the monolayer at 37 °C for 1 h.

6 Collect media from both the upper and bottom chambers and measure FITC–dextran with a fluorometer in a 3 ml cuvette (excitation 492 nm; emission 520 nm).^{b}

7 Calculate the unidirectional (apical→basolateral) flux as follows: divide the fluorescence intensity of a given sample of the bottom solution (arbitrary units) by the corresponding value of the upper solution diluted as indicated. The figure obtained corresponds to the flux (J_{DEX}) in ng cm^{-2} h^{-1}.

^{a} Dextran tracers of different molecular weights are available with several different fluorophores.

^{b} Upper chamber: 30 µl, dilute to 3.0 ml. Bottom chamber: 2.0 ml plus 1.0 ml of P buffer to complete 3.0 ml.

The study of the physiology of epithelia started more than a century ago with the frog skin which can be dissected and mounted between two chambers for tracer flux and electrical studies. It exhibits a TER of several thousand Ohms per square centimetre, a spontaneous electrical potential ($\Delta\Psi$) across, and a short circuit current (a measure of net Na$^+$ transport) of hundreds of microamperes per square centimetre (50). Under the same circumstances, epithelia such as the intestinal mucosa and the gall bladder showed instead a much lower TER, no spontaneous $\Delta\Psi$, and no short circuit current. These were considered to be too delicate to withstand dissection and mounting, and consequently were discarded. Yet, it was noticed that in spite of this 'miserable' condition, these epithelia were able to transport *in vitro* solutes and fluids very efficiently. Epithelia were thereafter classified into 'tight' and 'leaky' according to their TER. The use of impaling microelectrodes demonstrated that the plasma membrane of cells in leaky epithelia have a high electrical resistance, so that if a permeating ion had to cross the epithelium through a transcellular route and traverse the plasma membrane twice, the epithelium would have an overall electrical resistance comparable with that in the frog skin. This led to the demonstration that the lack of a sizeable TER, short circuit current, and spontaneous $\Delta\Psi$ was due to a predominant permeation route that avoided the cytoplasm. By passing an electric current perpendicular to the epithelium, and scanning the surface with a microelectrode to detect the points where the current was flowing, it was found that these correspond to intercellular spaces (4). In other words, 'leaky' epithelia were as healthy as 'tight' epithelia, only that their leakiness was a physiological property due to a predominant paracellular permeation route.

Protocol 7

Measurement of the transepithelial electrical resistance (TER)

Equipment and reagents

- Lucite Ussing chamber[a,b]
- Nitrocellulose filters (Millipore)
- CDMEM (see *Protocol 1*)
- Vacuum grease

Method

1 Plate cells at confluence on nitrocellulose filters.

2 Place filters with monolayers between the two Lucite chambers. To minimize leaks, it is advisable to use a thin coat of vacuum grease on the two contacting surfaces.

3 Fill chambers with medium or buffer (CDMEM, Ca-free MEM, PBS, etc.).

4 Deliver a brief pulse of direct current (I) of approximately 100 μA.

5 Measure the elicited voltage deflection (ΔV) in milliVolts[c].

6 Calculate *ter* as:

$$ter \text{ (in } \Omega \text{ cm}^{-2}) = \Delta V/I \qquad\qquad 1$$

Values of *ter* should be corrected, by subtracting the contribution of the filter and the bathing solutions. This value is in turn obtained from a similar measurement with an empty filter.

7 To compare *ter* values in monolayers under different experimental conditions measured on different days, it is necessary to normalize the data. This is usually

Protocol 8 continued

done by running a set of control monolayers and obtaining a mean value each time (m_i). Calculate M, the overall average of the values of m_i obtained on different occasions. Transform the values of *ter* into TER as follows:

$$\text{TER} = ter \, (M/m_i)^d \qquad\qquad 2$$

[a] A model is illustrated in *Figure 3B*. Each chamber, 8 ml capacity, has an exposed area of approximately 1 cm². Current is delivered via $Ag/AgCl_2$ electrodes placed far from the monolayer (~2 cm) and the voltage deflection measured with electrodes placed as close as possible (~1 mm) to the membrane.

[b] There are some forms commercially available equipment, such as Evom, that include chopstick electrode sets as well as an Ussing chamber, and Endohm, whose improved electrode set decreases error drastically (World Precision Instruments).

[c] To avoid leaks due to edge damage, each monolayer is used for a single measurement.

[d] To calculate the fraction of already sealed tight junctions (f_c) in a calcium switch study from the initial (TER_o), final (TER_∞) and intermediate (TER_t) values of the electrical resistance, use Equation 3:

$$f_c = \frac{\text{TER}_\infty}{\text{TER}_\infty - \text{TER}_o} \left(1\frac{\text{TER}_o}{\text{TER}_t} \right) \qquad\qquad 3$$

The paracellular route is controlled primarily by the TJ. However, when TER is particularly low (below, say, 25 Ω cm^{-2}), the length, narrowness, and tortuosity of the intercellular space may not be neglected (51).

5 The tight junction also functions as a fence

Membrane molecules may be anchored to the cytoskeleton, either directly or through an attachment to molecular complexes. Thus, the α-subunit of a K^+ channel is immobile due to the binding of its C-terminus to actin microfilaments, but the β-subunit is only fixed indirectly, because of its association with the α-subunit. Unbound molecules can diffuse freely in the plane of the membrane. Yet, in epithelial cells, even free diffusing molecules are confined to the apical or to the basolateral domain due to a fence imposed by the TJ (52–54). Its role as a fence can be investigated in several ways. One of the most common ways consists of adding a fluoresceinated lipid probe to the outer bathing medium and observing its distribution in transverse optical sections with confocal microscopy (45, 55, 56). Plate 2 illustrates that the lipid probe BODIPY® is confined to the apical domain of a monolayer of co-cultured LLC-PK$_1$ (a pig kidney epithelial cell line; red) and MDCK (unstained) cells, demonstrating that the role of a fence is even played by TJs established between two different cell types. Similar studies were performed in other experimental situations, such as varying the concentration of Ca^{2+}, using inhibitors of the synthesis of ATP, transfecting deletants of occludin, etc.

Protocol 8

Measurement of diffusion of membrane lipids

Equipment and reagents

- BSA, fatty acid-free (Sigma)
- BODIPY®FL-C$_{12}$-sphingomyelin (Molecular Probes)
- BODIPY®FL-C$_{12}$-sphingomyelin/BSA stock solution (250 mg ml^{-1} BODIPY®FL-C$_{12}$-sphingomyelin, 5 mg ml^{-1} BSA)[a]
- Transwell inserts of 4.5 cm^2 (see *Protocol 4*)[b]
- 12-well plastic cell culture plates (Corning Costar®)
- P buffer (see *Protocol 6*)
- Nail varnish
- Glass slides and coverslips (Fisher)

A. Preparation of BODIPY®FL-C$_{12}$-sphingomyelin/BSA complex

1 Dissolve BSA in P buffer to a final concentration of 5 mg ml^{-1}.

2 Dissolve 250 mg of BODIPY®FL-C$_{12}$-sphingomyelin in 1 ml of BSA solution. Mix by vortexing.[c]

B. Membrane lipid diffusion assay

1 Grow cells to confluence on Transwell inserts placed in 12-well plastic cell culture plates.

2 Wash the monolayers three times with P buffer in both compartments. Leave the inserts in P buffer.

3 Label with BODIPY®FL-C$_{12}$-sphingomyelin/BSA complex solution (5 nmol ml^{-1}) for 10 min on ice.[d]

4 Wash twice with cold P buffer (rapid).

5 Incubate on ice for 1 h.

6 Cut the membrane from the insert frame and mount in P buffer on a glass slide that has a drop of dried nail varnish in each corner.

7 Cover the samples with a rectangulat coverslip so that it rests on the drops of nail varnish.

8 Analyse the samples immediately (within 5–10 min) by confocal microscopy (vertical sections).

[a] Stock solution should have equivalent quantities of BODIPY®FL-C$_{12}$-sphingomyelin and defatted BSA.

[b] The insert membrane should be transparent.

[c] Keep small aliquots at $-20\,°C$ in light-protected vials.

[d] Add to the apical compartment 15.5 ml of BODIPY®FL-C$_{12}$-sphingomyelin/BSA stock solution per 1 ml of P buffer to reach a final concentration of 5 nmol ml^{-1}.

6 Analysis of second messenger regulation of tight junction sealing and activity

6.1 Ca^{2+} signalling

A widely used approach to study the assembly and sealing of the TJ is the so-called calcium switch. Here, epithelial cells are plated at confluence; 30–60 h later, the resulting monolayer is transferred to Ca-free medium and incubated overnight. Next, Ca^{2+} is added and junction formation is followed through the measurement of TER, diffusion of extracellular molecules, distribution of TJ-associated molecules by immunofluorescence analysis, or determination of the degree of TJ protein phosphorylation following extraction and electrophoretic separation, etc. Furthermore, calcium switching can be done in the presence of specific inhibitors to characterize the molecules and organelles involved. In this way, it was learned that Ca^{2+} triggers the sealing of the TJ (57) by acting on the extracellular side (8) of the E-cadherin molecule (58). Calcium triggers a cascade of signal transduction that includes G-proteins, phospholipase C, protein kinase C, and calmodulin (9, 59, 60). Phosphorylation of several molecular components is required not only for the sealing of a newly assembled TJ but also for the opening of an already sealed TJ provoked by Ca^{2+} removal (61, 62).

6.2 Changes in phosphorylation of TJ-associated molecules

As mentioned above, several TJ proteins are known to be phosphorylated. Thus ZO-1 and ZO-2 proteins in MDCK cells are phosphorylated on serine (63) and tyrosine residues (64, 65). Cingulin is phosphorylated on serine residues (61). Occludin has several levels of phosphorylation on serine and threonine so that it migrates as a series of bands (>10) between 62 and 82 kDa on sodium dodecyl sulphate–polyacrylamide gel electrophoresis (SDS–PAGE) (66–69) (*Figure 6*). During the sealing and re-opening of the TJ provoked by changes in Ca^{2+} concentration

OCCLUDIN MIGRATION PATERN

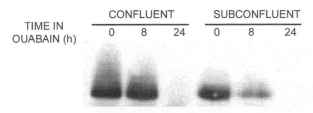

Figure 6 TJ molecules modify their level of phosphorylation in response to changes in the attachment state. Occludins from confluent monolayers of MDCK cells, which have several degrees of serine and threonine phosphorylation, migrate in acrylamide gels with a 'smeared' pattern due to bands of diverse molecular weight. In subconfluent cultures (i.e. without TJs), the amount of occludin is smaller. Addition of 10 μM ouabain, which stops the Na$^+$,K$^+$-ATPase pump and detaches cells from each other and from the substrate, progressively reduces the content of occludin, as well as the phosphorylated states (68).

or the use of inhibitors, the level of phosphorylation of these proteins is also changed (70). However, in spite of the fact that these correlations are well documented, we still have no dynamic model to explain how Ca^{2+} levels, small GTP-binding proteins, and TJ-associated molecules would be combined, and this is why the protocols described here can be useful to shed light on junctional phenomena.

Protocol 9

Measurement of cytosolic Ca^{2+} activity

Equipment and reagents

- Fluorometer (see *Protocol 6*), equipped with a XY recorder and a system for continuous agitation
- Orbital shaker (Hoeffer)
- Plastic fluorometric cuvettes (Sigma)
- 1 mM calcium indicator in culture medium: fura-2 or Indo-2 acetoxymethyl esters (Molecular Probes) first dissolved in dimethylsulphoxide (DMSO)
- Teflon stopper[a]

- 0.8×2.5 cm glass coverslips
- 60-mm tissue culture plate (Corning Costar®)
- DMSO
- Ringer buffer (140 mM NaCl, 5 mM KCl, 1 mM $MgCl_2$, 1.8 mM $CaCl_2$, 20 mM Tris, 10 mM dextrose, 0.05 mM probenedic)
- 2 mM ionomycin (Calbiochem®) in ethanol
- 2 M $MnCl_2$

Method

1 Plate MDCK cells on a tissue culture plate containing 4–5 glass coverslips.

2 After a 20 h incubation, transfer coverslips with confluent monolayers to new culture plates with fresh medium containing calcium indicator and incubate for 30 min at room temperature with continuous agitation in an orbital shaker.

3 Wash monolayers seven times with Ringer buffer.

4 Mount coverslips in the Teflon stopper of a plastic fluorometric cuvette containing 3.0 ml of Ringer buffer at a 30° angle with respect to the excitation beam.

5 Allow cells to stabilize for 5 min at room temperature.

6 Perform fluorescence measurements placing the cuvette in the cell compartment of the fluorometer under continuous agitation. For fura-2, fix excitation/emission wavelengths at 338 and 510 nm, respectively.

7 Register the fluorescence continuously until a stable value of basal fluorescence (F) is reached.

8 Add 10 ml of ionomycin solution with a thin pipette. Once it reaches a plateau (in ~5 min), read the F_{max}.

9 Add 10 ml of $MnCl_2$ to quench the fluorescence to a stable low value: F_{Mn}.

Protocol 9 continued

10 Calculate the cytosolic Ca^{2+} according to the following general equation:

$$[Ca^{2+}]_c = K_d(F - F_{min})/(F_{max} - F)$$ 4

where K_d is the dissociation constant for Ca^{2+} binding (224 nM for fura-2). F_{min} is the minimum fluorescence calculated according to the following equation:

$$F_{min} = 1/6 [F_{max} - F_{Mn}] + F_{Mn}$$ 5

[a] The Teflon stopper is necessary to hold the coverslips, from above, at a 30° angle with respect to the excitation beam to decrease light scattering. It must have a groove to insert the coverslip. A small piece of Parafilm in the slot helps to hold the coverslip. Make a couple of holes in the stopper to liberate pressure and introduce thin pipettes (see *Figure 7*).

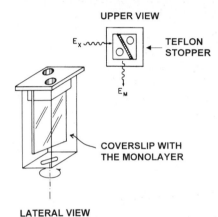

UPPER VIEW

E_x

TEFLON STOPPER

E_M

COVERSLIP WITH THE MONOLAYER

LATERAL VIEW

Figure 7 Measurement of cytosolic Ca^{2+} in monolayers of epithelial cells that were loaded with fura-2. The glass coverslip with the monolayer is inserted in the groove of the Teflon stopper. The cuvette is filled and emptied through a couple of holes in the stopper, and continuously stirred by a small magnetic bar. The groove is machined at 30° with respect to the emission beam (E_M). Cells face the incoming excitation beam (E_x).

Acknowledgements

We wish to acknowledge the efficient and pleasant assistance of C. Flores-Maldonado, L. Roldán, and A. Lazaro, and the financial support of the National Research Council of México (CONACYT).

References

1. Cereijido, M. (1991). In *Tight junctions* (ed. M. Cereijido). p. 1. CRC Press, Boca Raton, FL.
2. Cereijido, M., Shoshani, L., and Contreras, R. G. (2000). *Am. J. Physiol.*, **279**, G477.
3. Wright, E. M. and Diamond, J. M. (1968). *Biochim. Biophys. Acta*, **163**, 57.
4. Frömter E. and Diamond J. (1972). *Nature*, **235**, 9.

5. Balda, M. S. and Matter, K. (1998). *J. Cell Sci.*, **111**, 541.

6. Matter, K. and Balda, M. S. (1999). *Int. Rev. Cytol.*, **186**, 117.

7. Balda, M. S. (1991). In *Tight junctions* (ed. M. Cereijido). p. 121. CRC Press, Boca Raton, FL.

8. Contreras, R. G., Miller, J. H., Zamora, M., González-Mariscal, L., and Cereijido, M. (1992). *Am. J. Physiol.*, **263**, C313.

9. Balda, M. S., González-Mariscal, L., Contreras, R. G., and Cereijido, M. (1991). *J. Membr. Biol.*, **122**, 193.

10. Karnaky, K. J. (1991). In *Tight junctions* (ed. M. Cereijido), p. 175. CRC Press, Boca Raton FL.

11. Bentzel, C. J., Palant, C. E., and Fromm, M. (1991). In *Tight junctions* (ed. M. Cereijido). p. 151. CRC Press, Boca Raton, FL.

12. Schneeberger, E. E. and Lynch, R. D. (1994). In *Molecular mechanisms of epithelial cell junctions: from development to disease* (ed. S. Citi). p. 123. CRC Press, Boca Raton, FL.

13. Gottardi, C. J., Arpin, M., Fanning, A. S., and Louvard, D. (1996). *Proc. Natl Acad. Sci. USA*, **93**, 10779.

14. González-Mariscal, L. Islas, S., Contreras, R. G., García-Villegas, M. R., Betanzos, A., Vega, J., *et al.* (1999). *Exp. Cell Res.*, **248**, 97.

15. Cramer, E. B. (1991). In *Tight junctions* (ed. M. Cereijido). p. 321. CRC Press, Boca Raton, FL.

16. Byers, S. and Pelletier, R. M. (1991). In *Tight junctions* (ed. M. Cereijido). p. 279. CRC Press, Boca Raton, FL.

17. Willott, E., Balda, M. S., Fanning, A. S., Jameson, B., Van Itallie, C., and Anderson, J. M. (1993). *Proc. Natl Acad. Sci. USA*, **90**, 7834.

18. Marin, M. L., Greenstein, A. J., Geller, S. A., Gordon, R. E., and Aufses, A. H. (1983). *Am. J. Gastroenterol.*, **78**, 537.

19. Diamond, J. M. (1974). *Fed. Proc.*, **33**, 2220.

20. Reuss, L., Segal, Y., and Altenberg, G. (1991). *Annu. Rev. Physiol.*, **53**, 361.

21. Boulpaep, E. L. and Sackin, H. (1980). *Curr. Top. Membr. Transp.*, **13**, 169.

22. Madara, J. L., Nash, S., Moore, R., and Atisook, K. (1990). *Monogr. Pathol.*, **31**, 306.

23. Ribeiro, A. F. and David-Ferreira, J. F. (1996). *Cell Biol. Int.*, **20**, 513.

24. Fleming, T. P., McConnell, J., Johnson, M. H., and Stevenson, B. R.(1989). *J. Cell Biol.*, **108**, 1407.

25. Fleming, T. P. (1994). In *Molecular mechanisms of epithelial cell junctions: from development to disease* (ed. S. Citi). p. 67. CRC Press, Boca Raton, FL.

26. Chambard, M., Gabrion, J., and Mauchamp, J. (1981). *J. Cell Biol.*, **91**, 157.

27. Wang A. Z., Ojakian, G. K., and Nelson, J. (1990). *J. Cell Sci.*, **95**, 153.

28. Pitelka D. R., Taggart, B. N., and Hamamoto, S. T. (1983). *J. Cell Biol.*, **96**, 613.

29. Cereijido, M., Robbins, E. S., Dolan, W. J., Rotunno, C. A., and Sabatini, D. D. (1978). *J. Cell Biol.*, **77**, 853.

30. Salas, P. J., Vega-Salas, D. E., and Rodriguez-Boulan, E. (1987). *J. Membr. Biol.*, **98**, 223.

31. Ponce, A. and Cereijido, M. (1991). *Cell Physiol. Biochem.*, **1**, 13.

32. Rabito, C. A., Tchao, R., Valentich, J., and Leighton, J. (1980). *In Vitro*, **16**, 461.

33. Steele, R. E., Preston, A. S., Johnson, J. P., and Handler, J. S. (1986). *Am. J. Physiol.*, **251**, C136.

34. Taub, M. (ed.) (1985). *Tissue culture of epithelial cells.* Plenum Press, NY.

35. Grobstein, C. (1956). *Exp. Cell. Res.*, **10**, 424.

36. Bolívar, J. J. and Cereijido, M. (1987). *J. Membr. Biol.*, **97**, 43.

37. González-Mariscal, L., Chávez de Ramírez, B., Lazaro, A., and Cereijido, M. (1989). *J. Membr. Biol.*, **107**, 43.

38. Miller, F. (1960). *J. Biophys. Biochem. Cytol.*, **8**, 689.

39. González-Mariscal, L., Chavez de Ramirez, B., and Cereijido, M. (1985). *J. Membr. Biol.*, **86**, 113.
40. Peralta Soler, A., Miller, R. D., Laughlin, K. V., Carp, N. Z., Klurfeld, D. M., and Mullin J.M. (1999). *Carcinogenesis*, **20**, 1425.
41. Staehelin, L. A. (1973). *J. Cell Sci.*, **13**, 763.
42. Claude, P. and Goodenough, D. A. (1973). *J. Cell Biol.*, **58**, 390.
43. Claude, P. (1978). *J. Membr. Biol.*, **39**, 219.
44. Simon, D. B., Lu, Y., Choate, K. A., Velazquez, H., Al-Sabban, E., Praga, M., *et al.* (1999). *Science*, **285**, 103.
45. Balda, M. S., Whitney, J. A., Flores, C., Gonzalez, S., Cereijido, M., and Matter, K. (1996). *J. Cell Biol.*, **134**, 1031.
46. Cereijido, M., González-Mariscal, L., and Contreras, R. G. (1989). *News Physiol. Sci.* **4**, 72.
47. Bullivant, S. (1978). In *Electron microscopy* (ed. J. M. Sturgess). Vol. 3, p. 659. Imperial Press, Canada.
48. Morita, K., Furuse, M., Fujimoto, K., and Tsukita, S. (1999). *Proc. Natl Acad. Sci. USA*, **96**, 511.
49. Moreno, J. H. (1975). *J. Gen. Physiol.*, **66**, 97.
50. Koefoed-Johnson, V. and Ussing H. H. (1958). *Acta Physiol. Scand.*, **42**, 298.
51. Harris, P. J., Chatton, J. Y., Tran, P. H., Bungay, P. M., and Spring, K. R. (1994). *Am. J. Physiol.*, **266**, C73.
52. Dragsten, P. R., Blumenthal, R., and Handler, J. S. (1981). *Nature*, **294**, 718.
53. Van Meer, G. and Simons, K. (1986). *EMBO J.*, **5**, 1455.
54. Hannan, L. A., Lisanti, M. P., Rodriguez-Boulan, E., and Edidin, M. (1993). *J. Cell Biol.*, **120**, 353.
55. Pagano, R. E. and Martin, O. C. (1994). In *Cell biology: a laboratory handbook* (ed. J. E. Celis). p. 387, Academic Press, London.
56. Calderón, V., Lázaro, A., Contreras R. G., Shoshani, L., Flores-Maldonado, C., González-Mariscal, L., *et al.* (1998). *J. Membr. Biol.*, **164**, 59.
57. González-Mariscal, L., Contreras, R. G., Bolívar, J. J., Ponce, A., Chávez de Ramirez, B., and Cereijido, M. (1990). *Am. J. Physiol.*, **259**, C978.
58. Gumbiner, B., Stevenson, B., and Grimaldi, A. (1988). *J. Cell Biol.*, **107**, 1575.
59. Denker, B. M., Saha, C., Khawaja, S., and Nigam, S. K. (1996). *J. Biol. Chem.*, **271**, 25750.
60. Rosson, D., O'Brien, T. G., Kampherstein, J. A., Szallasi, Z., Bogi, K., Blumberg, P. M., *et al.* (1997). *J. Biol. Chem.*, **272**, 14950.
61. Citi, S. and Denisenko, N. (1995). *J. Cell Sci.*, **108**, 2917.
62. Citi, S. and Cordenonsi, M. (1998). *Adv. Mol. Cell Biol.*, **28**, 203.
63. Anderson J. M., Stevenson, B. R., Jesaitis, L. A., Goodenough, D. A., and Mooseker, M. S. (1988). *J. Cell Biol.*, **106**, 1141.
64. Takeda, H. and Tsukita, S. (1995). *Cell Struct. Funct.*, **20**, 387.
65. Van Itallie, C. M., Balda, M. S., and Anderson, J. M. (1995). *J. Cell Sci.*, **108**, 1735.
66. Cordenonsi, M., Mazzon, E., De Rigo, L., Baraldo, S., Meggio, F., and Citi, S. (1997). *J. Cell Sci.*, **110**, 3131.
67. Sakakibara, A., Furuse, M., Saitou, M., Ando-Akatzuka, Y., and Tsukita, S. (1997). *J. Cell Biol.*, **137**, 1393.
68. Contreras, R. G., Shoshani, L., Flores-Maldonado, C., Lazaro, A., and Cereijido, M. (1999). *J. Cell Sci.*, **112**, 4223.
69. Farshori, P. and Kachar, B. (1999). *J. Membr. Biol.*, **170**, 147.
70. Pérez-Moreno, M., Avila, A., Islas, S., Sánchez, S., and González-Mariscal, L. (1999). *J. Cell Sci.*, **111**, 3563.

Chapter 5
Leukocyte cell adhesion interactions

Carl G. Gahmberg, Leena Valmu, Tiina J. Hilden, Eveliina Ihanus, Li Tian, Henrietta Nyman, Ulla Lehto, Asko Uppala, Tanja-Maria Ranta and Erkki Koivunen

Department of Biosciences, Division of Biochemistry, Viikinkaari 5, 00014, University of Helsinki, Helsinki, Finland.

1 Introduction

1.1 General

During the past 20–25 years, cell adhesion has become a major object of study (1, 2) and, among its various subdisciplines, leukocyte adhesion certainly plays a pivotal role. There are two major reasons for the fact that research on leukocyte adhesion is flourishing:

(1) the various leukocyte subgroups and a number of leukocyte 'antigens' have been throughly studied and are now relatively well characterized by immunologists and scientists studying, for example, infectious diseases, leukaemias, and lymphomas; and

(2) it is now realized that many of the major diseases affecting mankind involve leukocytes.

Therefore, much of what we currently know about cell adhesion in general derives from studies on leukocyte adhesion molecules.

In contrast to most cells in mammals, the majority of leukocytes circulate in the body and accumulate in lymphoid organs and in altered tissues due to infections or for other reasons such as cancer. Therefore, leukocytes are usually not constitutively adhesive but need activation to attach. This fact is in contrast to most other cells where adhesion is not as strictly regulated or is regulated for other reasons. This has resulted in the mechanisms of activation of leukocyte adhesion becoming a major focus of study, and here we certainly have an advantage as compared with many other systems (3, 4).

The most important of the leukocyte-specific families of adhesion molecules are the integrins, notably the β_2-integrins, members of the immunoglobulin superfamily, and the carbohydrate-binding selectins (3–5).

1.2 The leukocyte β₂-integrins

Figure 1 shows members of the most important leukocyte adhesion molecule families drawn schematically. The integrins are complex proteins formed by two polypeptides (*Figure 1A*). Four β₂-integrins (CD11/CD18) are known, with a common β₂-chain (CD18) and four different α-chains (CD11a, CD11b, CD11c, and CD11d) (2, 4). They are heavily *N*-glycosylated (6). CD11a/CD18 (LFA-1, $\alpha_L\beta_2$) is enriched on lymphocytes but is also found on other leukocytes, albeit at lower levels. CD11b/CD18 (Mac-1, $\alpha_M\beta_2$) is characteristic of neutrophils but is found in smaller amounts on other leukocytes as well. CD11c/CD18 (P150/95) and CD11d/CD18 are enriched on monocytes and macrophages. The α-chains contain an approximately 200 amino acid long sequence called the I- (intervening) or A-domain. The I-domain is involved in ligand binding and its three-dimensional structure has been determined from CD11b and CD11a (7, 8).

1.3 The ICAMs

The β₂-integrins bind to the intercellular adhesion molecules (ICAMs) expressed on the surface of various cells. CD11b/CD18 and CD11c/CD18 bind to some soluble proteins as well, such as fibrinogen, complement factor C3bi, and factor X. The ICAMs are members of the immunoglobulin (Ig) superfamily containing 2–9 immunoglobulin domains (*Figure 1B*) (4).

ICAM-1 (CD54) is expressed on leukocytes, endothelial cells, and many other cells. It contains five Ig-domains, and characteristic of this ICAM is that it is induced easily by a number of cytokines. It is bound by CD11a/CD18, CD11b/CD18, and possibly by CD11c/CD18.

ICAM-2 (CD102) has only two Ig-domains and shows a more restricted distribution. It is expressed on leukocytes, platelets, and endothelial cells. In contrast to ICAM-1, it is not induced easily by commonly studied cytokines (9, 10).

ICAM-3 (CD50) is strongly expressed on leukocytes, it has five Ig-domains and it may be most important in signal transduction.

ICAM-4 has an interesting history. The Landsteiner–Wiener (LW) antigen is a classic blood group antigen exclusively expressed on red cells and their precursors. Originally, it was confused with the Rhesus (Rh) antigen, but when it was cloned and its cDNA sequenced, it was realized that it was related to the previously known ICAMs (11). Subsequently, it was shown to bind leukocytes through their β₂-integrins (12). It has been renamed ICAM-4 and, like ICAM-2, has two Ig-domains.

ICAM-5 (telencephalin) is a central neuron-specific ICAM (13, 14). It is, however, much more complex than the other described ICAMs with its nine Ig-domains. It binds β₂-integrins through its first Ig-domain, but the sixth domain also shows leukocyte binding activity (15). It may, in addition to leukocyte binding in the brain, be important in neuron–neuron interactions (16).

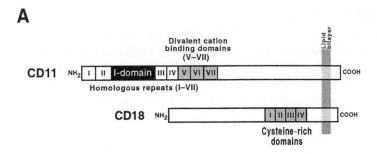

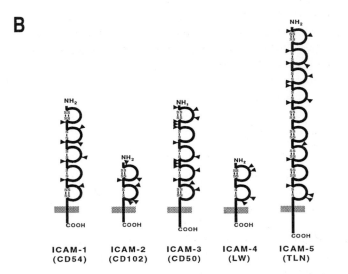

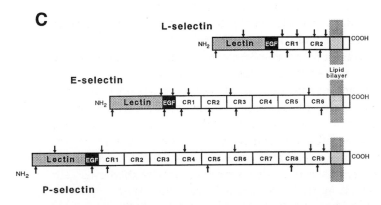

Figure 1 Schematic drawings of leukocyte adhesion molecules. (**A**) A leukocyte CD11/CD18 integrin. The α-chains (CD11) contain homologous repeats and the important ligand-binding intervening domain (I). The β_2-chain (CD18) is rich in cysteines near the membrane. (**B**) The ICAM family members. The immunoglobulin domains and the potential N-glycosylation sites (arrowheads) are shown. The CD nomenclature is given for ICAM-1, -2, and -3. LW = Landsteiner–Wiener antigen, TLN = telencephalin, a synonym for ICAM-5. (**C**) The selectins involved in leukocyte binding. The lectin and complement repeat domains are shown as well as the potential N-glycosylation sites (↑).

1.4 The selectins

Three carbohydrate-binding selectins have been described, L-selectin (CD62L), E-selectin (CD62E), and P-selectin (CD62P), their names reflecting the primary sites of expression: leukocytes, endothelial cells, and platelets, respectively (*Figure 1C*). These mammalian lectins bind to sialyl Le^x, sialyl Le^a and related oligosaccharides. L-selectin also binds to sulphatide glycolipids and is easily shed from leukocytes, which may be important for regulation of its activity. E-selectin is induced in activated endothelium, and P-selectin, which is also present in endothelial cells, becomes translocated from intracellular Weibel–Palade bodies to the cell surface upon activation. Characteristic of the selectins is their rapid response upon activation and their induction of 'rolling' of granulocytes and other haematopoietic cells. The rolling phenomenon is due to weak interactions characterized by high off rates. Importantly, during the rolling, leukocyte integrins become activated, resulting in firm adhesion.

Patients with genetic defects in the biosynthesis of selectin ligands have been described, and individuals with leukocyte adhesion deficiency type II (LAD II) show an increased incidence of microbial infections. The condition is not, however, as severe as that of patients lacking or having malfunctioning β_2-integrins (LAD type I).

There is an extensive literature on selectins (5) and they will therefore not be dealt with further here.

1.5 Other leukocyte adhesion molecules

Additional leukocyte adhesion molecules have been described, some of which are relatively important. These include β_1-integrins which are essential for leukocyte migration into tissues and interaction with extracellular matrix molecules such as fibronectin, laminins, and collagens (1, 2). These integrins are most important in non-leukocytic cells and will not be described further. Many of their central features certainly resemble those of their leukocytic counterparts.

The vascular cell adhesion molecule (VCAM) is a member of the immunoglobulin superfamily. It binds to the $\alpha_4\beta_1$-integrin and perhaps to other integrins as well. It seems to be most important in the binding of lymphoblasts to endothelium.

Several other leukocyte molecules exist which show weak binding to their target molecules. The definition of a 'real' adhesion molecule is therefore problematic. Many leukocyte cell surface proteins interact with neighbouring cells but still cannot be considered to be important in adhesion. They have other primary functions which, however, rely on protein–protein interactions and therefore result in some cellular binding.

A number of experimental protocols are used in leukocyte adhesion research, and obviously they vary in different laboratories. Those described in the following sections have been used successfully in our laboratory and certainly include the most common procedures but also some more unusual ones, reflecting the interest of our research group.

2 Activation of leukocyte adhesion

2.1 Methods to activate integrins

Most leukocytes are in suspension and are non-adhesive under normal conditions. When stimulated, however, they may become adhesive. To be able to study adhesion *in vitro*, we therefore have to activate the cells, and some useful methods are available. The 'classical' way is to use phorbol esters, molecules which penetrate the cell plasma membrane and bind to intracellular protein kinase C molecules, which then phosphorylate intracellular proteins. A more physiological means is to activate T lymphocytes through the T-cell receptor using antibodies to the CD3 molecules. Several other monoclonal antibodies reacting with various leukocyte surface molecules have been described which are able to induce leukocyte adhesion. Interestingly, it has been found recently that the ICAMs themselves also cause activation of the β_2-integrins, and a peptide derived from the first Ig-domain of ICAM-2 has turned out to be very effective (17, 18).

By inhibition of adhesion with monoclonal antibodies, potential adhesion molecules can be identified. An initial characterization is performed using cell surface radioactive labelling, immune purification, and sodium dodecyl sulphate–polyacrylamide gel electrophoresis (SDS–PAGE).

Protocol 1

Isolation and activation of T cells

Equipment and reagents

- Human buffy coats obtained from a local blood transfusion service
- Ficoll-Hypaque gradient medium (Pharmacia)
- 50-ml centrifuge tubes (Falcon)
- Nylon wool (Cellular Products Inc.)
- RPMI 1640 medium A: RPMI 1640 (Sigma) with 10% fetal calf serum (FCS), 1% L-glutamine, 1% streptomycin and penicillin
- RPMI 1640 medium B: RPMI 1640 (Sigma) with 10% FCS, 1% L-glutamine, 1% streptomycin and penicillin, 50 mM N-(2-hydroxyethyl) piperazine-N'-(2-ethanesulphonic acid) (HEPES), pH 7.4

- FCS
- Phorbol dibutyrate (PDBu; Sigma), 20 mM in dimethylsulphoxide (DMSO)
- Phosphate-buffered saline (PBS)
- 100-ml syringes
- CD3 antibody (e.g. OKT3 from ATCC clone CRL 8001)
- Integrin-activating peptide (P1) derived from ICAM-2 (ref. 17)
- Sorvall low-speed centrifuge
- Haemocytometer
- Cell microscope

A. Isolation of T cells

1 Dilute buffy coat with PBS (2:1; i.e. 25 ml of PBS to 50 ml of buffy coat)

2 Add 15 ml of Ficoll-Hypaque to each Falcon tube.

3 Carefully add 35 ml of diluted buffy coat on top of the Ficoll-Hypaque in each tube.

4 Centrifuge at room temperature at 2500 r.p.m. for 20 min. Accelerate and decelerate slowly without brakes.

5 Collect the buffer/gradient interphase layer.

6 Wash the mononuclear cell fraction in PBS four times using 1800, 1500, 1300, and 1200 r.p.m., 6 min each.

7 During the centrifugations, place the nylon wool in 100 ml syringes to form columns. Activate the columns by incubation at 37°C in RPMI 1640 medium A for 30 min.

8 Suspend the cell fraction in 5–10 ml of warm RPMI 1640 medium A and add to the columns. Leave the cell suspension in the columns for 30 min at 37°C.

9 Slowly elute the T cells with 100 ml of RPMI 1640 medium A.

10 Wash the collected fractions in PBS for use in *Protocol 1B* or *C*.

B. Activation of isolated T cells with phorbol esters

1 Suspend the cells in RPMI 1640 medium B to a density of 20–50×10^6 cells ml^{-1}.

2 Add PDBu to a final concentration of 200 nM.

3 Incubate the cells under gentle shaking for 20–60 min at 37°C.

4 Determine the percentage of aggregated cells in at least four aliquots of each sample by immediately counting single cells in the haemocytometer.

5 Treat control samples in the same way but omit the PDBu.

C. Activation with CD3 antibody

1 Prepare a cell suspension as in *Protocol 1B* step 1 and add 10 μg ml^{-1} CD3 antibody.

2 Incubate the cells with gentle shaking at 37°C for 60 min.

3 Count the proportion of single cells as in *Protocol 1B* step 4.

D. Activation with P1 activating peptide

1 Add the P1 peptide to a cell suspension of T cells to a final concentration of 40 μM.

2 Incubate the cells under gentle shaking at 37°C for 1–4 h.

3 Count the proportion of single cells as in *Protocol 1B* step 4.

3 Identification and preliminary characterization of leukocyte adhesion molecules

A useful way to find potential cell adhesion molecules is to use monoclonal antibodies to cell surface molecules. If the antibody is present during the activation and adhesion processes, it may inhibit adhesion. Obviously, monoclonal antibodies reacting with an adhesion molecule are not always inhibitory because they react with non-adhesive sites of the antigen distant from the adhesion site. Another possibility is that some adhesion molecules may be missed although the antibodies may block an adhesion site in a given molecule. The reason may be that the contribution of the molecule to adhesion at the cellular level is not that important, and other adhesion molecules are efficient enough to cover the effect of the reactive structure. To gain information about the adhesion molecule, the glycoproteins are surface-labelled (19), immunoprecipitated, and studied by SDS–PAGE (20).

Protocol 2

Identification of potential leukocyte adhesion molecules

Equipment and reagents

- Equipment, cells, and reagents as in *Protocol 1*.
- Control monoclonal antibody known not to react with cell surface molecules
- Monoclonal antibodies known to react with cell surface molecules (e.g. 7E4 antibody reacting with CD18 (21))

Method

1. Incubate T cells for 15 min at room temperature with 50 μg ml^{-1} 7E4 antibody. Incubate control cells with the same amount of control antibody.

2. Treat the cell suspension with one of the activating agents as described in *Protocol 1*.

3. Count the percentage of single cells after appropriate times (*Protocol 1B*, step 4).[a]

[a] If a substantial inhibition of aggregation is seen in the antibody-treated samples, the antigen may be an adhesion molecule.

Protocol 3

Immunoprecipitation of leukocyte extracts and analysis by SDS–PAGE

Equipment and reagents

- Sodium borohydride (NaB[^{3}H]$_4$; Amersham; 6–10 Ci mmol^{-1}) dissolved in 100 μl aliquots of 2 mCi in 0.01 M NaOH. Store at $-70\,°$C until use
- PBS, pH 7.4
- PBS, pH 8.0
- 1% Triton X-100 in PBS
- Freshly prepared 0.1 M Na periodate (21 mg ml^{-1}) in PBS, pH 8.0
- Reagents for SDS–PAGE (20)
- SDS–PAGE equipment
- Gel drying equipment
- X-ray film (Kodak BioMax)

- Fluorography reagent (Amersham NAMP 100 Amplify)
- 50–100 $\times$ 10^6 cells, for example T lymphocytes
- Monoclonal antibodies
- Rabbit anti-mouse IgG antiserum (several suppliers available)
- Protein A–Sepharose (Pharmacia)
- 1% Triton X-100 in PBS, pH 7.4
- 10% slurry of protein A–Sepharose in 1% Triton X-100 in PBS
- 1% SDS in water
- Lyophilizer

Protocol 3 continued

Method

1 Wash the cells in PBS, pH 7.4 by centrifugation and suspend in 1 ml of PBS in a plastic centrifuge tube.

2 Place the tube on crushed ice for approximately 15 min to cool down.

3 Add 20 μl of Na periodate solution to the tube, shake once, then put the cells in the dark (e.g. wrap the tube in aluminium foil) on ice for 10 min.

4 Wash the cells twice by centrifugation with PBS, pH 8.0 and suspend them in 0.5 ml of the same buffer.

5 Add 20 μl of NaB[^{3}H]$_4$ solution.

6 Incubate the cells for 30 min at room temperature in a well-ventilated hood.

7 Wash the cells three times by centrifugation in PBS, pH 8.0 and collect the supernatants as radioactive waste.

8 Add 1 ml of 1% Triton X-100 in PBS on ice, shake the tube, and leave on ice for 5 min.

9 Centrifuge the tubes at 5000 r.p.m. for 10 min and collect the supernatants for immunoprecipitation.

10 Take 100 μl of supernatant for each immunoprecipitation in microfuge tubes.

11 Add 5 μg of monoclonal antibody and incubate the samples for 2 h on ice.

12 Add 5 μl of rabbit anti-mouse IgG antiserum on ice and continue the incubation for an additional hour.

13 Add 100 μl of 10% slurry of protein A–Sepharose on ice and incubate for 15 min on ice.

14 Wash the protein A–Sepharose three times in 0.5 ml of 1% Triton X-100 in PBS. Wash once in cold water.

15 Elute the labeled antigen by boiling the protein A–Sepharose tubes in 200 μl of 1% SDS in water for 1 min.

16 Centrifuge and save the supernatants.

17 Lyophilize the supernatants.

18 Solubilize the supernatants in 10–25 μl of SDS–PAGE sample buffer. Boil for 1 min.

19 Run SDS–PAGE.

20 Incubate the gels with fluorography reagent.

21 Wash the gels in water for 1 h.

22 Dry the gels using a gel dryer.

23 Apply Kodak X-ray film in the dark in direct contact with the gel, pack the sample to exclude light, and incubate the package at $-70\,^{\circ}$C for 1–14 days, depending on the radioactivity.

24 Develop the film to visualize the radioactive antigen.

4 Phosphorylation of adhesion molecules

Activation by phorbol esters involves the protein kinase C family of serine/threonine protein kinases. Alternatively, activation through the T-cell receptor involves, in addition, tyrosine protein kinases. Therefore, it has become important to be able to study phosphorylation of integrins and other molecules in detail. Tyrosine phosphorylation is usually detected by immunoblotting with phosphotyrosine-specific antibodies, whereas serine/threonine phosphorylation requires labelling with radioactive phosphate. Recently, however, commercial phosphoserine and phosphothreonine antisera have become available (New England Biolabs). In the following protocol, we describe analysis of ^{32}P-labelled proteins (22, 23).

Protocol 4

^{32}P-labelling and analysis of T lymphocytes

Equipment and reagents

- T cells (see *Protocol 1*)
- Phosphate-free minimal essential medium (MEM; Sigma) supplemented with 5% dialysed FCS, 10 mM HEPES, 1% L-glutamine and 1% antibiotics
- [^{32}P]Orthophosphate: 10 mCi ml^{-1}, 5000 Ci mmol^{-1} in aqueous solution (Amersham, UK)
- 10 mM ethylenediamine tetraacetic acid (EDTA):PBS, pH 7.4 (1:1)
- 2 mM EDTA:PBS, pH 7.4 (1:1)
- Lysing buffer: 1% Triton X-100, 10 mM sodium phosphate, 10 mM sodium pyrophosphate, 50 mM sodium fluoride, 1 mM sodium orthovanadate, 10 mM EDTA, 350 mM NaCl, and protease inhibitors, pH 7.4. The protease inhibitors are: 5 mM iodoacetamide, 1 mM phenylmethylsulphonyl fluoride (PMSF), 10 μg ml^{-1} aprotinin (Sigma), 10 μg ml^{-1} leupeptin (Sigma), 0.1 mM quercetin (Sigma), 0.1 mM TLCK (N-p-tosyl-lysine chloromethyl ketone) (Sigma), 0.1 mM TPCK (N-p-tosyl-phenylalanyl chloromethyl ketone) (Sigma), 0.1 mM ZPCK (N-benzyloxycarbonyl-L-phenylalanine chloromethyl ketone) (Sigma)
- Protein G–Sepharose (Pharmacia)
- 0.05% Triton X-100 in 0.15 M NaCl
- Monoclonal antibody
- 1% SDS and reagents for SDS–PAGE (20)
- SDS–PAGE equipment
- Fractionation buffer 1: 1% Triton X-100, 10 mM EDTA, 50 mM NaCl, 50 mM NaF, 10 mM sodium phosphate, pH 7.4 and protease inhibitors as above
- Fractionation buffer 2: as buffer 1 above but with 300 mM NaCl and also containing 5 mM sodium pyrophosphate, 1 mM sodium orthovanadate, and 10 mM ATP
- Pyrophosphate solution: 5 mM sodium pyrophosphate, 300 mM NaCl, 1 mM sodium orthovanadate, 10 mM ATP
- Polyvinylidene difluoride (PVDF) membranes (Millipore)
- 6 M HCl
- Lyophilizer
- Phosphoserine, phosphothreonine, and phosphotyrosine (Sigma)
- Two-dimensional (2D) thin-layer electrophoresis equipment and reagents
- pH 1.9 buffer: 50 ml of 88% formic acid, 156 ml of acetic acid, 1794 ml of deionized water
- pH 3.5 buffer: 100 ml of acetic acid, 10 ml of pyridine, 1890 ml of deionized water

Protocol 4 continued

- Thin-layer chromatography (TLC) cellulose plates (Merck)
- Alkylating reagents: 100 mM dithiothreitol (DTT) in 30 mM EDTA; 200 mM iodoacetic acid
- Polyvinylpyrrolidone 360 (Sigma)
- Phosphoimager or autoradiography (*Protocol 3*, steps 20–24)
- Refrigerated microfuge (Eppendorf or other company)
- Ultracentrifuge
- Protease-free DNase I (Boehringer, Germany)
- Sequencing grade modified trypsin (activity >2.5 U mg^{-1}) (Promega)

A. *In vivo* phosphate labelling and activation of leukocytes

1 Wash the cells with phosphate-free MEM and resuspend them in the same medium at a density of 50×10^6 cells ml^{-1}

2 Pre-incubate the cells for 45–60 min at 37 °C.

3 Add 1–4 mCi of [^{32}P]orthophosphate to the cells and incubate overnight.

4 Divide the cell suspension into equal aliquots and activate with adhesion-inducing agents, for example PDBu (*Protocol 1B*), adhesion-activating antibodies (*Protocol 1C*), or ICAM-derived ligands (*Protocol 1D*). Leave the control samples untreated.

5 Stop the activation by adding ice-cold 10 mM EDTA:PBS and wash cells once with ice-cold 2 mM EDTA:PBS.

B. Extraction of adhesion molecules without fractionation of the cells

1 Lyse the cells in lysing buffer at 0 °C for 15 min. Adjust the amount of buffer to 1 μl 10^{-6} cells.

2 Centrifuge the cell lysate at 60 000 *g* for 30 min at 2 °C.

3 Pre-clear the lysate with 25 μl of protein G–Sepharose (or protein A–Sepharose) for 1 h at 4°C and immunoprecipitate the adhesion molecules with monoclonal antibody covalently coupled to protein G (or A)–Sepharose for 2.5 h at 4 °C under constant rolling.

4 Wash the Sepharose beads extensively three times with 0.05% Triton X-100 in 0.15 M NaCl.

5 Elute the adhesion molecules with 1% SDS at 40°C for 40 min and subject the sample to SDS–PAGE.

6 Expose the gels to autoradiography or to a phosphoimager.

C. Preparation of subcellular fractions from phospholabelled cells

1 Lyse the cells at 0°C for 15 min in fractionation buffer 1.

2 Centrifuge the cell lysate in the refrigerated microfuge at 4000 r.p.m. for 5 min at 4 °C. The insoluble pellet is the nuclear fraction.

3 Centrifuge the supernatant further at 100 000 *g* for 2 h at 2 °C. The pellet is the cytoskeletal fraction and the supernatant the soluble fraction.

4   Solubilize the nuclear and cytoskeletal fractions in fractionation buffer 2. Cyto-skeletally bound proteins are released by incubation with 2 U of protease-free DNase I for 50 min at 22 °C. The solubilized proteins are obtained after centrifugation and subjected to immunoprecipitation as described in *Protocol 3*.

5   Add 1 ml of pyrophosphate solution to 100 μl of the soluble fraction. The fraction is subjected to immunoprecipitation as above (*Protocol 3*).

D. Phosphoamino acid analysis of phosphorylated adhesion molecules

1   Electrophorese immunoprecipitated adhesion molecules from ^{32}P-labelled cells on SDS–PAGE and blot them onto PVDF membranes, which are subjected to autoradio-graphy or to phosphoimaging.

2   Excise the region containing the desired protein.

3   Hydrolyse the protein on the membrane with 6 M HCl for 2 h at 110 °C.

4   Lyophilize the hydrolysate, add 100 μl of water, and lyophilize again. Repeat this several times. Add 20 μg each of standard phosphoserine, phosphothreonine, and phosphotyrosine.

5   Analyse phosphoamino acids by 2D thin-layer electrophoresis (pH 1.9 in the first and pH 3.5 in the second dimension) on cellulose plates (22, 23).

6   Visualize the standard amino acids by ninhydrin staining (*Figure 2A*).

7   The radioactive amino acids are detected by autoradiography or phosphoimaging (*Figure 2B*).

E. Determination of phosphorylation sites in adhesion molecules by phosphopeptide mapping

1   Lyophilize the ^{32}P-labelled immunoprecipitated adhesion molecules.

2   Add 8 ml of water and 8 ml of SDS–PAGE sample buffer without 2-mercaptoethanol. Then add 2 ml of DTT/EDTA alkylating reagent and boil for 1 min.

3   Add 2 ml of iodoacetic acid alkylating reagent and incubate for 15 min at room temperature.

4   Electrophorese the proteins by SDS–PAGE.

5   Blot the electrophoresed proteins onto PVDF membranes and localize them by autoradiography or phosphoimaging.

6   Excise the region containing the desired adhesion molecule.

7   Saturate the filter pieces with polyvinylpyrrolidone 360 at 37 °C for 1 h and wash several times with water.

8   Digest the filter pieces with trypsin at 37 °C overnight.

9   Wash the filters four times with water.

10 Separate the resulting peptides in the first dimension by electrophoresis at pH 8.0 and in the second dimension by ascending chromatography (22, 23)

Protocol 4 continued

11 Visualize the phospholabelled peptides by autoradiography or phosphoimaging.

12 Phosphorylation sites in the adhesion molecules are identified by simultaneous separation of synthetic phosphopeptides visualized by ninhydrin staining.

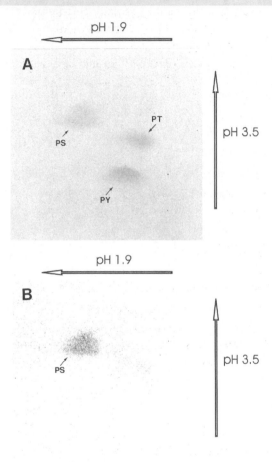

Figure 2 Analysis of phosphorylated amino acids. The phosphoamino acids have been separated by two-dimensional electrophoresis at pH 1.9 and 3.5. (**A**) Ninhydrin-stained phosphoamino acids. (**B**) Radioactive phosphoamino acids. PS = phosphoserine, PT = phosphothreonine, PY = phosphotyrosine.

5 Isolation of active β_2-integrins

To study cell–cell interactions *in vitro*, it is often essential to use purified, active adhesion molecules. To acquire active integrins, they must be rapidly isolated from fresh cells in the presence of divalent cations (24).

Protocol 5

Isolation of leukocyte integrins in an active form

Equipment and reagents

- 10–100 buffy coats obtained fresh from a local blood transfusion service
- Lysing buffer: 1% Triton X-100, 2 mM $MgCl_2$ in PBS, pH 7.4
- pH 11.5 buffer: 0.1 M $NaHCO_3$–NaOH, 2 mM $MgCl_2$, 1% octyl glucoside
- 1 M acetic acid
- Monoclonal anti-integrin antibodies (e.g. anti-CD18 antibody 7E4) coupled to Sepharose in column (21)
- Low-speed centrifuge (e.g. Sorvall RC5C)
- High-speed centrifuge (e.g. Beckman L4)

Method

1 Wash the buffy coat cells in PBS.

2 Lyse the cells at 0°C in lysing buffer.

3 Centrifuge in the low-speed centrifuge at 10 000 r.p.m. for 10 min.

4 Recover the supernatant and centrifuge at 100 000 g for 60 min.

5 Recover the supernatant and add it to the monoclonal antibody–Sepharose column at 4°C.

6 Wash the column with 10 vols of cold lysing buffer.

7 Elute the integrins with pH 11.5 buffer.

8 Collect 1–2 ml fractions and check the pH with pH paper.

9 Neutralize the first 3–4 alkaline fractions (which contain the integrins) immediately with 1 M acetic acid.

10 Check the purity of the fractions using SDS–PAGE and Coomassie blue or silver staining.

6 Preparation of I-domains

The I-domains in the α-chains of the β_2-integrins are directly involved in adhesion. Both ICAMs and soluble proteins bind to this part of the polypeptides. The I-domains can be expressed in bacteria, which makes their preparation relatively easy. High expression may, however, result in precipitation in inclusion bodies.

Protocol 6

Expression and purification of the CD11a I-domain

Equipment and reagents

- *Escherichia coli* strain BL21 containing the pGex-5X-2 and I-domain constructs (obtained from Dr D. Altieri, Yale University)
- LB plates containing 100 μg ml^{-1} ampicillin
- LB media containing 100 and 50 μg ml^{-1} ampicillin

- Spectrophotometer
- 0.2 M isopropyl-β-D-thiogalactopyranoside (IPTG) solution
- Sorvall RC5C centrifuge, rotors GSA and SS34
- Shaker for bacterial growth at 37°C
- Lysing buffer: 50 mM Tris–HCl, pH 8.0, 10% sucrose, 0.5% Triton X-100, 5 mM EDTA, 1 mM DTT, 1 mM PMSF, 5 μg ml^{-1} aprotinin, 5 μg ml^{-1} leupeptin (Sigma)
- Lysozyme powder (Sigma)
- Sonicator (Sonifier Cell Disruptor B-30, Branson Sonic Power Co.)
- 2% Triton X-100, 50 mM Tris–HCl, pH 8.0, 5 μg ml^{-1} aprotinin, 5 μg ml^{-1} leupeptin

- SDS–PAGE equipment
- Urea solution: 7 M urea, 5 mM MnCl$_2$, 10 mM Tris–HCl, pH 7.5, 5 mM 2-mercaptoethanol
- Buffer 1: 5 mM MnCl$_2$, 10 mM Tris–HCl, pH 7.5, 5 mM 2-mercaptoethanol
- Buffer 2: 50 mM Tris–HCl, pH 8.0
- Buffer 3: 50 mM Tris–HCl, pH 8.0, 1% Triton X-100, 0.5 M NaCl
- Buffer 4: 50 mM Tris–HCl, pH 8.0, 20 mM glutathione
- Glutathione–Sepharose 4B (Pharmacia)
- Biogel P-10 column made in 20 mM Tris–HCl, pH 8.0
- Factor Xa (optimal; Boehringer Mannheim)

A. Protein expression

1   Grow the recombinant *E.coli* on LB-ampicillin plates at 37°C.

2   Pick up four single colonies from the LB plate and put each in 100 ml of LB medium containing 100 μg ml^{-1} ampicillin, shake overnight at 37°C.

3   Add 50 ml of the *E.coli* suspension to 500 ml of LB medium containing 50 μg ml^{-1} ampicillin, making eight incubation flasks. Shake for about 2–3 h at 37°C.

4   Check the optical density (OD) at 600 nm every 30 min. When the OD is 0.5–2, start IPTG induction.

5   Add IPTG to the cultures, making the final concentration 0.2 mM. Shake for 3–4 h at 37°C and then take a 50 μl sample of the culture for SDS–PAGE.

6   Harvest the *E.coli* by centrifugation at 10000 r.p.m. for 10 min at 4°C using the Sorvall rotor GSA.

7   Combine the pellets and wash with PBS. Centrifuge once more at 10000 r.p.m. for 10 min at 4°C. Store the pellet at −70°C.

B. Lysis of *E.coli*

1   Melt the frozen bacterial pellet at room temperature. After melting, keep the sample on ice. Lyse the cells by adding 20–30 ml of cold lysing buffer to the pellet. Transfer it to a 50 ml Sorvall tube.

2   Add lysozyme to a final concentration of 350 μg ml^{-1}.

3   End-to-end shake the sample for 1 h at 4°C.

4   Incubate the sample at 37°C for 10 min.

5   Sonicate the sample for 10 s (OUTPUT CRTL 6, duty cycle 60%), put on ice for 30–60 s. Repeat 10 times. Avoid foaming of the sample.

6   Add 5 ml of the 2% Triton X-100 solution to 10 ml amounts of lysate.

7 Sonicate the sample for 10 s as above, put on ice for 30–60 s. Repeat five times. Avoid foaming of the sample.

8 Centrifuge the sample at 13 000 r.p.m. for 45 min at 40 °C using rotor SS34.

9 Collect the supernatant. Keep a small amount of the supernatant and pellet for SDS–PAGE. Store both the pellet and supernatant at −70 °C.

C. Solubilization of the fusion protein from the pellet (optional)

1 Dissolve the pellet with 10 ml of urea solution. End-to-end shake for 2 h at room temperature.

2 Centrifuge at 13 000 r.p.m. for 45 min at 4 °C using rotor SS34. Collect the supernatant. Keep a small amount of the supernatant and pellet for SDS–PAGE.

3 To renature the protein, remove urea by dialysing the supernatant step by step into buffer 1. If there is a precipitate after the dialysis, centrifuge as above. Keep a small amount of the supernatant and pellet for SDS–PAGE. Save the supernatant and discard the pellet.

D. Purification of the soluble fusion protein from the supernatants

1 Follow the instructions of Pharmacia Biotech for preparation of a 2-ml bed volume of glutathione–Sepharose 4B gel required for 4000 ml of *E.coli* culture. Use a bench centrifuge at 2000 r.p.m. for 10 min at 4 °C to sediment the gel matrix.

2 Before use, equilibriate the matrix by washing three times with buffer 3.

3 Add the supernatant to the tube containing the equilibrated glutathione–Sepharose 4B gel. End-to-end shake for 2 h at 4 °C. Centrifuge at 2000 r.p.m. for 10 min.

4 Remove the supernatant and take a 50 μl sample of both the supernatant and the sedimented matrix for SDS–PAGE. Store the supernatant at −70 °C.

5 Wash the glutathione–Sepharose 4B gel as follows: buffer 2, about 50 ml; buffer 3, 50 ml; buffer 2, about 80 ml. Collect the washing solutions in separate tubes and take a sample of each for SDS–PAGE.

6 Pack the gel into a suitable column. Elute the bound material with buffer 4. Collect 8–10 fractions of 1 ml and check them by SDS–PAGE to detect the fusion protein. Run the SDS–PAGE and include all the samples taken so far.

7 Pool the fractions containing the fusion protein. To remove the glutathione and further purify the protein, run the pooled fractions on the Biogel P-10 column with 20 mM Tris–HCl, pH 8.0.

8 Pool the fractions containing the protein. Measure the protein concentration using a protein assay reagent kit and concentrate the sample if necessary.

9 Cleave off the glutathione *S*-transferase (GST) from the fusion protein if necessary with overnight treatment with factor Xa protease at a substrate ratio of 10% (w/w). Check with SDS–PAGE.

10 To remove GST, repeat the affinity chromatography and collect the flow through or run ion exchange chromatography using high-performance liquid chromatography (HPLC).

7 Construction and purification of Fc chimeric proteins

Fusion proteins consisting of the extracellular region of the molecules of interest and the Fc region of human IgG_1 have been used successfully in biochemical experiments to identify unknown ligands or receptors. The Fc chimeric proteins have several advantages. First, because these proteins are produced in mammalian cells (e.g. COS cells), they undergo appropriate post-translational processing. Secondly, they become water-soluble when released from the transfected cells. Thirdly, the Fc chimeric proteins can be purified easily from culture supernatants by protein A affinity chromatography. They are detected by anti-human IgG antibodies conjugated with enzymes or fluorescent tags.

There are also disadvantages with the Fc chimeric proteins. In some cases, the Fc region binds to cellular receptors. The method is usually applicable only to type I integral membrane proteins and secretory proteins that have hydrophobic signal peptides in their N-termini. Otherwise, an artificial signal peptide must be added to the N-terminus of the protein.

The vector used for the construction of Fc chimeric plasmids in this protocol is pEF-Fc (6.8 kb). The backbone of pEF-Fc is pEF-BOS (25), a mammalian expression vector containing the cloning sites *Bst*XI or *Xba*I, elongation factor-1× promoter, granulocyte colony-stimulating factor polyadenylation signal, and pUC119 backbone (ampicillin resistance). pEF-Fc is constructed by inserting into pEF-BOS the multicloning sites and the Fc region (hinge, CH2, CH3 domains) of human IgG_1 genomic DNA (14, 26).

Since the Fc region of human IgG_1 in pEF-Fc is a genomic DNA fragment containing introns, an artificial splicing donor signal (GTGAGT) must be added at the 3′ terminus of the cDNA insert. In addition, because the first codon of the hinge region contains two nucleotides (GA), the last codon of the extracellular region must be made by adding one nucleotide, preferably G to be translated to glycine (GGA). T should never be used as the last nucleotide of the extracellular region because an opal stop codon (TGA) then results. A default design of the polymerase chain reaction (PCR) primer is the following:

```
                                          Splicing RE recognition
            20–25 nucleotides of the 3′ region   donor site
  5′-XXX XXX XXX XXX XXX XXX XXX XXX G GTGAGT XXX XXX CGG-3′
  3′-XXX XXX XXX XXX XXX XXX XXX XXX C CACTCA XXX XXX GCC-5′
```

Protocol 7

Construction of Fc chimeric plasmids and preparation of Fc-containing protein

Equipment and reagents

- Primers: forward (sense) primer, usually a 5'-non-coding sequence of the cDNA with *Sal*I or *Cla*I sites at the 5' terminus; reverse (antisense) primer, 3'-sequence of the extracellular region with an artificial splicing donor signal and an *Spe*I or *Bam*HI site at the 5'-terminus[a]
- PCR thermocycler
- PCR mixture: 1–10 ng of template, 0.5–1 μM sense primer, 0.5–1 μM antisense primer, 100–200 μM dNTPs, 1 U of *Pfu* DNA polymerase, 5–10 μl of 10× buffer, water up to 50–100 μl
- Digestion solution: purified PCR product (~500 ng) or pEF-Fc vector (~1 μg), 5 μl of 10× buffer, 1–2 μl of 5' restriction endonuclease, 1–2 μl of 3' restriction endonuclease, water up to 50 μl
- Ligation solution: digested pEF-Fc vector, digested PCR product as insert (vector: insert ~1:4), 1 μl of 10× buffer, 0.5 μl of T4 DNA ligase, water up to 10 μl
- Competent *E.coli* (DH5× or JM109)
- LB agar plates containing 100 μg ml[-1] ampicillin
- COS7 mammalian cells

- Serum-free Dulbecco's modified Eagle's medium (DMEM): DMEM, 10 mM Tris–HCl, pH 7.4, 1% L-glutamine, 1% penicillin + streptomycin[b]
- Serum-free DMEM/diethylaminoethyl (DEAE)-dextran: DMEM, 10 mM Tris–HCl, pH 7.4, 1% L-glutamine, 1% penicillin + streptomycin, 600 mg ml[-1] DEAE-dextran (0.15 ml per 10 ml of 40 mg ml[-1] stock stored at −20 °C)[b]
- DNA solution: 1:5:6 mixture of plasmid DNA in TE, pH 7.5; serum-free DMEM; serum-free DMEM/DEAE-dextran[b,c]
- 10 mM Tris–HCl, pH 7.4 (TBS)[b]
- TBS/glycerol: TBS with 2 mM CaCl$_2$, 2 mM MgCl$_2$, 20% glycerol
- DMEM with serum: DMEM with 10% FCS, 1% L-glutamine, 1% penicillin + streptomycin[b]
- Centricon concentrator (Amicon)
- 1 M and 100 mM Tris–HCl, pH 8.0
- Protein A column
- 1 M glycine buffer, pH 3.0
- 1 M Tris–HCl, pH 9.5

A. PCR

1 Perform PCR by standard methods. Use purified cDNA (fragment of plasmid) or reverse-transcribed crude cDNA from tissues as a template for PCR.

2 Set PCR cycles as: (i) 95 °C, 5 min, followed by (ii) 25–35 cycles: 95 °C, 30–60 s; 50–65 °C, 30 s; 72 °C, 30–90 s; and 72 °C, 3–5 min.

B. Digestion and ligation of the PCR product and the pEF-Fc vector

1 Purify the PCR product and digest it at 37 °C for 1 h with the digestion solution containing two appropriate restriction endonucleases which cleave the 5' and 3' ends of the product, respectively. Transfer to 65 °C for 20 min to stop the reaction.

2 Digest the pEF-Fc vector with the same two enzymes as above.

Protocol 7 continued

3 Purify the digested DNAs and ligate them at 16 °C overnight using the ligation solution.

C. Transformation and screening

1 Transform the ligated plasmid into competent *E.coli* (DH5× or JM109).

2 Grow the *E.coli* on LB agar plates containing 100 μg ml^{-1} ampicillin.

3 Select appropriate clones by restriction mapping.

4 Confirm the sequence by DNA sequencing as PCR errors may cause production of recombinant proteins with no or altered functions.

D. Transient transfection of cDNA into mammalian cells (DEAE-dextran method)

1 For small-scale (SS) production, seed COS7 cells at a density of 1×10^4–2×10^5 cells per well (6-well plate), or for large-scale (LS) production, plate cells at a density of 4×10^6 cells per 15-cm dish (total 30×15-cm dishes).[d]

2 Begin transfection the following day by washing cells three times with 2 ml (SS) or 10 ml (LS) of serum-free DMEM.

3 Add 600 μl of DNA solution (2 μg of plasmid DNA per well for SS production) or 6.5 ml of DNA solution (10 μg of plasmid DNA per dish for LS production).

4 Incubate at 37 °C for 5 h in 5% CO_2.

5 Remove the DNA solution and add 1 ml (SS) or 5 ml (LS) of TBS/glycerol and incubate at room temperature for 90 s.

6 Wash the cells twice with 2 ml (SS) or 10 ml (LS) of serum-free DMEM.

7 Add 2 ml (SS) or 15 ml (LS) of DMEM with serum and incubate overnight at 37 °C in 5% CO_2.

8 Change medium the next day to remove dead cells. Wash cells three times for 5 min at room temperature with serum-free DMEM.

9 Add 2 ml (SS) or 15 ml (LS) of serum-free DMEM and incubate for 2 days (SS) or 3 days (LS) at 37 °C in 5% CO_2.

10 For SS production only, after 2 days take 2 ml of supernatant medium, concentrate by Centricon, and apply 1/10 vol. to SDS–PAGE and Western blotting.

11 For LS production only, after 3 days harvest the culture supernatant (total 450 ml) into 9×50 ml tubes, centrifuge at 3000 r.p.m. for 20 min at 4 °C, recover the supernatant, and add 1/10 vol. (50 ml) of 1 M Tris–HCl (pH 8.0).

E. Purification of Fc fusion proteins

1 Immediately apply the supernatant onto a pre-washed protein A column at 4 °C overnight.

2 Wash the column with 100 mM Tris–HCl (pH 8.0).

3 Elute the bound proteins with glycine buffer into 10 fractions, 1 ml per fraction, and neutralize with 100 μl per fraction of 1 M Tris–HCl (pH 9.5).

4 Regenerate the column.

[a] Do not add a restriction site that is present in the PCR-amplified region of the cDNA.

[b] Prepare transfection media and reagents on the day of transfection.

[c] DNA solution should be mixed well; see method for concentration of plasmid DNA.

[d] Cell density depends on cell line and purpose, e.g. for staining of COS7 cells, 1×10^4–5×10^4; for Western analysis, 5×10^4–2×10^5.

8 Cell adhesion assay

Cell adhesion using purified reagents is usually performed using plastic dishes onto which the protein ligand is coated. Quantitation of adherent cells can be either by counting several microscopic fields, by enzyme assays, or by using radioactive cells. One of the most difficult steps to standardize is the washing because adherent cells are often only weakly bound. Therefore, we have found it very useful to perform the washing by turning the plates upside down in buffer.

Protocol 8

Cell adhesion assays

Equipment and reagents

- 96-well plates (Nunc, Roskilde, Denmark)
- Binding medium: RPMI 1640 or DMEM containing 50 mM N-(2-hydroxyethyl) piperazine-N'-(2-ethanesulphonic acid) (HEPES), 2 mM MgCl₂, 0.5% bovine serum albumin (BSA)
- 1% BSA
- Coating buffer: 25 mM Tris, pH 8.0, 150 mM NaCl, 2 mM MgCl₂
- Lysing buffer: 1% Triton X-100 in 50 mM sodium acetate, pH 5.0, containing 3 mg ml⁻¹ p-nitrophenyl phosphate (Sigma)
- Spectrophotometer

Method

1 Coat various amounts (0.1–1 μg) of purified ICAM proteins, leukocyte integrins, or control proteins on plastic 96-well plates in coating buffer at 4°C overnight.

2 Block the wells with 1% BSA by coating for an additional 1–2 h at room temperature.

3 Wash the wells 2–3 times with binding medium.

4 For blocking experiments, pre-treat the cells or protein-coated wells for 10–20 min with inhibitory monoclonal antibodies at final concentrations of 10–40 μg ml⁻¹.

5 Add the cells (leukocytes, red cells, or adhesion molecule-transfected cells) in a total volume of 100 μl and incubate for 20 min at room temperature for leukocytes or transfected cells and for 60 min at 37°C for red cells.

Protocol 8 continued

6 Fill the plates carefully with binding buffer.

7 To remove non-adherent cells, place the microplates to float upside down for 1 h in PBS.

8 After washing, lyse the bound leukocytes or transfected cells by adding 100 μl per well of lysing buffer and incubate at 37 °C for 30 min.

9 Terminate the reaction by adding 50 μl per well of 1 M NaOH.

10 Measure the absorbance at 405 nm. Calculate the percentage of bound cells as:

$$\% \text{ bound cells} = \frac{A_{405} \text{ bound cells/well}}{A_{405} \text{ total amount of cells/well}} \times 100$$

11 Quantify the binding of red cells by counting bound cells in four randomly chosen fields from duplicate wells, omitting steps 8–10.

9 Detection of ICAMs

It has become apparent that ICAMs are released from cells in soluble form. Either the whole extracellular portion or parts of it are cleaved from the cells. Evidently, this often occurs during inflammatory conditions, perhaps due to increased expression of ICAMs and high protease activity. Because ICAM assays may be very important for diagnostic purposes, we describe below the assay for soluble ICAM-5. The TL-1 and TL-3 monoclonal antibodies to ICAM-5 have been described recently (15).

Protocol 9

Detection of soluble ICAM-5 (sICAM-5) in the cerebrospinal fluid using sandwich enzyme-linked immunosorbent assay (ELISA)

Equipment and reagents

- 96-well microtitre plate (flat bottom ELISA plate, Greiner)

- Microtitre plate reader

- Recombinant soluble Fc fusion proteins: ICAM-5–Fc protein which contains domains 1–2 of human ICAM-5 fused with the Fc region of human IgG1; ICAM-1–Fc recombinant protein which contains the extracellular part of human ICAM-1 fused with the Fc region of human IgG1

- Monoclonal antibodies: TL-3 which recognizes domain 1 of human ICAM-5 (15); biotinylated TL-1 which recognizes domains 1–2 of human ICAM-5 (15)[a]

- Washing buffer: 0.05% Tween-20 in PBS

- Blocking buffer: 0.05% Tween-20, 1% BSA, in PBS (make fresh for each ELISA)

- Dilution series of ICAM-5–Fc in blocking buffer: 10, 20, 40, 80, 160, 320, 640, and 1280 pg ml^{-1}.

- Plating buffer: PBS, pH 7.4
- ICAM-1–Fc diluted to 1280 pg ml^{-1} in blocking buffer
- Biotinylated TL-1 diluted to 5 g l^{-1} in blocking buffer
- Horseradish peroxidase (HRP)-conjugated streptavidin (1 mg ml^{-1}, Pierce) diluted 1:1000 in blocking buffer

- Citrate buffer: 0.1 M Na$_2$HPO$_4$, 0.1 M citric acid, pH 5.0
- Substrate solution: dissolve 10 mg of o-phenylenediamine in 10 ml of citrate buffer and add 1 μl of 30% H$_2$O$_2$ just before use
- Stop solution: 10% H$_2$SO$_4$

Method

1 Coat a 96-well microtitre plate with the first monoclonal antibody, TL-3, diluted to 10 μg ml^{-1} in plating buffer with 50 μl per well for 3 h at 37 °C.

2 Wash the wells three times with washing buffer, block with blocking buffer for 1 h at 37 °C, and then wash the wells again three times with washing buffer.

3 Add 50 μl of each dilution of ICAM-5–Fc in blocking buffer into wells in triplicate to act as protein standard. Add ICAM-1–Fc diluted to 1280 pg ml^{-1} in blocking buffer as a negative control and blocking buffer alone as a blank control.

4 Dilute the cerebrospinal fluid samples from patients[b] in blocking buffer if needed, and add to the wells as duplicates.

5 Allow the antigens to bind to the antibody-coated wells overnight at 4 °C, after which the wells of the plate are washed in washing buffer as before.

6 Add 50 μl per well of the second antibody, biotinylated TL-1, in blocking buffer and incubate the plate for 1 h at 37 °C.

7 Wash the wells of the plate as above.

8 Add 50 μl per well of HRP-conjugated streptavidin solution and allow it to bind to the biotin molecules in the second antibody for 1 h at 37 °C.

9 Wash the wells of the plate as before and prepare the substrate solution.

10 Add 50 μl of substrate solution to each well and incubate the plates at room temperature until the yellowish colour appears (5–7 min).

11 Add 50 μl of 10% H$_2$SO$_4$ to each well to stop the reaction.

12 Quantify the binding assay using a microtitrr plate reader at 492 nm.

[a] The biotinylation of TL-1 was carried out using the EZ-Link Sulfo-NHS Biotinylation Kit (Pierce). Both monoclonal antibodies were produced by a hybridoma cell line in our laboratory.

[b] Cerebrospinal fluid is obtained by lumbar puncture at 8–9 a.m. following overnight fasting and is immediately frozen for storage at −70 °C.

10 Use of phage display libraries for isolation of peptide ligands to β$_2$-integrins

Libraries of random peptides displayed on filamentous phage provide an important source of potential ligands to integrins and other cell surface receptors (27).

Many integrins recognize short peptides of 4–9 amino acid residues in length, and phage display libraries have become a convenient means to characterize the peptide-binding specificity of those integrins (28). This has been accomplished with several integrins including $\alpha_5\beta_1$ (29, 30), $\alpha_v\beta_3$ (31), $\alpha_{IIb}\beta_3$ (31, 32), and $\alpha_v\beta_6$ (33) which recognize peptides containing the RGD tripeptide motif. By development of specific phage binding and selection procedures, it has been possible to discover relatively selective peptide ligands to each of these integrins. However, it has been more cumbersome to isolate binders to other members of the integrin family that do not recognize the RGD motif. This is probably due to the fact that those integrins bind large surface areas of proteins such as immunoglobulin folds, or laminin or collagen repeats. Thus, the binding determinants may not be present in libraries of short peptides and an individual peptide may exhibit such a weak affinity that it is not enriched during phage library panning. So far, there have been only a few reports of phage library screening with non-RGD-directed integrins, one report focusing on the $\alpha_6\beta_1$ integrin (34) and one on the $\alpha_M\beta_2$ integrin (35).

We describe here a procedure for isolation of β_2 integrin-binding peptides from phage display libraries. We found two parameters to be important for a successful binding and enrichment of phage. First, the leukocyte integrin preparation needs to be used within a couple of days after its isolation; integrin activity is lost during storage and freezing–thawing for reasons we do not know. Secondly, if short peptide inserts containing 10 or fewer random amino acids are screened, the phage-displayed peptides should carry conformation-restricting determinants such as four cysteine residues making disulphide bridges. The β_2-integrins recognize immunoglobulin-like domains, thus having poor binding activity for short linear peptides. We observed that when at least two conformation-stabilizing disulfide bonds were present on the peptides, the phage readily bound to the β_2-integrins. When longer peptide inserts of compact domains of ICAM molecules are displayed on the phage, it is not necessary to include such conformation constrains. However, an important point here is to use phage libraries possessing as much peptide diversity as possible. The current phage display methods allow construction of libraries of peptides of 5–8 amino acids carrying up to 10^9–10^{10} different members. However, in the case of 10 or more variable amino acid residues, such high peptide diversity may be tedious, if not impossible, to obtain (36, 37).

Protocol 10

Identification of integrin-binding sequences by phage display

Equipment and reagents

- 96-well microtitre plates (Greiner, Costar, or Nunc)
- TBS: Tris-buffered saline, pH 7.4

- 1 M MnCl$_2$ stock solution in water
- TBS, 25 mM *n*-octyl-β-D-glucopyranoside (TBS/OGP; Calbiochem)

Protocol 10 continued

- TBS, 1 mM MnCl$_2$ (TBS/MnCl$_2$)
- Leukocyte integrin preparation
- TBS, 5% BSA (TBS/BSA)
- TBS, 5% BSA, 1 mM MnCl$_2$ (TBS/BSA/MnCl$_2$)
- Phage library
- Pilus-positive bacterial strain K91kan, for amplification of filamentous phage (30)
- Terrific Broth medium (see ref. 38)
- LB medium
- TBS, 0.5% Tween-20 (TBS/Tween)

- 0.1 M glycine-HCl, pH 2.2, 0.1% BSA, 0.01% phenol red (Gly/BSA)
- Resistance antibiotic (tetracycline)
- Polyethylene glycol
- TBS, 0.02% NaN$_3$ (TBS/NaN$_3$)
- Sequencing primers for PCR
- Spectrophotometer
- PCR thermocycler
- Dye Terminator Cycle Sequencing Core KIT and ABI 310 Genetic analyzer (PE Applied Biosystems)

Method

1 Purify the desired species of the β$_2$-integrin by immunoaffinity chromatography (see *Protocol 5*).

2 Coat the integrin on 96-well microtitre plates that possess high protein-binding capacity. Pipette 1 μg of the integrin per well in TBS/OGP.

3 Add 200 μl of TBS/MnCl$_2$ per well to dilute the OGP, thus inducing better coating efficiency for the integrin.[a] Incubate the plate overnight at 4°C.

4 Remove the integrin solution and saturate the wells in TBS/BSA for 1 h at 22°C. Wash five times with TBS/BSA and empty the wells by shaking against a paper towel.

5 For the first panning, incubate each phage library in a different well in 100 μl of TBS/BSA/MnCl$_2$ with constant shaking at 4°C. Use at least 100 equivalents of the diversity library, thus approximately 10^{11}–10^{12} transducing units (tu) of phage. To improve the chance of finding good peptide ligands, try different types of peptide inserts.[b] The same day, start an overnight 2 ml culture of bacteria such as K91kan.

6 The following day, dilute a 100 μl aliquot of the bacteria in 10 ml of Terrific Broth media and let the bacteria grow for 3–5 h until 1/10 dilution of the culture reaches an OD of 0.1–0.25 at 600 nm.

7 Remove the unbound phage by washing 10 times with TBS/Tween. Press the plate against paper towels after each wash.

8 Elute the bound phage with 100 μl of Gly/BSA for 10 min. Transfer the eluate to a 50 ml volume tube containing 100 μl of the competent bacteria as described above. Neutralize the pH. Wash the well and the pipette tip with another 100 μl of the bacteria to collect any remaining phage, and combine the bacterial solutions.

9 Incubate for 15–30 min at 22°C and add 20 ml of 0.2 μg ml^{-1} tetracycline antibiotic, against which the phage carries a resistance gene, for a period of 30–60 min at 37°C. Then, adjust the tetracycline concentration to 20 μg ml^{-1}. At this point, titre the bound phage by making several dilutions on agar plates containing tetracycline.[c]

10 Grow the bacteria for 24 h at 37 °C with constant shaking. Centrifuge for 10 min at 10 000 r.p.m. to pellet the bacteria. To ensure pelleting of all cell remnants, repeat the centrifugation. Precipitate the phage from the supernatant by adding polyethylene glycol and NaCl as described (28, 37). Repeat the precipitation after the phage is dissolved in TBS. Finally, dissolve the phage in 100 μl of TBS/NaN$_3$.

11 Perform a second round of panning with the amplified phage. To determine the extent to which integrin-binding phage is enriched, add the same amount of phage to a BSA-coated well as that added to an integrin-coated well. Use duplicate or triplicate wells if necessary. If desired, examine the effect of different coating concentrations of the integrins as well as different dilutions of phage such as 10^8, 10^9, 10^{10}, 10^{11}, and 10^{12} tu per well. After the binding, washing, and elution steps as described above, titre the phage bound to each well and prepare agar plates.

12 Perform up to five rounds of pannings to enrich the integrin-binding phage further, if necessary.

13 Sequence the phage directly from a bacterial colony in an agar plate without further purification of the phage. Pick up each bacterial colony with a sterile pipette tip and mix with 10 μl of TBS in a microtitre well, thereby allowing up to 96 samples to be stored in one 96-well plate at −20 °C. To sequence the FUSE5 phage, thaw the samples and take a 1 μl aliquot to subject to PCR with 10 pmol each of the forward primer 5'-TAATACGACTCACTATAGGGCAAGCTGAATAAACCGATACAATT-3' and the reverse primer 5'-CCCTCATAGTTAGCGTAACGATCT-3'. The PCR conditions are 92 °C for 30 s, 60 °C for 30 s, and 72 °C for 60 s, and the cycle number is 35. Apply 1 μl of the PCR for direct sequencing. Use either one of the primers above (15 pmol per reaction).[d]

14 To amplify a phage clone for which a DNA sequence has been obtained, take a 1 μl aliquot from the stock stored in the microtitre plate, and grow in K91kan bacteria overnight. Verify the phage binding to the integrin. For further studies, the binding peptide can be reproduced either as a synthetic peptide, or made as part of recombinant GST fusion protein in *E.coli* as described (38).

[a] The Mn^{2+} cation stimulates a high affinity state of the integrin and improves the peptide-binding capacity.

[b] We have experience of the FUSE5 phage vector from several years and, for example, we have constructed and employed the libraries CX$_7$C, CX$_8$C, CX$_9$C, CX$_{10}$C, CX$_3$CX$_3$CX$_3$C, CX$_3$CX$_4$CX$_2$C, and CXCX$_2$CX$_3$CX$_2$CXC, where C is a cysteine residue and X is any of the 20 amino acids.

[c] Alternatively, we try to collect all phage after the first round of panning and therefore usually titre the bound phage only after the second and subsequent pannings when the phage is already amplified.

[d] We usually sequence the DNA in both directions.

Acknowledgements

We thank Yvonne Heinilä for expert secretarial assistance. The scientific work on which this chapter is based was supported by the Sigrid Jusélius Foundation, the Academy of Finland, and the Finnish Cancer Society.

References

1. Ruoslahti, E. (1991). *J. Clin. Invest.*, **87**, 1.
2. Hynes, R. O. (1992). *Cell*, **69**, 11.
3. Springer, T. A. (1990). *Nature*, **346**, 425.
4. Gahmberg, C. G., Tolvanen, M., and Kotovuori, P. (1997). *Eur. J. Biochem.*, **245**, 215.
5. Westweber, D. (ed.) (1997). *The selectins*. Harwood Academic Publishers, Amsterdam.
6. Asada, M., Furukawa, K., Kantor, C., Gahmberg, C. G., and Kobata, A. (1991). *Biochemistry*, **30**, 1561.
7. Lee, J.-O., Rieu, P., Arnaout, M. A., and Liddington, R. (1995). *Cell*, **80**, 631.
8. Qu, A. and Leahy, D. J. (1995). *Proc. Natl Acad. Sci. USA*, **92**, 10277.
9. Nortamo, P., Li, R., Renkonen, R., Timonen, T., Prieto, J., Patarroyo, M., *et al.* (1991). *Eur. J. Immunol.*, **21**, 2629.
10. de Fougerolles, A. D., Stacker, S. A., Schwarting, R., and Springer, T. A. (1991). *J. Exp. Med.*, **174**, 253.
11. Bailly, P., Hermand, P., Callebaut, I., Sonneborn, H. H., Khamlichi, S., Mornon, J.-P., *et al.* (1994). *Proc. Natl Acad. Sci. USA*, **91**, 5306.
12. Bailly, P., Tontti, E., Hermand, P., Cartron, J.-P., and Gahmberg, C. G. (1995). *Eur. J. Immunol.*, **25**, 3316.
13. Yoshihara, Y., Oka, S., Nemoto, Y., Watanabe, Y., Nagata, S., Kagamiyama, H., *et al.* (1994). *Neuron*, **12**, 541.
14. Tian, L., Yoshihara, Y., Mizuno, T., Mori, K., and Gahmberg, C. G. (1997). *J. Immunol.*, **158**, 928.
15. Tian, L., Kilgannon, P., Yoshihara, Y., Mori, K., Gallatin, W. M., Carpén, O., *et al.* (2000). *Eur. J. Immunol.*, **30**, 810.
16. Tian, L., Nyman, H., Kilgannon, P., Yoshihara, Y., Mori, K., Andersson, L. C., *et al.* (2000). *J. Cell Biol.*, **150**, 243.
17. Li, R., Nortamo, P., Kantor, C., Kovanen, P., Timonen, T., and Gahmberg, C. G. (1993) *J. Biol. Chem.*, **268**, 21474.
18. Kotovuori, A., Pessa-Morikawa, T., Kotovuori, P., Nortamo, P., and Gahmberg, C. G. (1999). *J. Immunol.*, **162**, 6613.
19. Gahmberg, C. G. and Andersson, L. C. (1977). *J. Biol. Chem.*, **252**, 5888.
20. Laemmli, U. K. (1970). *Nature*, **227**, 680.
21. Nortamo, P., Patarroyo, M., Kantor, C., Suopanki, J., and Gahmberg, C. G. (1988). *Scand. J. Immunol.*, **28**, 537.
22. Valmu, L., Fagerholm, S., Suila, H., and Gahmberg, C. G. (1999). *Eur. J. Immunol.*, **29**, 2107.
23. Valmu, L., Hilden, T. J., van Willigen, G., and Gahmberg, C. G. (1999). *Biochem. J.*, **339**, 119.
24. Dustin, M. L. and Springer, T. A. (1989). *Nature*, **341**, 619.
25. Mizushima, S. and Nagata, S. (1990). *Nucleic Acids Res.*, **18**, 5322.
26. Nishimura, Y., Yokoyama, M. Araki, K., Ueda, R., Kudo, A., and Watanabe, T. (1987). *Cancer Res.*, **47**, 999.
27. Smith, G. P. and Scott, J. K. (1993). In *Methods in enzymology* (ed. R. Wu). Vol. 217, p. 228. Academic Press, London.

28. Koivunen, E., Arap, W., Rajotte. D., Lahdenranta, J., and Pasqualini, R. (1999). *J. Nucl. Med.*, **40**, 883.

29. Koivunen, E., Gay, D. A., and Ruoslahti, E. (1993). *J. Biol. Chem.*, **268**, 20205.

30. Koivunen, E., Wang, B., and Ruoslahti, E. (1994). *J. Cell Biol.*, **124**, 373.

31. Koivunen, E., Wang, B., and Ruoslahti, E. (1995). *Bio/Technology*, **13**, 265.

32. O'Neill, K. T., Hoess, R. H., Jackson, S. A., Ramachandran, N. S., Mousa, S. A., and DeGrado, W. F. (1992). *Proteins*, **14**, 509.

33. Kraft, S., Diefenbach, B., Mehta, R., Jonczyk, A., Luckenbach, A., and Goodman, S. L. (1999). *J. Biol. Chem.*, **274**, 1979.

34. Murayama, O., Nishida, H., and Sekiguchi, K. (1996). *J. Biochem.*, **120**, 445.

35. Feng, Y., Chung, D., Garrard, L., McEnroe, G., Lim, D., Scardina, J., *et al.* (1998). *J. Biol. Chem.*, **273**, 5625.

36. Koivunen, E., Wang, B., Dickinson, C. D., and Ruoslahti, E. (1994). In *Methods in enzymology* (ed. E. Ruoslahti and E. Engvall). Vol. 245, p. 346. Academic Press, London.

37. Koivunen, E., Restel, B. H., Rajotte, D., Lahdenranta, J., Hagedorn, M., Arap, W., *et al.* (1999). *Methods Mol. Biol.*, **129**, 3.

38. Sambrook, J., Fritsch, E. F., and Maniatis, T. (1989). *Molecular cloning: a laboratory manual.* Cold Spring Harbor Laboratory Press. Cold Spring Harbor, NY.

Chapter 6

Studying cell interactions during development of the nervous system in *Drosophila*

David Shepherd, Louise Block, James Folwell and Darren Williams

School of Biological Sciences, University of Southampton, Bassett Crescent East, Southampton SO16 7PX, UK

1 Introduction

1.1 Why study *Drosophila*?

The fruit fly *Drosophila melanogaster* is one of the key model organisms for studying both the genetic and molecular mechanisms of cell interactions. The versatility of the fly as a model experimental organism means that by working with *Drosophila* it is possible to combine molecular, genetic, and micromanipulation techniques to address fundamental questions in embryology, cell biology, and neurobiology. This versatility has attracted many research workers to *Drosophila* and has generated a vast literature. The volume of work done and the high degree of evolutionary conservation have meant that *Drosophila* has had a huge impact on many areas of biological research. The completed sequencing of the *Drosophila* genome further adds to its power as a research organism. The combination of the genome database and the available experimental tools mean that the humble fruit fly will play a key role in the post-genomic era in establishing the functions of the gene sequences held in the databases. This will ensure that *Drosophila* will remain one of the key model organisms.

Within the scope of this chapter it would be impossible to provide a comprehensive review of the techniques and strategies used to study *Drosophila*. This chapter, therefore, focuses specifically on techniques used to study cell interactions during nervous system development and function. Many of these techniques can be applied to other areas of *Drosophila* research. It is our aim that the chapter will be useful to newcomers and experienced research workers alike. The chapter is divided into two sections. The first describes techniques and strategies for visualizing cells in the nervous system. The second details strategies for experimentally manipulating cells. Each describes the methodologies used in our laboratory and gives full details of the techniques.

1.2 For the absolute beginner

This chapter assumes that the reader has some familiarity with the essentials of handling and rearing of *Drosophila*. If you are a novice, you should consult some of the resources on the basic aspects of using *Drosophila*. Whilst the best way to get started in *Drosophila* research is to spend time in a fly laboratory, there are books that give an excellent introduction to the fundamentals of *Drosophila* husbandry and genetics. For the basics of how to look after flies, and to make and handle food, the introductory chapter of '*Drosophila*: a practical approach' by Roberts (1) is excellent. This book also provides practical guides to key *Drosophila* methodologies and makes an excellent companion to this chapter. For an introduction to *Drosophila* genetics and genetic nomenclature 'Fly pushing' by Greenspan (2) is a practical option. For descriptions of the anatomy and development of *Drosophila*, the book edited by Bate and Martinez Arias (3) gives a definitive overview of the vast literature on *Drosophila* and is a must for anybody who wishes to work seriously with flies. Another resource is the laboratory manual edited by Sullivan *et al.* (4) which provides a practical guide to methods used in *Drosophila* research. The Internet is a vast reserve of information. The most useful is Flybase (http://flybase.bio.indiana.edu:82/), a database of genetic and molecular data for *Drosophila* which include information on the 64 000+ alleles of more than 23 000 genes, their expression, transcripts, and functions. Flybase is the first port of call to find mutations, clones, fly stocks, publications, and people. There is a *Drosophila* bulletin board, which is a forum to answer specific questions (http://www.bio.net/hypermail/dros/). Another useful database is Flybrain (http://flybrain.uni-freiburg.de/) which is an excellent resource dedicated to the nervous system of *Drosophila* and is excellent for unravelling the mysteries of the three-dimensional structure of the brain. The Interactive Fly web site (http://flybase.bio.indiana.edu:82/allied-data/lk/interactive-fly/aimain/1aahome.htm) provides a good textual guide to *Drosophila* genes and their functions. All these sites have links to other useful resources. There are a number of stock centres from which *Drosophila* stocks can be ordered, the largest of which is the Bloomington stock centre (http://flystocks.bio.indiana.edu/). Their website also provides advice on working with *Drosophila*, including the handling of stocks, various recipes for media, and problems such as mite infestation.

2 Preparing tissues for study

In the early embryo, it is possible to examine neuronal cell interactions with the nervous system *in situ*. At all other stages of development, this is not possible, and, to apply the techniques described in this chapter, it is necessary to dissect the nervous system from the animal. Whilst the basic skills for doing this are generic, the details differ depending on the stage in the life cycle one wishes to study. Before detailing the various approaches that can be used to examine cell–cell interactions within the nervous system, we shall give an overview of the dissections at different developmental stages and the equipment required.

2.1 Equipment required

2.1.1 Dissecting instruments

Dissections are usually performed under a physiological saline called *Drosophila* saline. A 10× concentrate comprises 0.15 g of KCl, 7.48 g of NaCl, 0.81 g of MgCl$_2$, 0.26 g of CaCl$_2$·2H$_2$O, 12.3 g of sucrose, 1.19 g of N-(2-hydroxyethyl) piperazine-N'-(2-ethanesulphonic acid) (HEPES) in 1 litre of water, with pH adjusted to 7.4 with NaOH. This can be aliquoted and stored at −20°C and diluted in distilled H$_2$O before use. Alternatively, phosphate-buffered saline (PBS) can be used. The saline is placed in 35-mm plastic Petri dishes filled to half depth with sylgard silicone rubber (Sylgard Elastomer 184, Dow Corning). Top quality dissecting instruments are essential. Dissecting instruments must be well maintained and kept in top condition. The basic dissecting kit includes: two pairs of Dumont INOX No 5 forceps, one pair of Moria extra delicate mini-Vannas spring scissors with straight tips, and a pin holder with a stainless steel minutien pin. Animals are pinned with stainless steel minutien pins (0.1 mm diameter, 10 mm long). All dissecting tools are obtained from Fine Science Tools. The small size of the tissues necessitates that all dissections are done under a stereomicroscope. You will also need a means of anaesthetizing flies. The best option is CO$_2$; details of anaesthetizer designs can be found in ref. 5.

2.1.2 Lighting

The quality of the lighting is crucial; it is best to use a cold light source (Schott 1500 W) with a double swan neck fibre optic light pipe. Dissections are carried out against a black background with transmitted rather than incident light. This can be achieved by placing the fibre optic light pipes either side of the Petri dish.

2.2 Handling tissues

The tissues are delicate and must be kept in aqueous solutions, any air–water interface will be very destructive. Tissues are transferred between solution and containers either by using Pasteur pipettes or by leaving tissues in containers whilst changing solutions. Ideal containers for tissues are: glass embryo dishes, 10-ml scintillation vials, or 1.5-ml microfuge tubes. All changes of solution and tissue transfers are observed directly under the microscope to ensure that no tissues are mistakenly discarded. It is wise to place all wastes into a small container, which can be checked for stray tissues before being disposed of.

2.3 Dissections

2.3.1 Embryo

In most cases, it is not necessary to dissect the central nervous system (CNS) from embryos because it can be readily observed *in situ*. Embryos can be processed whole (Section 3.3).

2.3.2 Larva

Dissection of the larval CNS is simple. The CNS can be dissected as a single mass that includes all neuromeres. Immerse larvae in cold saline in a dissecting dish and identify the anterior and posterior ends. The CNS is located in the anterior third of the larva. Simply cut the larva in half transversely; the internal organs and CNS will spill out of the cut end of the anterior section. Draw up the anterior end into a pipette and transfer to a glass vial containing 4% paraformaldehyde fixative. Fix at room temperature for 30 min with rotation. After fixation, transfer the tissue to a dissecting dish and immerse in 4% paraformaldehyde fix. Under the stereomicroscope, locate the CNS and tease from the body with forceps and mounted pin. It is important that the CNS is dissected completely free of all extraneous tissues. Transfer the dissected CNS to another glass vial containing 4% paraformaldehyde and fix for a further 90 min. After fixation, wash the tissue in PBS-Tx (PBS + 0.4% Triton X-100) and process as required.

2.3.4 Pupa

Select a pupa of the required age and immerse in saline in a dissecting dish. To identify pupal stages, consult Bainbridge and Bownes (6). Examine under the stereomicroscope and identify anterior and posterior ends. The anterior is easily recognized because of the pair of spiracles that protrude with the flattened region of cuticle (operculum) between them. The operculum is on the dorsal surface. Pin the pupa to the dish, dorsal uppermost, by placing a pin through the operculum. Cut off the posterior-most tip of the abdomen and make a single cut along the dorsal midline towards the anterior. Pin out the body wall with four minutien pins, placing one at each corner. Remove loose debris by gently squirting with saline from a pipette; this is vital because if you do not clear the debris it will be impossible to see the CNS. Locate the CNS, which should be found on the ventral surface. Remove the complete CNS with a mounted minutien pin. In early pupae, the cephalic neuromeres and ventral nerve cord will be removed as a single piece. In later pupa (>50 h old), the head separates from the thorax and it is necessary to remove the ventral nerve cord and cephalic neuromeres separately. Draw up the dissected CNS into a pipette, transfer to 4% paraformaldehyde fix in a glass vial, and fix for 2 h.

2.3.5 Adult

The dissection used to isolate the adult CNS depends upon which part of the nervous system one wishes to study.

2.3.5.1 Cephalic neuromeres

In adults, the cephalic neuromeres are separated from the ventral nerve cord, and the dissection of these two regions are different. Anaesthetize the flies and dip in 95% ethanol to reduce surface tension in aqueous medium. Transfer to 4% paraformaldehyde fix in a dissecting dish. Pin through the abdomen (ventral up) with the head towards you. Remove the front four legs close to their insertion.

Tilt the head towards you and extend the proboscis and remove by pinching it and pulling. You will see tracheae in the head; remove these with forceps. With two pairs of forceps, grab the outer edges of the hole made by removing the proboscis and peel the head cuticle away. It should 'shell' the eye cuticle away from the optic lobe. Break the cuticle behind the brain and remove the head capsule. If you now grasp the cuticle near the 'neck' (avoiding the neck), and tear away from the fly, the cuticle should bring the CNS with it. This cuticle can be removed by grasping it with forceps at two points and tearing the cuticle between these points with all of the force applied at a tangent to the brain. Do not touch the brain. Always hold the cuticle/tracheae attached to the brain and manipulate these. If you see the brain distorting, stop and try pulling at two different pieces of cuticle. Normally, the gut will be attached, this can be pulled out. The last (small) pieces of cuticle and tracheae should be removed, although the risk increases the more you touch the brain, so indirect tearing is always recommended. Fix overnight in 4% paraformaldehyde at 4°C. Practice is needed.

2.3.5.2 Ventral nerve cord

Anaesthetize flies, decapitate, and remove abdomen. Transfer the thorax to a microfuge tube containing 4% paraformaldehyde fix and shake vigorously for 20 s. Fix overnight at 4°C with rotation. Transfer tissues to a dissecting dish in fixative. With irridectomy scissors, remove the dorsal thorax (notum). Turn the thorax ventral up and, whilst holding the tissue by the posterior-most piece of cuticle, use a second pair of forceps to take hold of the prothoracic legs and peel them anteriorly. The legs should break away cleanly to reveal the ventral nerve cord. With a mounted needle or tips of forceps, tease the ventral nerve cord free. Transfer the ventral nerve cord with a pipette to PBS-Tx.

3 Revealing neurons and glia

The essence of understanding nervous system function and development is detailed knowledge of the anatomical organization of neurons and glia. Key to this is the ability to visualize neurons within the nervous system. Neuron visualization techniques enable researchers to describe the functional organization of the nervous system that underlies its synaptic connectivity. Similarly, studies of axon guidance and synapse formation rely on the ability to visualize neurons. Such observations suggest interactions with components of the surrounding environment, which can be tested. In short, we need to know the structural organization of neurons in order to study their interactions. In this respect, *Drosophila* offers a vast array of resources and strategies for visualizing neurons.

3.1 Reporter genes

The introduction of transgenic techniques revolutionized our ability to analyse the cellular interactions that underlie the development and function of the nervous system. Key amongst these techniques has been the development of reporter gene constructs. Reporter genes are now used routinely to produce insights into the cell biology of the nervous system in a number of important model systems, e.g. *Drosophila* (7, 8), mouse (9), zebrafish (10), and *Caenorhabditis elegans* (11). The value of transgenic techniques is that they are non-invasive, applicable to living material, and can provide reliable and repeatable visualization of defined neurons or populations of neurons in either a mutant or wild-type background. In *Drosophila*, perhaps more than other organisms, reporter genes have been used to great effect, and consequently there are many different reporter gene constructs available, each with its particular application and value.

3.1.1 Driving reporter gene expression

The key to reporter gene studies is the methodology used to drive expression of the reporter construct in specific cells or subsets of cells. In *Drosophila*, the two most common methods of driving gene expression are enhancer trapping (12, 13) and promoter fusions (see below).

3.1.2 Choice of reporter genes

The first widely used transgenic reporter gene was the bacterial β-galactosidase (*lacZ*) gene that was incorporated into the p*LacZ* enhancer trap construct. This provided a readily detectable reporter gene that could be visualized using histochemical techniques (*Protocol 1*) or immunocytochemistry (*Protocols 4* and *5*). Whilst p*lacZ* enhancer trap lines were widely used as cell markers, their use in neurons was limited due to the protein labelling only nuclei. The concurrent development of cytoplasmic *lacZ* reporters and a more versatile method of enhancer trapping (GAL4 techniques (see below)) circumvented this problem (13) and revealed detailed neuronal structures (7), providing transgenic reporters that labelled neurons in a way that enabled study of growth, interactions, and anatomy (*Figure 1*). Although hugely important, these early reporters were simple in design and relied on diffusion of the protein to reveal neuronal structures. Whilst these reporters show much of the structure of neurons, they do not display their full structure and, in some cases, they reveal little more than the cell body. These original reporters, therefore, provided a novel, but limited, set of tools for studying neurons.

Reporter genes have now been developed that are designed to improve visualization of neurons. In most cases, these reporters work by producing proteins that utilize transport systems, such as axonal transport, to distribute the protein. Reporter genes can encode native proteins targeted to axon transport, for example inactive tetanus toxin (IMPTNT (14)) and the microtubule-associated protein tau (15). Both these reporters encode proteins not normally expressed in *Drosophila* and rely on the use of specific antibodies raised to the reporter protein. Of these, IMPTNT is by far the best, but its use has been limited by the lack of a commercially available anti-tetanus antibody. Alternatively, chimeric fusion proteins can be created, e.g. kinesin–*lacZ* (16) and tau–*lacZ* (17) which contain the functional domains of transported proteins in-frame with a reporter gene. In these cases, the reporter gene *lacZ* can be detected either histochemically (*Protocol 1*) or immunocytochemically (*Protocols 4* and *5*).

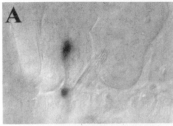

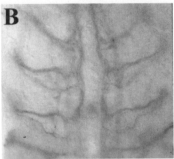

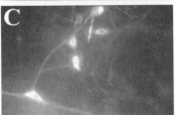

Figure 1 Revealing neurons with reporter genes. (**A**) LacZ expression detected histochemically in a single element in the lateral pentascolopale of a third instar larva. (**B**) Central afferent projections of proprioceptive sensory neurons in the CNS of a third instar larva revealed with an anti-β-galactosidase antibody. (**C**) A GFP reporter revealing sensory neurons in the periphery of a live third instar larva.

Protocol 1

Histochemical detection of *lacZ* gene expression in isolated CNS

Equipment and reagents

- Dissecting instruments and CO_2 anaesthetizer
- Bench-top centrifuge
- X-gal staining solution: 10 ml of $K_3Fe(CN)_6$ [31 mM], 10 ml of $K_4Fe(CN)_6$ [31 mM], 5 ml of NaH_2PO_4 [0.1 M], 5 ml of Na_2HPO_4 [0.1 M], 15 ml of NaCl [1.0 M], 200 ml of $MgCl_2$ [0.5 M]; make up to 100 ml with distilled water
- 4% paraformaldehyde fixative: 4 g of paraformaldehyde dissolved in 17 ml of 0.63 M NaOH, then add 83 ml of 0.188 M NaH_2PO_4
- Embryo dishes (cavity blocks or solid watch glasses)
- Mounting medium: add 7 g of gelatin to 42 ml of distilled H_2O, stir, leave for 5 min, boil until gelatin dissolves. Add 63 g (50 ml) of glycerol and store at 4°C. Warm to 50°C for use.
- 70% glycerol
- Microscope slides and coverglasses
- 10-ml glass vials
- Rotator/agitator
- Vortexer
- Pasteur pipettes
- PBS-Tx (PBS + 0.1% Triton X-100)
- Parafilm

Protocol 1 continued

Method

1 Fix and isolate the required material (see above).

2 Transfer tissues with Pasteur pipettes to a solid watch glass.

3 Wash in PBS-Tx for 1 h with rotation/agitation.

4 Remove X-gal staining solution from the freezer and warm to 37 °C.

5 Vortex the X-gal solution.

6 Centrifuge the X-gal solution at 13 000 r.p.m. for 5 min.

7 Replace PBS-Tx with X-gal supernatant.

8 Cover with Parafilm to prevent evaporation.

9 Incubate at 37 °C for 2–4 h, checking the progress of the staining every 30 min.

10 Wash tissues in PBS-Tx.

11 Clear in 70% glycerol.

12 Warm mountant to 50 °C and transfer the tissue to a microscope slide in a small volume of glycerol.

13 Immerse in mountant, cover with a coverglass, and view under the microscope.

The most widely used reporter genes are based on green fluorescent protein (GFP) because of the ability to visualize cells in live material (*Figure 1*). A number of GFP fusion genes have been produced specifically for studying neurons. These include tau–GFP (18), gap–GFP (A.Chiba, personal communication), and CD8–GFP (19) that target GFP to the plasma membrane. Of these, tau–GFP is the most widely used (20). Although in most cases GFP is detected using fluorescence microscopy, there are occasions when immunocytochemistry is required. Anti-GFP antibodies are available commercially (Boehringer Mannheim (http://biochem.boeringer-mannheim.com)).

3.1.3 Potential problems with reporter genes

Although reporter genes are used routinely, they can disrupt expressing cells. Kinesin–*lacZ* for example is embryonic lethal when expressed in neurons (15). Tau–*lacZ* interferes with mitosis and can be lethal (20). Furthermore, all tau reporter genes cause neurodegeneration (21) and it is wise to avoid tau-based reporter genes. Whilst these are isolated examples, it is evident that reporter genes can influence the functions of cells. It is important to ensure that the selected reporter gene does not have deleterious effects and it is advisable to carry out control experiments.

3.2 Dye injections

In addition to reporter genes, classical methods of physically introducing dyes into neurons are an important tool. Numerous dyes have been used to stain neurons in *Drosophila*. Staining neurons with metals such as cobalt has a long history and works well in *Drosophila*. Cobalt can either be injected into neurons or introduced into the cut ends of axons or dendrites. Protocols describing the use of cobalt and other dyes are available elsewhere (22). Whilst cobalt is a good neuron label, its use has been limited. There are several reasons for this. First, it is toxic to neurons and, secondly, the processing required to visualize staining is noxious (H_2S containing), protracted, and capricious. Neurobiotin and horse-radish peroxidase (HRP) can also be used in a similar way to cobalt, with the same drawbacks (23).

3.2.1 Carbocyanin dyes (DiI and DiO)

Lipophilic dialkylcarbocyanin fluorescent dyes such as DiI (Molecular Probes) are widely used to reveal neuron morphology and are without doubt the best dyes to use in *Drosophila*. These dyes label neurons by migration through the plasma membrane and work in living and fixed material. The rate of migration depends upon the state of the tissue; in fixed tissue it is 0.2–0.6 mm per day and in living neurons up to 6 mm per day. The dyes are stable and can be used for up to several years to trace neuron projections. Used successfully, they reveal detailed structures of neurons. There are several dyes in the DiI family varying in their emission spectra and mobility. In our work, we use DiI and fast DiI, a modified variant of DiI with a faster rate of mobility.

Carbocyanin dyes can be applied in a number of ways. The simplest is to apply crystals directly to cut axons or intact neurons. More typically, the dyes are used as a paste mixed with oil or silicon grease and diluted to the desired consistency with absolute ethanol before being applied to cells or cut axons. DiI introduced in this way has been used in a number of studies in *Drosophila* to map the positions of motor neurons (24) and to reveal clones of neurons in the CNS (25–27). It is possible to inject the dyes into neurons but this is rarely done in *Drosophila*. For this, it is recommended that you sonicate, centrifuge, or filter the dye to remove undissolved crystals.

When labelling fixed tissues, tissues are fixed in 4% paraformaldehyde in 0.1 M phosphate buffer, pH 7.4, at room temperature. Other fixatives, e.g. gluta-raldehyde, should be avoided. Materials stored during long-term labelling can be kept in fixative at room temperature or 4°C. Higher temperatures increase the rate of mobility but also increase the probability of transneuronal labelling. Per-meabilization agents and detergents such as Triton X-100 and Tween-20 should be avoided. Mounting media such as glycerol should be avoided.

3.2.2 Handling DiI

Fast DiI is provided by Molecular Probes (order no. D3899) and comes in the form of an oil. On receipt, dilute the oil with 100 μl of 100% ethanol, aliquot into 50 μl volumes, and store at -20°C. When using an aliquot, add another 100 μl of 100% ethanol to a 50 μl aliquot and keep at -20°C. To use, warm to room temperature and transfer a few microlitres to a micro-cup. The micro-cup is a container designed for easy access for the micropipette when applying the dye and is made by cutting the bottom off a 1.5-ml microfuge tube with a razor blade. After use, return to $-20\,°C$.

In our hands, DiI oil is apparently air reactive and changes its properties with prolonged and repeated exposure to air. When the oil loses its oily consistency and begins to form 'toffee strings', discard and use a new aliquot.

Protocol 2

Labelling adult sensory neurons with DiI

Equipment and reagents

- Epifluorescence microscope with Zeiss Rhodamine filter set 15
- Dental wax[a] (The Hygenic Corporation)
- Dissection instruments and CO_2 anaesthetizer
- Stereomicroscope and light source
- DiI (Molecular Probes)
- Glass micropipettes

- Coverglasses
- Depression slides
- 4% paraformaldehyde fixative in PBS
- PBS
- *Drosophila* saline (see Section 2.1.1)
- 24-well plate

Method

1 Anaesthetize flies with carbon dioxide.
2 Isolate the thorax by cutting off the head and the tip of the abdomen.
3 Transfer to 4% paraformaldehyde fixative.
4 Shake vigorously for 30 s.
5 Fix overnight at $4\,°C$.
6 Remove the fly from the fixative.
7 Transfer to a blob of dental wax on a small circular coverglass, 13 mm diameter. Orient to the correct position.
8 Break off the required sensory bristle by rocking it gently back and forth with a mounted pin.[b]
9 Dry the structure and apply DiI with a broken glass micropipette. A small drop of DiI oil should be applied.
10 Transfer each preparation to paraformaldehyde fixative in a 24-well plate and leave at room temperature for 5 days.
11 Ensure only the intended area/structure is labelled and dissect out the CNS in *Drosophila* saline.
12 Wash in PBS.
13 Mount in PBS on a depression slide and view under fluorescence using a rhodamine filter set.

[a] Called tacky wax; it is extremely sticky and excellent for holding flies down temporarily.
[b] Breaking is necessary for success, do not remove the bristle by pulling.

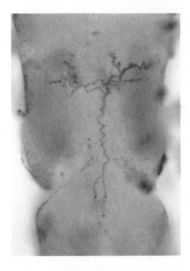

Figure 2 The central afferent projection of the posterior scutellar macrochaete of adult *Drosophila* labelled with DiI applied peripherally and photoconverted with DAB.

3.2.3 Photoconversion of DiI

Although DiI fluorescence is bright and long lasting, it is often desirable to produce an optically dense product for long-term storage and detailed analysis. In these cases, it is easy to photoconvert the dye into an optically dense substrate using diaminobenzidine (DAB) as a substrate (*Figure 2*).

Protocol 3

Photoconversion of DiI[a]

Equipment and reagents

- Epifluorescence microscope with Zeiss rhodamine filter set 15
- DAB solution: 150 mg of diaminobenzidine (DAB; 0.2%) added to 75 ml of 50 mM Tris buffer, pH adjusted to 7.4–7.6 after adding DAB
- PBS

- Microscope depression slides
- Coverglasses
- Bench-top centrifuge
- Alcohol series
- Xylene
- Mounting medium (permount)

Method

1 Wash the CNS in PBS.

2 Defrost the DAB solution and centrifuge at 13 000 r.p.m. for 5 min.

3 Transfer the supernatant to a new tube and repeat centrifugation.

4 Transfer the CNS to a depression slide.

5 Incubate the CNS in DAB solution in the dark for 5 min.

6 Cover the preparation with a coverglass.

Protocol 3 continued

7 Expose to green light until the reaction is complete.[b]

8 Dehydrate through alcohol series, clear in xylene, and mount in permount.

[a] Modified from David Merritt (personal communication) and refs 28 and 29.

[b] This is done empirically, but typically takes 20 min.

3.3 Using antibodies to reveal specific neurons

Immunocytochemistry can also be used to reveal the structure and identity of neurons. Much of the pioneering work on axon guidance and patterning in the ventral nerve cord has relied on the ability to visualize axons within the CNS, and to identify neuroblasts using antibodies. Antibodies can be used to reveal neuron structure through the native expression of neuron-specific antigens or by driving reporter gene expression in subsets of neurons. The aim of this section is to highlight immunocytochemical procedures used to visualize cells of the CNS.

Primary antibodies can be detected using either fluorescent or biotinylated secondary antibodies. Fluorescent antibodies are useful for co-localization studies as the signals from the different fluorescent secondary antibodies can be over-laid to determine co-localization. The disadvantage of fluorescent antibodies is that the material only lasts for a short time and that for double-labelling studies two primary antibodies raised in the same species cannot be used. Biotinylated secondary antibodies produce material that can last for years, and can be re-examined later. These can also be used for co-localization by varying the sub-strates to produce different coloured reaction products for each antibody. As each antibody reaction is developed sequentially, it is possible with biotinylated antibodies to use primary antibodies raised in the same species. However, this technique is limited to antibodies that label different cellular compartments, otherwise the signals cannot be distinguished readily. Whilst the first choice for many laboratories is to use fluorescent secondary antibodies, we prefer biotinyl-ated antibodies. For all of this work, we use Vectastain ABC kits which are robust and reliable, but alternatives are available.

3.3.1 Embryos

Antibody staining of embryos is easy. It is possible to examine large numbers of embryos *en masse* without dissection or manipulation. There are many procedures and short cuts available for collecting embryos and processing them. Most are variations of techniques developed by Wisechaus and Nusslein-Volhard (30). Here we present the variants that work well in our hands.

3.3.2 Collecting and handling embryos

Embryos can be collected with a simple egg collecting kit constructed from common laboratory items. This chapter does not aim to cover these techniques in detail; for more details of these procedures, consult Roberts (5).

A problem to overcome is the outer protective layers of the egg, which are impenetrable to aqueous fixatives and antibodies, and must be removed. The chorion can be removed by washing embryos in bleach. To remove the vitelline membrane, the embryos must be partly fixed. First stage fixation is achieved in a mixture of aqueous fixative (4% paraformaldehyde) and heptane. These solutions are not miscible and separate into two layers. The upper layer is heptane and the embryos will sink in the heptane and lie at the interface. The heptane will saturate with fixative allowing it to penetrate the vitelline membrane. After fixation, the vitelline membrane can be removed with methanol.

Some antigens are sensitive to methanol and, for these, the embryos must be de-vitellinized by hand. Embryos for hand de-vitellinization are collected and fixed as above up to the de-vitellinization step. The fixative layer is replaced by heptane and the embryos transferred to a basket and air dried. The embryos are transferred to a 35-mm Petri dish lined with double-sided sticky tape and covered with 4% paraformaldehyde. Embryos that adhere to the tape are teased out of the vitelline membranes using a sharp needle, fixed for a further 20 min, and can then be processed as normal.

Protocol 4

Antibody staining of embryos

Equipment and reagents

- Egg collection basket[a] and paintbrush
- Apple juice agar plates[b]
- 1.5-ml microfuge tubes
- 35-mm plastic Petri dishes
- Bleach
- Heptane
- 4% paraformaldehyde in PBS
- Methanol series (30, 50, 70%, and absolute) prepared in PBS-Tw
- PBS-Tw (PBS + 0.4% Tween-20)
- Vectastain ABC kit
- DAB stock solution (10 mg ml^{-1} in PBS)
- H$_2$O$_2$ (40% v/v)
- Primary antibody
- Rotator
- Pasteur pipettes
- Distilled water
- Mounting materials[c]

Method

1 Collect embryos overnight on apple juice agar plates.

2 Transfer embryos to the egg basket with the paintbrush.

3 Dechorionate by washing in bleach for no more than 2 min.[d]

4 Wash thoroughly in distilled water.[b]

5 Transfer the embryos with a paintbrush from the egg basket to a 1.5-ml microfuge tube containing a 1:1 mixture of heptane:4% paraformaldehyde.

6 Fix for 20 min with rotation.

7 Remove the lower aqueous fixative with a Pasteur pipette.

Protocol 4 continued

8 Add an equal volume of absolute methanol and shake vigorously for 20 s to remove the vitelline membrane.

9 Wash embryos in absolute methanol, and rehydrate in a series of washes in 70% methanol/PBS-Tw, 50% methanol/PBS-Tw, and then three times in PBS-Tw.[e]

10 Incubate embryos for 30 min in 2.5% normal horse serum (NHS, from Vectastain kit; 25 μl in 1 ml of PBS-Tw), or whichever serum is appropriate for the secondary antibody used.

11 Remove 0.5 ml and replace with an equal volume of PBS-Tw (final concentration 1.25% NHS).

12 Add the appropriate volume of primary antibody.

13 Incubate overnight at 4 °C with rotation.

14 Wash three times for 20 min and three times for 1 min in PBS-Tw.

15 Add the secondary antibody in PBS-Tw.

16 Incubate for 3 h at room temperature.

17 Wash three times for 20 min and three times for 1 min in PBS-Tw.[f]

18 Mix 10 μl each of Vectastain A and B reagents in 1 ml of PBS-Tw for 30 min prior to use to make the ABC solution.

19 Incubate embryos in ABC solution for 30–60 min.

20 Wash six times for 1 min in PBS-Tw.

21 Transfer embryos to 1 ml of PBS-Tw + 30 μl of DAB stock (final concentration 0.3 mg ml^{-1}).

22 Add 10 μl of 1% H_2O_2 to the DAB solution (final concentration 0.01%).

23 Observe the colour reaction to determine the optimal end point.

24 Wash embryos twice for 1 min in PBS-Tw to stop the reaction.

25 Clear and mount.[g]

[a] A small section of 1.5 cm diameter polypropylene tubing stuck to a piece of plankton net with epoxy adhesive to create a small sieve.

[b] A 35-mm Petri dish containing apple juice-laced agar designed to induce flies into laying eggs. To make, boil 3 g of sucrose in 33 ml of apple juice. In a separate container, boil 3 g of agar in 100 ml of distilled water. Mix the two solutions, pour into Petri dishes, and allow to cool. Usually primed for use by the addition of yeast paste.

[c] 50 and 70% glycerol, gelatin mountant, clear nail varnish, microscope slides, and coverglasses.

[d] Solutions are kept in 35-mm Petri dishes and embryos are moved between solutions in the basket, blotting on absorbant blue paper between each change. The blue paper acts as an indicator for when all the bleach has been washed off.

[e] It is possible to freeze and store the embryos at −20 °C in methanol at this stage.

[f] For fluorescent secondary antibodies, transfer embryos to 70% glycerol for mounting after step 17.

[g] Clear in 50% glycerol for 30 min, transfer to 70% glycerol for at least 1 h, and mount in a small volume of 70% glycerol. To avoid crushing specimens, support the coverglass by placing two coverglasses either side of the tissue. Seal with clear nail varnish.

3.3.3 Antibody staining of larval and adult tissues

Immunocytochemistry can be applied to the nervous system at all stages of development. To examine the nervous system of later stages, it is necessary to dissect the nervous system out of the body as discussed earlier (Section 2).

Protocol 5

Antibody staining of the larval, pupal, and adult nervous system

Equipment and reagents

- Glass vials
- Pasteur pipettes
- Embryo dishes
- PBS-Tx (PBS + 0.1% Triton X-100)
- 2 M HCl in PBS

- Vectastain ABC kit
- Appropriate primary antibody
- DAB (10 mg ml^{-1} stock solution)
- Hydrogen peroxide (40% v/v)
- Mounting materials (see *Protocol 3*)

Method

1. Dissect the nervous system.
2. Wash twice for 15 min in PBS-Tx.
3. Wash in 2 M HCl in PBS for 30 min.
4. Wash four times for 15 min in PBS-Tx.
5. Incubate with NHS (from the Vectastain kit) or appropriate serum for selected secondary antibody for 30 min (30 μl of NHS in 1 ml of PBS-Tx).
6. Replace 600 μl of the 1 ml of PBS-Tx containing NHS with 600 μl of PBS-Tx.
7. Incubate with the appropriate primary antibody overnight at 4°C.
8. Wash four times for 15 min in PBS-Tx.
9. Incubate overnight in secondary antibody.
10. Wash four times for 15 min in PBS-Tx.
11. Incubate for 2 h in ABC solution (Vectastain ABC kit)
12. Wash four times for 15 min in PBS-Tx.
13. Incubate in DAB for 15 min (30 μl of stock in 1 ml of PBS) in the dark.
14. Develop the reaction by adding 10 μl of a 1:100 dilution of H_2O_2.
15. Stop the reaction by washing with PBS-Tx.
16. Wash four times for 15 min in PBS-Tx.
17. Clear and mount (see *Protocol 3*).

3.3.4 Double staining with two different antibodies

It is often necessary to determine if two proteins are co-localized. This is done with two different fluorescent tags on the secondary antibodies. Typical markers are: fluorescein isothiocyanate (FITC), tetramethylrhodamine isothiocyanate (TRITC), Texas red, Cy2, Cy3, and Alexa dyes. It is also possible to double label with biotinylated antibodies.

3.3.4.1 Double staining with different fluorescent secondary antibodies

Double staining with fluorescent tags is simple because both sets of antibodies can be processed simultaneously. For examining the nervous system, the simplest procedure is to detect the protein of interest using the appropriate primary antibody with a TRITC-tagged secondary antibody combined with an FITC-conjugated anti-HRP primary antibody (labels all neurons). In this case, the anti-HRP antibody and the TRITC secondary antibody can both be added after primary antibody processing.

3.3.4.2 Double staining with different biotinylated secondary antibodies

Double staining with biotinylated antibodies requires repeating the process for each antibody. With the first antibody, the protocol is exactly as above, but 60 μl of 1% NiCl is added to the DAB/H_2O_2 reaction to produce a black precipitate, which can be distinguished from the normal brown reaction product. For the second label, on completion of the first reaction, embryos should be washed in PBS-Tw (PBS with 0.1% Tween-20) and the antibody protocol repeated for the second antibody reaction but without nickel (*Figure 3*). A helpful tip is to ensure that the black reaction product is generated first for the antibody that stains the more internal cellular compartment (e.g. nucleus), otherwise the black reaction product can obscure the brown product. Various other colour combinations are also possible, and for details see Patel (31).

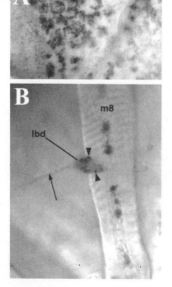

Figure 3 Examples of lineage tracing. (**A**) BrdU labelling of neurons in the cephalic neuromeres of a third instar larva. (**B**) Biotinylated dextran (BDA) labelling of embryonically derived cells in the body wall of a newly eclosed adult. The labelling reveals a single sensory neuron (lbd) and muscle 8 (m8), a larval muscle that persists through metamorphosis.

3.4 Lineage tracing and birthdating of cells

There are techniques available to determine the developmental origins of a particular cell type, which include lineage studies and birthdating of cells. The most widely used method is with bromodeoxyuridine (BrdU). BrdU is a thymidine analogue that becomes incorporated into newly synthesized DNA during DNA replication in cells that are undergoing mitosis. All those cells that are born during the period of BrdU exposure will be differentially labelled and those cells that have incorporated BrdU can be detected using a monoclonal antibody against BrdU (32; *Figure 3*). By limiting exposure to BrdU to short periods, cells can be birthdated accurately.

3.4.1 Birthdating neurons using BrdU

To label neurons generated during larval life is easy since BrdU can be introduced in the diet via a BrdU-laden yeast paste added to the surface of food media (33, 34). To label neurons produced during the non-feeding pupal stages is not so easy because the BrdU has to be injected. Injections are done with a glass micropipette, injecting approximately 50 nl of 3 mg ml^{-1} BrdU into the thorax of pupae of the required age. The injection is in the region of the thoracic musculature and does not cause serious damage to the animal. Culture pupae to the required age and process as described below (*Protocol 6*).

Protocol 6

BrdU labelling of neurons

Equipment and reagents

- Dried baker's yeast
- Distilled water
- Food tubes
- Red food dye
- Dissection instruments
- Hotplate
- Beaker
- Stirrer
- Alcohol series

- Relevant antibodies
- Carnoy's fixative (60 ml of absolute ethanol, 30 ml of chloroform, 10 ml of glacial acetic acid)
- 2 M HCl in PBS
- 3 mg ml^{-1} BrdU in PBS
- Anti-BrdU antibody
- 1.5-ml microfuge tubes

Method

1 Dissolve bakers' yeast in water to make a paste.

2 Simmer for 10 min to denature yeast.

3 Mix three parts yeast paste to one part 3 mg ml^{-1} BrdU solution to give a final BrdU concentration of 0.75 mg ml^{-1}.

4 Add a small portion of red food dye.

Protocol 6 continued

5 Spread yeast paste on the surface of the food medium.

6 Feed the animals in food tubes for 1–12 h (according to experimental design), starting at the required developmental stages.

7 Confirm ingestion by the presence of red food dye in the larval gut.

8 Dissect out the CNS (see Section 2).

9 Fix with Carnoy's fixative for 30 min in 1.5-ml microfuge tubes.

10 Wash twice with 95% ethanol.

11 Rehydrate in the alcohol series.

12 Pre-treat with 2 M HCl in PBS for 20 min to denature the DNA necessary to enable antibody to detect incorporated BrdU.

13 Incubate with anti-BrdU primary antibody and process according to *Protocol 4* or *5*.

3.4.2 Other lineage markers

The injection of a lineage tracer such as HRP or biotin-labelled 70 kDa dextran (BDA) can be used to label cells and to trace their developmental fate.

The tracer is injected into syncytial embryos and incorporated into all cells. As cells divide, the label becomes progressively diluted, leading to differential labelling of cells that have undergone relatively few divisions. In tissues where large numbers of cell divisions have taken place (imaginal discs), the label will no longer be detectable. Cells that have undergone only a few rounds of mitosis (embryonic sensory neurons) will contain sufficient labelled molecule to be detectable. Whilst HRP was the lineage tracer used by Tix *et al.* (35), the switch to BDA by Williams *et al.* (36) enabled signal amplification using avidin and biotin complexes, improving the signal to noise ratio and resulting in better resolution (*Figure 3*).

Protocol 7

Lineage tracing with BDA labelling

Equipment and reagents

- Apple juice agar plates (see *Protocol 4*)
- Yeast paste (see *Protocol 6*)
- Microscope slides
- Double-sided sticky tape
- Blunt forceps (size 4)
- Coverglasses
- 10S Voltalef fluorocarbon oil (ELF Atochem)
- Picoinjector (Narashige US Inc.)
- Lysine-fixable BDA (Molecular Probes): 25 mg ml^{-1} BDA in 0.2 M KCl
- Micromanipulator (Narashige MX-Z)
- 4% paraformaldehyde in PBS
- PBS-Tw (PBS + 0.4% Tween-20)
- Vectastain ABC kit
- DAB (10 mg ml^{-1} stock solution in PBS)
- Hydrogen peroxide (40% v/v)

Protocol 7 continued

Method

1 Set up wild-type *Drosophila* to lay eggs on apple juice agar plates.

2 Collect embryos every 20 min.

3 Dechorionate embryos by fixing them to a slide coated with double-sided sticky tape and gently popping the embryo out of the chorion using blunt forceps.

4 Transfer dechorionated embryos to a coverglass coated in double-sided sticky tape and cover with Voltalef fluorocarbon oil.

5 Inject approximately 2 nl of BDA solution into the centre of the embryo at approximately 50% egg length using a picoinjector and a micromanipulator to target the injection.

6 Remove excess oil and place the coverglass on a fresh apple juice agar plate.

7 Place in a humid chamber at 18 °C for 24 h.[a]

8 Add a small amount of yeast paste to the apple juice agar plate and transfer to 25 °C to continue development.

9 Dissect (see Section 2) and fix in 4% paraformaldehyde for 30 min once individuals have reached the desired stage.

10 Wash four times in PBS-Tw.

11 Incubate in 1 ml of PBS-Tw containing 10 _l of Vectastain reagent A and 10 μl of Vectastain reagent B for 30 min.

12 Wash four times in PBS-Tw.

13 Incubate for 20–30 min in 1 mg ml^{-1} DAB in PBS.

14 Develop the reaction with 0.003% hydrogen peroxide.

15 Stop the reaction using PBS-Tw.[b]

[a] This is necessary to improve survival rates.

[b] The end point has be determined empirically.

4 Manipulation of cells

This section describes techniques to manipulate cells experimentally by cellular ablation, mosaic analysis, and transgene misexpression. We focus on strategies for ablation of specific subsets of cells and for achieving spatial and temporal control of ectopic gene activation.

4.1 Methods of ectopic gene activation

Ectopic expression can be a powerful way to elucidate a gene's role in determining cell fate. It provides a counter approach to the classical genetic approach of examining the effect of loss-of-function mutations. Such approaches are invaluable where there are no loss-of-function mutations available, particularly as until very recently there has been no method for creating targeted gene knockouts in *Drosophila*. In addition, there are instances where loss-of-function mutations generate no discernible phenotype due to redundancy or where a gene is essential for embryonic development but also plays a role later in development.

Traditionally, there are two techniques for achieving ectopic expression: promoter-driven expression and heat shock activation. We shall consider both approaches.

4.1.2 Promoter-driven gene expression

This technique uses a defined promoter sequence to activate gene expression in the same spatial and temporal pattern as the gene from which the promoter is derived. Theoretically, any gene sequence can be fused to this promoter sequence and introduced into the genome. This permits expression of that gene in a clearly defined temporal and spatial pattern, providing a high degree of control over ectopic expression. Whilst extremely powerful as an ectopic expression approach, the main drawback to this technique is the limited availability of characterized promoters. There are few promoter constructs available which recapitulate the pattern of expression of a known gene. Consequently, promoter-driven ectopic gene expression can only be directed within a few tissues, which severely limits the utility of this approach.

4.1.3 Heat shock induction of gene expression

A second and more versatile method of ectopic expression is to activate gene expression from the *Drosophila* heat-shock promoter, hsp70. Transgenic *Drosophila* can be created that contain the gene of interest coupled to the heat-shock promoter. By providing a heat shock stimulus, the heat-shock promoter is activated, initiating transcription of the target gene. This enables gene expression to be switched on at a precise point in development, providing good temporal control over ectopic activation. The flexibility of *Drosophila* as a transgenic organism means that many lines exist with a wide variety of genes coupled to the hsp70 promoter, and the ease and cheapness of making transgenics means that any gene can be activated in this manner. The major drawback to this technique is that there is no spatial control of the resulting expression and that it is transient.

Protocol 8

Embryonic heat shock protocol

Equipment and reagents
- Apple juice agar plates (see *Protocol 4*)
- Yeast paste (see *Protocol 6*)
- Parafilm

Method
1   Put a dab of yeast paste on an apple juice agar plate.

Protocol 8 continued

2   Collect embryos on apple juice agar plates either overnight or for 1 h timed collections.

3   Seal the plates with Parafilm and place in a water bath at 35 °C for 20 min.

4   Transfer plates to either 18 or 25 °C and allow to recover for at least 1 h prior to examining.

5   Process embryos for antibody staining to detect gene activation.[a]

[a] If *lacZ* is being used as a reporter gene, embryos can be processed for X-gal staining (*Protocol 1*) or antibody staining (*Protocol 4*).

4.2 The GAL4 enhancer-trap technique

The GAL4 technique (18) is one of the most powerful and widely used tools for studying cell–cell interactions in *Drosophila*. It enables cloned genes to be activated ectopically in a cell- or tissue-specific manner, and allows temporal control of that expression. GAL4 has been useful for a variety of studies with respect to gene function and cellular mechanisms. The technique involves crossing two lines of flies. One line carries the yeast transcriptional activator, GAL4, under the control of an endogenous promoter or enhancer which directs GAL4 expression in a distinct spatial or temporal pattern. The second line contains the coding sequence for the gene of interest coupled to the GAL4 target promoter, the upstream activating sequence (UAS). Crossing both lines generates progeny that express the gene of interest in the same temporal and spatial pattern as GAL4. The system provides a convenient means of selectively misexpressing a variety of genes, simply by crossing a suitable GAL4 line with different UAS target gene lines (*Figure 4*).

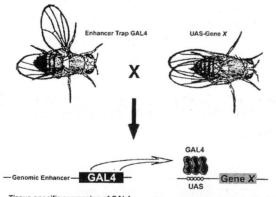

Figure 4 (**A**) The GAL4 technique requires crossing of flies of two different genotypes. One line contains the GAL4 reporter gene driven, in this example, by a genomic enhancer. The second genotype is transformed with the gene of interest (gene X) coupled to the GAL4-binding upstream activation sequence (UAS). (**B**) The offspring from this cross contain both transgenes and the expressed GAL4 binds to the UAS and initiates expression of gene X in the cells that express GAL4.

4.2.1 Applications

GAL4 has been applied to a diverse array of problems in *Drosophila* nervous system function and development. It has been used as a neuroanatomical tool to study functional organization (7, 37) and axon growth (8). GAL4-driven toxins (diphtheria, ricin, and tetanus) can be used to ablate chemically specific groups of cells (see below) or disrupt specific cellular functions (38). Whilst GAL4 was designed for creating misexpression phenotypes, the technique can also generate loss-of-function effects. GAL4 can drive expression of double-stranded RNA (39) and of dominant-negative forms of a gene. This can overcome problems of early lethality resulting from loss of function or a requirement for a gene product at several stages of development of a single structure (40).

The key to success is to have the correct GAL4 and responder lines. It is essential to obtain lines that express GAL4 in the required cells. Whilst it is possible to make/isolate such a line, the preferred route is to use Internet resources to find existing lines. There are many UAS transgenes available, but if the gene you wish to study is unique to your laboratory then it will be necessary to make your own lines. For a detailed description of how to generate new GAL4 and UAS target gene lines, see ref. 20. The pattern of GAL4 expression should be assayed with a suitable reporter gene (see above).

4.3 Laser gene activation in single cells

It is possible to target gene expression to single cells using a laser. Any gene coupled to the heat-shock promoter can be activated by using the laser to provide a heat shock stimulus (41). This results in activation of gene expression in individually targeted cells. One drawback to this technique, however, is that the resulting gene expression is transient. We have, therefore, modified this basic technique to combine laser-activated gene expression with the flippase (FLP)

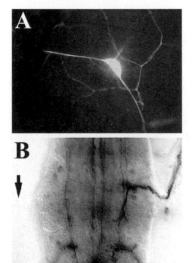

Figure 5 Using laser to induce expression and to ablate neurons selectively. (**A**) A single larval sensory neuron expressing a GFP transgene following focal application of a laser to a single neuron to induce local gene activation. (**B**) The CNS of a pupa following unilateral ablation of a persistent embryonic sensory neuron. Note the absence of axons entering the CNS on the ablated side (arrow).

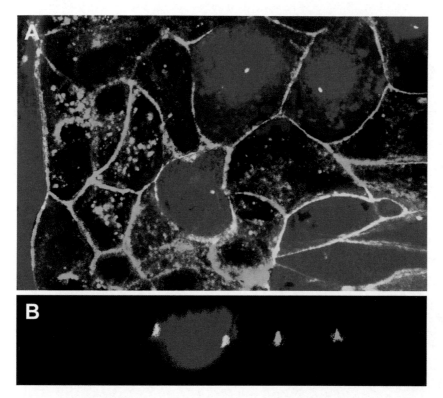

Plate 1 TJs can be formed between epithelial cells of different types. (**A**) Co-culture of CMTMR-labelled LLC-PK$_1$ cells (pig, red) and MDCK cells (dog, dark) were stained with an antibody against occludin, followed by an FITC-labelled secondary antibody (green). The image was obtained by confocal laser scanning microscopy. (**B**) Confocal lateral optical slice of the same region of monolayer shown in (A). Notice that occludin is expressed in homologous (dark/dark) and heterologous (dark/red) borders, indicating that TJs can be assembled by cells of different animal species.

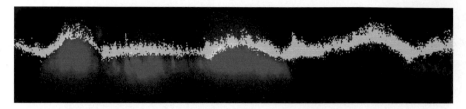

Plate 2 TJs act as fences, restricting the diffusion of membrane molecules from the apical to the basolateral domain. A monolayer of LLC-PK$_1$ cells which were stained with CMTMR (red) beforehand and co-cultured with MDCK cells (unstained) is shown with confocal microscopy in a transverse section. The lipid fluorescent probe BODIPY®FL-C$_{12}$-sphingomyelin added to the apical solution incorporates into the apical membrane of all cells, but is stopped at the level of the TJs and cannot reach the basolateral domain.

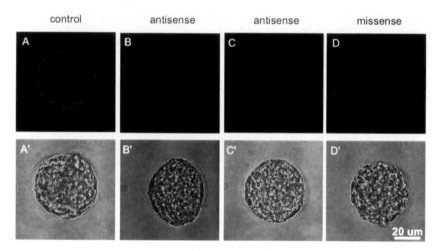

Plate 3 Detection of Kit receptors on rat oocytes using immunofluorescence and anti-Kit antibodies. Kit is detected on the plasma membrane of uninjected (**A**) or missense oligonucleotide-injected oocytes (**D**), but little or no Kit is evident after microinjection of antisense oligonucleotides (**B** and **C**). **A′**, **B′**, **C′**, and **D′** are transmitted light photomicrographs of the oocytes pictured in A, B, C, and D, respectively.

recombinase site-specific recombination system to create a method that results in constitutive gene activation in targeted cells (see Section 4.4; *Figure 5*).

Laser attachments to microscopes were at one time the reserve of the laser specialist. Now there are several purpose-built 'plug and play' laser systems available for this type of work that are extremely easy and safe to use (even for novices). For our work, we have used a Micropoint Laser Ablation System manufactured by Laser Photonics Inc.

Protocol 9

Laser gene activation in embryos

Equipment and reagents

- Flies of the appropriate genotype[a]
- Apple juice agar plates (see *Protocol 4*)
- Yeast paste (see *Protocol 6*)
- Microscope slides
- Double-sided sticky tape
- Blunt forceps
- Glue-coated coverglasses[b]
- Voltalef 10S fluorocarbon oil
- Heptane

- 4% paraformaldehyde in PBS
- 1.5-ml microfuge tubes
- Egg basket
- Pasteur pipettes
- Dissecting pins
- PBS-Tw (PBS + 0.4% Tween-20)
- Compound microscope
- Laser[c]
- Epifluorescence source

Method

1   Put a dab of yeast paste on an apple juice agar plate.

2   Set up a cross to lay for 2 h at 18 °C on apple juice agar plates.

3   Keep plates at 18 °C until the embryos reach the desired stage.

4   Dechorionate embryos manually.[d]

5   Transfer the embryos to a glue-coated coverglass and cover with Voltalef oil to prevent desiccation.

6   Target cells to be lasered using the 100× microscope objective.

7   Attenuate the laser beam using a series of neural density filters (exact settings vary for individual lasers and need to be determined empirically) and apply a 75–120 s laser burst of 4-ns pulses at a frequency of 3–5 Hz to targeted cells.[e]

8   Return embryos to apple juice agar plates and allow to recover under oil at 18 or 25 °C for 2–4 h.

9   Wash embryos off the coverglass with heptane.

10 Transfer embryos to a microfuge tube containing a 1:1 mixture of heptane:4% paraformaldehyde and fix for 20 min.

11 Remove embryos from the heptane–paraformaldehyde interface using a Pasteur pipette and transfer to an egg basket to air-dry before manually devitellinizing them under 4% paraformaldehyde (see Section 3.3.2).

Protocol 9 continued

12 Fix for a further 20 min in 4% paraformaldehyde.

13 Wash three times, 15 min each, in PBS-Tw.

14 Continue with the embryonic antibody staining protocol (Protocol 4) from step 10.

[a] Example parental genotypes: (virgin females) Hs-GAL4 × UAS-GFP (males)
 F_1 genotype: Hs-GAL4; UAS-GFP

[b] Dissolve the adhesive from double-sided sticky tape in heptane, spot onto coverglasses with a Pasteur pipette and air dry.

[c] Micropoint laser ablation system (Laser Photonics Inc.)

[d] Adhere embryos to a microscope slide coated with double-sided sticky tape and gently pop the embryo out of the chorion with a pair of blunt forceps.

[e] It is essential to monitor constantly the effectiveness of the laser system. It is vital to calibrate the laser before each day of use.

Protocol 10

Laser gene activation in larvae

Equipment and reagents

- Flies of the appropriate genotype[a]
- Food plates
- Diethyl ether
- Glass dish with heavy lid for ether
- Cotton
- Microscope slides
- PBS

- Masking tape
- Razor blade
- Coverglasses
- Compound microscope with epifluorescence
- Laser

Method

1 Grow larvae at 18 °C on plates of standard *Drosophila* food until the second larval instar.

2 Place a small amount of cotton and a stack of slides inside the glass dish and pour ether into the cotton.

3 Place larvae on the stack of slides under PBS and etherize larvae individually for approximately 5 min.[b]

4 Mount larvae singly in PBS on a slide and cover with a coverglass.[c]

5 Screen larvae for GFP expression using epifluorescence prior to laser treatment.

6 Target cells to be lasered using the 63× or 100× microscope objectives.

7 Attenuate the laser beam using a series of neural density filters and apply a 75–120 s laser burst of 4-ns pulses at a frequency of 3–5 Hz to targeted cells.[d]

8 Allow larvae to recover on food plates at 18 or 25 °C for at least 48 h.

9 Re-examine larvae for GFP activation in targeted cells using epifluorescence.

10 Larvae can be processed for antibody staining, if required (*Protocol 5*)

[a] Example parental genotypes: (virgin females) Hs-FLP × FRT-GAL4-UAS-GFP (males)
F_1 genotype: Hs-FLP; FRT-GAL4-UAS-GFP

[b] Observe until movements cease; the slides are used to prevent direct contact with liquid ether; must be performed in a fume cupboard.

[c] This is done by creating a small depression slide by cutting a rectangle out of a piece of masking tape which is adhered to the slide.

[d] Exact settings will vary for individual lasers and must to be determined empirically. It is vital that the laser is calibrated at the start of each session.

4.4 Mosaic analysis

Mosaic analysis has a long history for studying cell interactions in *Drosophila*. The essence of mosaic analysis is to create clones of genetically distinct cells within an otherwise wild-type organism so that the organism becomes 'mosaic' for a particular mutation. In this way, it has been possible to examine the role of particular genes within specific tissues, and to explore fundamental biological questions such as autonomy of gene action and restriction of cell fate. The best examples of this approach are the seminal studies of eye development in *Drosophila* (42).

Mosaics are made by inducing mitotic recombination between homologous chromosomes in heterozygous individuals. Traditionally, X-rays have been used to cause chromosomal breaks, which can result in exchange between homologous chromosome arms. Daughter cells homozygous for a given mutation or marker gene can be formed after mitosis, creating a somatic clone of mutant cells. Whilst the use of ionizing radiation to generate mosaics has been highly successful, it has limitations. The most obvious is that induction of mitotic recombination by ionizing radiation occurs at a low frequency, and achieving an acceptable frequency of clones often causes cell death and can result in developmental abnormalities unrelated to mosaicism (43). As a result, the full power of the mosaic approach has not been fully realized.

4.4.1 FLP–FRT

To create a more efficient method of generating mosaics, the FLP–flippase recombinase target (FRT) site-specific recombination system of the yeast 2 μm plasmid has been adapted for use in *Drosophila*. Initial studies placed the FLP recombinase gene under the control of the hsp70 promoter and used the *white* gene as a marker (44). Heat shock results in the induction of FLP recombinase, which catalyses recombination between the two FRT sites in those cells responding to the stimulus. This leads to the excision of the FRT-flanked *white* gene,

A

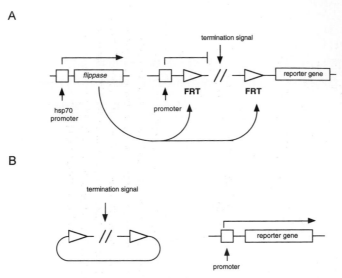

B

Figure 6 Using FLP–FRT to control expression of transgenes. (**A**) In the absence of FLP activity, transcription is initiated at the promoter but is stopped by the presence of transcriptional termination signals shortly before the reporter gene coding sequence. Heat shock induces the FLP recombinase which catalyses recombination between two FLP recombinase target (FRT) sites. (**B**) This leads to the excision of the stop signal that lies between the promoter and the reporter gene coding sequence, resulting in reporter gene expression in those cells responding to the stimulus.

making cells mosaic for this marker and enabling mosaic cells within the eye to be recognized by the absence of the marker gene (*Figure 6*).

4.4.2 Identifying clones in mosaics.

One limitation of classical mosaic techniques was how to distinguish clones of cells from the surrounding non-mosaic cells. For this reason, most of the mosaic studies were limited to cuticular structures or eyes, both of which had pigment-ation, loss of which could be used as a marker to label clones negatively. FLP–FRT site-specific recombination, however, can also be used to label mosaic clones positively. A modification of the technique enables genes to be switched on only in cells where recombination has taken place. FLP-mediated site-specific recom-bination catalyses the excision of transcriptional termination signals that lie between the promoter and the gene to be expressed, removing the block to gene transcription and resulting in gene expression (45). This creates positively labelled clones expressing a marker (46). The versatility of GAL4 has been incorporated into this technique by including GAL4 as the gene whose expression is switched on by FLP-mediated site-specific recombination (47).

4.4.3 MARCM (*mosaic analysis with a repressible cell marker*)

The drawback to mosaic analysis is the inability to control in which cells re-combination events will occur. These limitations have been largely overcome by the MARCM technique developed by Lee and Luo (19). MARCM uses GAL4 to

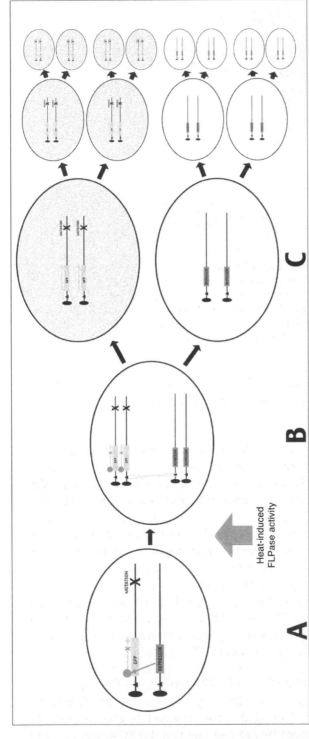

Figure 7 The MARCM technique described. (**A**) A transgene encoding a constituitively active repressor (Ga180) of GAL4 is placed distal to an FRT site ({invtri}). On the homologous chromosome arm from the repressor is placed a 'neuron-specific' GAL4 driver with a UAS–GFP reporter gene (or other required UAS transgene) and the desired mutation (if wanted). In the heterozygous situation, no GFP is expressed due to the repressor, and the mutant phenotype is also not expressed. (**B**) After heat shock activation of FLP recombinase, a small number of mitotically active cells will undergo recombination at the FRT sites. (**C**) As a result, the daughter cells will either be homozygous for the repressor or homozygous for the GFP and mutation. In this situation, the GFP reporter is no longer repressed and GFP is expressed. The GFP-expressing cells are also homozygous for the mutation. All other cells remain heterozygous and do not express GFP. These genotypes are inherited by all daughter cells, resulting in labelling of all clonal descendants of the GFP-expressing cells.

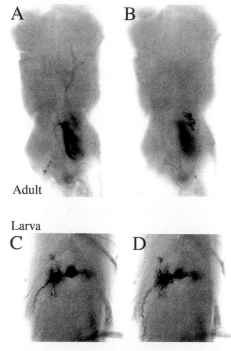

Figure 8 Examples of MARCM-induced labelling. (**A** and **B**) Two different focal planes of an adult thoracicoabdominal complex illustrating anti-GFP labelling of a single neuron clone after heat shocking as a first instar larva. The labelling reveals a distinct cluster of labelled cell bodies in the metathorax and their central arborizations. (**C** and **D**) A single motor neuron in the larval CNS is shown as an example. This preparation was heat shocked as an embryo and resulted in the labelling of a single neuron with an axon leaving the peripheral nerve.

restrict mosaicism to only cells in which the GAL4 is expressed. To achieve this, all cells also express GAL80, a transcriptional repressor of GAL4, so that in the normal situation there is no GAL4 activity in any cell. GAL4 methodology is again combined with FLP–FRT site-specific recombination but in this case induction of the FLP recombinase removes GAL80 repression and enables GAL4, and those genes under GAL4 transcriptional control, to be expressed (*Figure 7*). Cells mosaic for the gene of interest are positively marked by introducing a visible maker, such as GFP, which is also under GAL4 control. The system can also be used to make cells homozygous for a recessive mutation in a gene of interest and, again, those cells where recombination has occurred can be visualized using GFP (*Figure 8*).

To create mosaicism in the nervous system, the elav-GAL4 driver which is expressed in all neurons, is used. Since the FLP recombinase is under the control of a heat-shock promoter, by using appropriate heat shock conditions to induce recombination (and remove GAL80 repression), it is possible to recover single mosaic neurons which can be identified by expression of GFP (*Figure 8*). Thus, MARCM provides a means of targeting gene expression to single identified neurons. While laser-induced heat shock can also be used to activate gene expression in individual neurons, the advantage of MARCM is that it enables both loss- and gain-of-function analysis of genes of interest within single neurons.

4.4.4 Generating clones using MARCM

The flies needed for MARCM are available from Bloomington. To use MARCM to label clones, you need to cross a *Drosophila* line containing the ubiquitously expressed GAL80 construct, distal to the FRT site, with a *Drosophila* line that contains a FRT site in the same position but no GAL80 construct, thus enabling recombination to remove the GAL80 repressor. The resulting offspring must also contain a suitable GAL4 driver (elav-GAL4, for nervous system-specific clones), a marker gene (UAS CD8–GFP) to visualize clones, and a heat shock-inducible FLP recombinase. If you wish to generate clones of cells homozygous for a particular mutation, the mutation must also be introduced distal to the FRT site in the line that does not contain the GAL80 construct. The example below demonstrates the parental genotypes required to label positively clones within the nervous system (*Figure 8*).

Parental genotypes: (virgin females)   elav-GAL4, hs-FLP; FRT G13 GAL80/CyO

$$\times$$

(males)   FRT G13, UAS-mCD8::GFP/CyO

F$_1$ genotype: elav-GAL4, hs-FLP; FRT G13 GAL80/FRT G13, UAS-mCD8::GFP

Induction of the FLP recombinase via activation of the heat-shock promoter can result in somatic clones that are of the genotype elav-GAL4, hs-FLP; FRT G13 UAS-mCD8::GFP, i.e. these clones have lost the GAL80 transcriptional repressor and become homozygous for the UAS-mCD8::GFP transgene, resulting in a high level of GFP expression within the clone.

Protocol 11

MARCM

Equipment and reagents

- Food plates
- Yeast paste (see *Protocol 6*)
- Parafilm
- 39°C water bath

Method

1 Add a small dab of yeast paste to a food plate.

2 Collect virgin females of one parental genotype and males of the other.

3 Set up a parental cross on the food plate.

4 Collect eggs for 1 h at 25°C.

5 Allow embryos to develop until the required stage.[a]

6 Seal the plates with Parafilm and immerse in a water bath at 39°C for 40 min.

7   Remove Parafilm and return embryos to 25 °C to continue development.

8   Dissect and process for antibody staining[b] (*Protocol 4*).

[a] Clones can also be generated post-embryonically; adapt the timings and schedule to suit your needs.

[b] Once individuals have reached the stage of development at which you wish to analyse the clones.

4.5 Methods of targeted cellular ablation

Cell ablation, ranging from the crude to the refined, has been used extensively to study cell interactions in a wide variety of organisms. In *Drosophila*, cell ablation has become a powerful technique due to the level of control that can be achieved through either conditional genetic cell ablation or laser microsurgery. Genetic ablation strategies have used a variety of different effectors. These include the toxic polypeptides diphtheria toxin (DTA; 48) and ricin (49) and the cell death genes *ced3* and *Ice* (50). The key to chemical cell ablation is again GAL4 (13). GAL4 enables toxin expression to be targeted to selected cells using toxins that are coupled to the UAS promoter and for stable toxin gene transgenic lines to be created. For example, UAS-*ricin A* (51) enabled all GAL4-expressing cells to be ablated specifically upon crossing the UAS-*ricin A* flies with GAL4-expressing flies. Since the ricin B chain, which enables the toxin to cross cell membranes, has been removed, the toxin is cell specific and not taken up by neighbouring cells, resulting in an efficient and cell-autonomous method of cell ablation (51).

Cell ablation can also be accomplished using the programmed cell death genes *ced-3* and *Ice* which can overcome potential problems of non-specific cell damage to surrounding cells when cells are either ablated physically or undergo necrosis as the result of toxin expression (52).

Protocol 12

Laser ablation in embryos

Equipment and reagents

- Food plates
- Yeast paste (see *Protocol 6*)
- Diethyl ether
- Glass dish with heavy lid for ether
- Cotton
- Microscope slides
- PBS
- Masking tape
- Razor blade
- Coverglasses
- Compound microscope
- Laser

Method

1   Grow larvae on plates of standard *Drosophila* fly food to the desired stage.

Protocol 12 continued

2   Etherize each larva individually (*Protocol 10*).

3   Mount larvae singly in PBS on a slide and cover with a coverglass.[a]

4   Target cells to be lasered using the 63× or 100× microscope objective and administer 20–30 pulses of 4 ns of laser light at a frequency of 2 Hz.[b]

5   Allow larvae to recover on food plates and add a dab of yeast paste to each plate.

[a] Create a cavity slide by cutting a rectangle in the middle of a piece of masking tape placed on the slide.

[b] The exact settings will vary between lasers and need to be determined empirically. It is vital that the laser is calibrated at the start of each session.

Laser microsurgery provides a refined method of cell ablation. Using a laser microbeam enables an extremely high level of control to be achieved, and enables individual cells to be targeted accurately (53, 54; *Figure 6*). Whilst this gives accurate control on which cells are ablated, it is not easy to use this system to perform ablations in large numbers of individuals whereas with GAL4 toxin ablations, it is easy to generate many hundreds of preparations in one go.

Acknowledgements

We would like to thank the various people who have worked in our laboratory and in so doing have made contributions to the development of the many techniques we use. Work in this laboratory is supported by awards from the BBSRC and Wellcome Trust.

References

1. Roberts, D. B. (ed.) (1998). *Drosophila: a practical approach*. Oxford University Press, Oxford.
2. Greenspan, R. J. (1997). *Fly pushing: the theory and practice of Drosophila genetics*. Cold Spring Harbor Laboratory Press, Cold Spring Harbor, NY.
3. Bate, M. and Martinez-Arias, A. (ed.) (1993). *The development of Drosophila melanogaster*. Cold Spring Harbor Laboratory Press, Cold Spring Harbor, NY.
4. Sullivan, W., Ashburner, M., and Hawley, R. S. (2000). *Drosophila protocols*. Cold Spring Harbor Laboratory Press, Cold Spring Harbor, NY.
5. Roberts, D. B. and Standen, G. N. (1998) In *Drosophila: a practical approach* (ed. D. B. Roberts). p. 1. Oxford University Press, Oxford.
6. Bainbridge, S. P. and Bownes, M. (1981). *J. Embryol. Exp. Morphol.*, **66**, 57.
7. Smith, S. A. and Shepherd, D. (1996). *J. Comp. Neurol.*, **364**, 311.
8. Shepherd, D. and Smith, S. A. (1996). *Development*, **122**, 2375.
9. Mombaerts, P., Wang, F., Dulac, C., Chao, S. K., Nemes, A., Mendelsohn, M., *et al.* (1996). *Cell*, **87**, 675.
10. Higashijima, S., Hotta, Y., and Okamoto, H. (2000). *J. Neurosci.*, **20**, 206.
11. Lundquist, E. A., Herman, R. K., Shaw, J. E., and Bargmann, C. I. (1998). *Neuron*, **21**, 385.
12. O'Kane, C. J. and Gehring, W. J. (1987). *Proc. Natl Acad. Sci. USA*, **84**, 9123.

13. Brand, A. H. and Perrimon, N. (1993). *Development*, **118**, 401.

14. Sweeney, S. T., Broadie, K., Keane, J., Niemann, H., and O'Kane, C. J. (1995). *Neuron*, **14**, 341.

15. Ito, K., Sass, H., Urban, J., Hofbauer, A., and Schneuwly, S. (1997). *Cell Tissue Res.*, **290**, 1.

16. Giniger, E., Wells, W., Jan, L. Y., and Jan, W. N. (1993). *Wilhelm Roux's Arch. Dev. Biol.*, **202**, 112.

17. Callahan, C. A. and Thomas, J. B. (1994). *Proc. Natl Acad. Sci. USA*, **91**, 5972.

18. Brand, A. H. and Dormand, E. L. (1995). *Curr. Opin. Neurobiol.*, **5**, 572.

19. Lee, T. and Luo, L. Q. (1999). *Neuron*, **22**, 451.

20. Phelps, C. B. and Brand, A. H. (1998). *Methods—a companion to Methods in enzymology*, **14**, 367.

21. Williams, D. W., Tyrer, M., and Shepherd, D. (2000). *J. Comp. Neurol.*, **428**, 630.

22. Strausfeld, N. J. and Miller, T. A. (1980). *Neuroanatomical techniques: insect nervous system.* Springer-Verlag, New York.

23. O'Carroll, D. C., Osorio, D., James, A. C., and Bush, T. (1992). *J. Comp. Physiol. A*, **171**, 447.

24. Landgraf, M., Bossing, T., Technau, G. M., and Bate, M. (1997). *J. Neurosci.*, **17**, 9642.

25. Bossing, T. and Technau, G. M. (1994). *Development*, **120**, 1895.

26. Bossing, T., Udolph, G., Doe, C. Q., and Technau, G. M. (1996). *Dev. Biol.*, **179**, 41.

27. Schmid, A., Chiba, A., and Doe, C. Q. (1999). *Development*, **126**, 4653.

28. Grillenzoni, N. vanHelden, J., DamblyChaudiere, C., and Ghysen, A. (1998). *Development*, **125**, 3563.

29. Whitlock, K. E. and Palka, J. (1995). *J. Neurobiol.*, **26**, 189.

30. Wieschaus, E. and Nusslein-Volhard, C. (1998). In *Drosophila: a practical approach* (ed. D. B. Roberts). p. 179. Oxford University Press, Oxford.

31. Patel, N. H. (1994). *Methods Cell Biol.*, **44**, 445.

32. Gratzner, H. G. (1982). *Science*, **218**, 474.

33. Truman, J. W. and Bate, M. (1988). *Dev. Biol.*, **125**, 145.

34. Ito, K. and Hotta, Y. (1992). *Dev. Biol.*, **149**, 134.

35. Tix, S., Bate, M., and Technau, G. M. (1989). *Development*, **107**, 855.

36. Williams, D. W. and Shepherd, D. (1999). *J. Neurobiol.*, **39**, 275.

37. Armstrong, J. D., de Belle, J. S., Wang, Z. S., and Kaiser, K. (1998). *Learn. Mem.*, **5**, 102.

38. Reddy, S., Jin, P., Trimarchi, J., Caruccio, P., Phillis, R., and Murphey, R. K. (1997). *J. Neurobiol.*, **33**, 711.

39. Misquitta, L. and Paterson, B. M. (1999). *Proc. Natl Acad. Sci. USA*, **96**, 1451.

40. Freeman, M. (1996). *Cell*, **87**, 651.

41. Halfon, M. S., Kose, H., Chiba, A., and Keshishian, H. (1997). *Proc. Natl Acad. Sci. USA*, **94**, 6255.

42. Ready, D. F., Hanson, R. E., and Benzer, S. (1976). *Dev. Biol.*, **53**, 217.

43. Xu, T. and Harrison, S. D. (1994). *Methods Cell Biol.*, **44**, 655.

44. Golic, K. G. and Lindquist, S. L. (1989). *Genetics*, **122**, S36.

45. Struhl, G. and Basler, K. (1993). *Cell*, **72**, 527.

46. Brewster, R. and Bodmer, R. (1995). *Development*, **121**, 2923.

47. Ito, K., Awano, W., Suzuki, K., Hiromi, Y., and Yamamoto, D. (1997). *Development*, **124**, 761.

48. Lin, D. M., Auld, V. J., and Goodman, C. S. (1995). *Neuron*, **14**, 707.

49. Moffat, K. G., Gould, J. H., Smith, H. K., and O'Kane, C. J. (1992). *Development*, **114**, 681.

50. Shigenaga, A., Kimura, K., Kobayakawa, Y., Tsujimoto, Y., and Tanimura, T. (1997). *Dev. Growth Differ.*, **39**, 429.

51. Hidalgo, A., Urban, J., and Brand, A. H. (1995). *Development*, **121**, 3703.
52. Shigenaga, A., Kimura, K., Kobayakawa, Y., Tsujimoto, Y., and Tanimura, T. (1997). *Cell Death Differ.*, **4**, 371.
53. Cash, S., Chiba, A., and Keshishian, H. (1992). *J. Neurosci.*, **12**, 2051.
54. Chiba, A., Hing, H., Cash, S., and Keshishian, H. (1993). *J. Neurosci.*, **13**, 714.

Chapter 7

Tight junction protein expression in early *Xenopus* development and protein interaction studies

Sandra Citi

Department of Biology, University of Padova, Viale G. Colombo 3, Padova 35131, Italy and Department of Molecular Biology, University of Geneva, 30 Quai Ernest Ansermet, 1211 Geneva 4, Switzerland

Fabio D'Atri

Department of Molecular Biology, University of Geneva, 30 Quai Ernest Ansermet, 1211 Geneva 4, Switzerland

Michelangelo Cordenonsi, Pietro Cardellini

Department of Biology, University of Padova, Viale G. Colombo 3, Padova 35131, Italy

1 Introduction

This chapter is divided into two sections. The first section describes protocols to produce eggs and embryos of *Xenopus laevis* to investigate the organization and composition of tight junctions (or other structures). The second part describes approaches to study interactions between tight junction (or other) proteins using recombinant proteins and lysates of cultured epithelial cells (from *Xenopus* or mammalian organisms), reticulocytes, and insect cells. Combining studies on protein expression *in vivo* with studies on protein interactions *in vitro* can help to elucidate the molecular mechanisms involved in tight junction biogenesis. The reader is referred to review articles (1–5) for a suitable introduction to the structure, function, and molecular composition of epithelial tight junctions.

2 Tight junction formation during *Xenopus laevis* early development

Tight junction formation in the early stages of *X.laevis* development has been investigated for many years exclusively by electron microscopic and functional approaches. The characteristic morphology of sectioned tight junctions (membrane 'kisses') and fractured tight junctions (fibrils and grooves) could not be established unequivocally before the 32-cell stage (6, 7). In addition, microelectrode impalement studies (8) could not provide reliable results before the blastula stage, because of the small size of the incipient blastocoel. However, a few functional studies indicated that the 'barrier' and 'fence' functions of tight junctions could be detected following the first embryo cleavage, e.g. at the 2-cell stage (9, 10). As specific protein components of tight junctions were discovered, antibody tools have become available to investigate the biogenesis of *Xenopus* tight junctions at the protein level. The first report on the distribution of a tight junction protein in early *Xenopus* development showed that cingulin is maternally provided, is localized at junctions from the 2-cell stage onwards, and is localized at the boundary of distinct membrane domains even in embryos raised in low calcium, which reduces cadherin-dependent cell–cell adhesion (11). Further studies have shown that *Xenopus* homologues of several other tight junction proteins are maternally provided in the *Xenopus* egg (12–15). Characterization of the precise spatial and temporal pattern of expression and junctional assembly of some of these proteins in early embryos is providing interesting clues towards understanding the molecular mechanisms of generation of cell polarity *in vivo* (11, 13). A novel surface biotinylation method has been used to examine the permeability of tight junctions in *Xenopus* embryos, and thus correlate protein localization with development of a functional barrier (13, 14, 16). Further exploitation of *X. laevis* as an experimental system to modulate expression of maternal tight junction proteins will undoubtedly provide key information about the role of these proteins in early development.

2.1 Preparation of *Xenopus laevis* eggs and embryos

Xenopus laevis species originates from South Africa, and animals for laboratory use can be purchased from commercial suppliers (NASCO, Xenopus Express, and others) and then raised in the laboratory (17). Once females reach the age of 2 years, they are mature for egg production. Frogs are reared in plastic or glass tanks (volume 50 l) containing dechlorinated water (~4 l per animal), which is either filtered continuously, or changed every 2–3 days. The room should have controlled temperature (18–22°C) and light (12 h light:12 h darkness cycle). Animals are fed twice a week with beef heart cut (not ground) into small pieces, or with fish food pellets. Ovulation is induced only once every 3–6 months. Many factors can influence egg production: temperature, light cycle, season, genetic variety, recycling times, and diseases. Sometimes eggs are produced but are not viable or usable for experiments. Once fertilized eggs are obtained (see *Protocols 1* and *2*), embryos are allowed to develop either in the collection tray, or in a Petri dish, after gently stripping them from the tray using a plastic blade. To study early phases of development, it is essential to obtain freshly laid eggs. Embryos are staged according to Nieuwkoop and Faber (18) by observation with a stereomicrosope.

Protocol 1

Induction of egg production and normal fertilization in *Xenopus*

Equipment and reagents

- Tank with aerated, temperature-controlled dechlorinated water
- Light-tight metal box (30 cm each side) with lid, gridded base (1 cm mesh), and removable tray underneath
- Follicle-stimulating hormone (FSH) and human chorionic gonadotropin (hCG) (Sigma)

Method

1 One week before mating, inject FSH (7.5 IU) into a female. Make all injections into the dorsal lymphatic sac, using an insulin syringe and a volume of 0.5–1.0 ml.

2 At 48 h before mating, inject hCG (300 IU) into a male.

3 At 24 h before mating, inject hCG into the pre-stimulated female (400 IU) and male (300 IU).

4 In the evening, place the couple in the metal box inside the tank. Programme the temperature to shift from 20 to 25 °C starting from 8 h before ovulation.

5 Collect eggs in the morning, using the removable plastic tray located under the gridded floor of the box. To ensure that eggs are freshly laid, place a new tray under the grid every hour.

Protocol 2

Artificial fertilization of *Xenopus* eggs

Equipment and reagents

- MMR (100 mM NaCl, 2.0 mM KCl, 2.0 mM MgCl$_2$, 2.4 mM CaCl$_2$, 20 mM *N*-(2-hydroxyethyl) piperazine-*N*′-(2-ethanesulphonic acid) (HEPES, pH 7.8)
- Surgical dissection tools
- MMR1/4 (25 mM NaCl, 0.5 mM KCl, 0.5 mM MgCl$_2$, 0.6 mM CaCl$_2$, 5 mM HEPES, pH 7.8)

Method

1 Stimulate a male and a female as described in *Protocol 1*, but keep them in separate metal boxes.

2 Gently squeeze eggs out of the female by applying pressure to the abdomen while keeping legs apart. Distribute the eggs as a layer on a Petri dish. Fertilize eggs immediately or store for a few hours in MMR. In this case, just before fertilization, rinse with MMR1/4 and drain.

Protocol 2 continued

3 Anaesthesize a male by hypothermia (immersion in ice for 40 min). Dissect out the testicles. Use immediately or store at 4°C for up to 5 days in a Petri dish containing MMR, 10% fetal calf serum, 50 μg ml^{-1} gentamicin.

4 Smear small fragments of testicles directly on top of the eggs, or gently homogenize them in MMR1/4 and then pipette onto the eggs.

5 After 5 min, add dechlorinated water or MMR1/4 and allow to develop. The first cleavage of fertilized eggs occurs 60–80 min after fertilization.

2.2 Whole-mount immunolocalization of tight junction proteins in *Xenopus laevis* eggs and early embryos

Tight junctions surround the apical regions of polarized epithelial cells as narrow continuous belts, which extend for less than 1 μm along the apicobasal axis of the cell. Early *Xenopus* embryos are made of a few large cells, and after they are sectioned it is difficult to identify tight junctions, or immunolocalize tight junction proteins. On the contrary, at the blastula or gastrula stage, when the size of the embryo is the same but many small cells and abundant junctions are present, it is easier to localize tight junction proteins in sections of fixed or frozen embryos. In our experience, whole-mount immunolabelling (*Protocol 4*) is the protocol of choice to immunolocalize tight junction proteins in eggs or early embryos (11–13). Antibody penetration is good, and extensive washes reduce background staining. The presence of yolk particles can be a problem if fluorescent secondary reagents are used, because of yolk autofluorescence. Since the vitelline membrane also tends to give non-specific background staining, we prefer to remove it prior to immunolabelling (*Protocol 3*), even though this can lead to slight separation of blastomeres and deformation of the embryo during cell division. After removal of the vitelline membrane, eggs and embryos are fragile and should be handled very gently and transferred from one solution to the other only with specially adapted glass Pasteur pipettes. If the secondary antibodies are labelled with an enzyme, and a colour reaction is developed, it is preferable to use eggs from albino animals, which lack pigment granules. Gold-labelled secondary antibodies are used for immunoelectron microscopy (13). Once the specimens have been stained by the whole-mount procedure, they can either be observed directly *in toto* by light or fluorescence microscopy (preferably confocal, to reduce out-of-focus staining), or sectioned and analysed by electron, light, or immunofluorescence microscopy. If immunofluorescently stained embryos are sectioned, they are post-fixed in 3.7% formaldehyde in phosphate-buffered saline (PBS), dehydrated, and embedded in glycolmethacrylate before sectioning (13). For immunogold localization, embryos are post-fixed in 2% glutaraldehyde in PBS, stained with 1% uranyl acetate, embedded in 2% low-melting agarose, dehydrated, and embedded in Spurr's resin (13). Direct analysis of unsectioned embryos allows a better three-dimensional observation of protein distribution, whereas analysis of sections is preferred to obtain a precise spatial definition of protein distribution boundaries, and is the only possible method for electron microscopic examination. The success of the whole-mount immunolabelling protocol described here depends primarily on the quality of the primary antibodies. If the antibodies work for immunoblotting, it is a good idea to check beforehand that they recognize specifically only one polypeptide of the expected electrophoretic mobility in lysates of *Xenopus* embryos (*Protocol 6*). Good results have been reported with this protocol using antibodies to cingulin and occludin (11–13), zonula occludens 1 (ZO-1), β_1-integrin, and β-catenin (13).

Protocol 3

Removal of jelly coat and vitelline membrane from *Xenopus* eggs/embryos

Equipment and reagents

- Dissection microscope
- Watchmaker tweezers (Dumont no. 5) with tips thinned by fine abrasive paper (grade 6/0)
- Cysteine solution (165 mM L-cysteine hydrochloride in MMR1/4 (see *Protocol 2*), adjusted to pH 8.9 with 10 M NaOH)
- MMR (see *Protocol 2*)

Method

1   Collect eggs/embryos in a 6-cm Petri dish, and pour off all or most of the water. Add approximately 10 ml of cysteine solution.

2   Incubate eggs in the cysteine solution at room temperature for about 3 min, swirling gently and continuously. Using the dissection microscope, check that the jelly coat has dissolved.

3   Rinse three times in MMR, and leave eggs/embryos in MMR1/4 solution.

4   Remove the vitelline envelope manually, using two pairs of tweezers. Grab the vitelline membrane from the top with the first pair of tweezers and gently pull, then grab the resulting fold with the second pair of tweezers and pull the envelope apart in opposite directions laterally.

Protocol 4

Whole-mount immunostaining of *Xenopus* eggs/embryos

Equipment and reagents

- Glass slides with a lens-shaped cavity (15 mm diameter, 0.7 mm depth)
- Rocker platform
- Dent's solution (80% methanol, 20% dimethylsulphoxide (DMSO), stored at −20 °C)

Protocol 4 continued

- Wide-mouth glass Pasteur pipettes: cut the tip with a diamond knife to obtain an opening diameter of about 1 mm, polish the edge, and curve the pipette approximately 2 cm from the opening by flaming

- *Xenopus* PBS (XPBS: 110 mM NaCl, 1.9 mM KCl, 8 mM Na_2HPO_4, 2 mM KH_2PO_4, pH 7.4)

- Blocking medium (2 mg ml^{-1} of bovine serum albumin (BSA) or 1% fish skin gelatine (Sigma) in XPBS)

- First antibody/antiserum diluted in blocking medium, or undiluted hybridoma culture supernatant

- Secondary antibodies: we have used fluorescein- or rhodamine-conjugated antibodies (Jackson Laboratories, diluted 1:100 in blocking medium), or peroxidase-conjugated antibodies (Calbiochem, diluted 1:1000)

- Peroxidase substrate for colour reaction (1.3 mM diaminobenzidine (DAB), 42 mM $CoCl_2$, 42 mM $NiCl_2$, 0.015% H_2O_2)

- Murray's solution (benzyl alcohol:benzyl benzoate 1:2)

- Mounting medium (XPBS containing 50% glycerol and 2.5% diazabicyclo [2.2.2] octane)

Method

1 Transfer eggs/embryos into pre-cooled Dent's solution in a glass vial (at least 1 ml of solution per five embryos). Incubate at $-20\,^{\circ}$C (not a frost-free freezer) for at least 16 h, but no more than 1 month (best: 1 week).

2 Rehydrate embryos by incubating in a descending ethanol series (100, 90, 75, and 50% in XPBS), each for 10 min, using wells of a 24-well tissue culture plate (4–5 embryos per well). Incubate for 15 min at room temperature in XPBS.

3 Optional: incubate embryos twice for 20 min in blocking medium at room temperature.

4 Incubate with first antibody overnight (16–20 h) at 4°C, with gentle rocking. Use enough antibody solution to submerge the embryos completely.

5 Wash embryos three times with blocking medium, each for 30 min, at room temperature.

6 Incubate with secondary antibody overnight at 4°C in the dark, with gentle rocking.

7 Wash once with blocking medium and twice with XPBS, each for 30 min at room temperature and in the dark.

8 If peroxidase-conjugated secondary antibodies are used, develop the colour reaction with substrate for 5 min, wash twice with PBS for 10 min, and dehydrate in methanol for 1 h at room temperature. Make specimens transparent by immersion in Murray's solution for 30 min (or until clear) and place embryos in the well of a special slide using Murray's solution as mounting medium.

9 If fluorescently labelled antibodies are used, place embryos in the special slide using approximately 100 µl of mounting medium, trying to avoid air bubbles.

10 Observe by light or fluorescence microscopy.

2.3 Scanning electron microscopic analysis of junctions in *Xenopus* embryos

Scanning electron microscopy of *Xenopus* embryos allows direct visualization of blastomere membrane surfaces coated with electron-dense metal. Since tight junction assembly in early *Xenopus* embryos involves the fusion of membrane vesicles (containing ZO-1 and occludin (13)) with the plasma membrane in the junctional area, scanning electron microscopy analysis can be useful to study the distribution of vesicular structures in the fractured tight junction membrane. In addition, the steps in the formation of the adhesion crest can be visualized at high magnification and with a comprehensive view, which is impossible to achieve with other techniques.

Protocol 5

Preparation of *Xenopus* samples for scanning electron microscopy

Equipment and reagents

- Critical point dryer (Edwards)
- Sputter coater (Edwards 5150B)
- Fixation solution (2.5% glutaraldehyde in 0.1 M cacodylate, pH 7.3)
- Wash solution (0.1 M cacodylate, pH 7.3)
- Post-fixation solution (1% osmium tetroxide in 0.1 M cacodylate, pH 7.3)
- Liquid CO_2

Method

1 Dejelly embryos (*Protocol 3*) and allow to develop in MMR1/4 (see *Protocol 2*) until the desired stage.

2 Incubate in fixation solution for 1 h at room temperature in a sealed glass container.

3 Wash (wash solution) twice for 15 min.

4 Remove vitelline membrane manually (*Protocol 3*).

5 Wash four times for 30 min each.

6 Post-fix (post-fixation solution) for 1 h at room temperature in a sealed glass container.

7 Wash four times for 30 min each.

8 Dehydrate by sequential immersion in 30, 50, 60, 70, 85, 95, and 100% ethanol (30 min each).

9 Completely dehydrate using the critical point dryer with at least seven washes in liquid CO_2 at 4 °C. Remove CO_2 by evaporation at 40 °C.

10 Fracture samples (usually into two halves) manually, using a scalpel under microscopic examination.

Protocol 5 continued

11 Place samples onto holders using double-sided sticky tape, and a droplet of silver to achieve electrical connection.

12 Coat samples with gold–palladium in an argon atmosphere at 1.6 kV for 4 min.

13 Observe with a scanning electron microscope.

2.4 Biochemical analysis of *Xenopus* eggs and embryos

Lysates of *X.laevis* eggs and embryos are used for immunoblotting detection of proteins at different stages of development, and for a variety of other purposes, including immunoprecipitation, phosphorylation assays, and glutathione S-transferase (GST) pull-downs. The composition of the lysis buffer depends on the subsequent use of the lysate and the extractability of the protein(s) under examination. Addition of sodium dodecyl sulphate (SDS) to lysis buffers can be useful to extract insoluble proteins, but leads to solubilization of yolk particles, which creates problems in running SDS–polyacrylamide gels of lysates. Non-ionic detergents such as Triton X-100 or Nonidet P-40 (NP-40; 0.1%) do not cause such problems, and can be used to extract membrane proteins (12). Two simple protocols are described here to prepare lysates for immunoblotting analysis, and to fractionate embryo lysates into cytoplasmic and membrane fractions (19). Lysates prepared using this protocol were used to identify phosphorylation-dependent shifts in the electrophoretic mobility of occludin in *Xenopus* eggs and embryos (12).

Protocol 6

Preparation of lysates of *Xenopus* eggs or embryos

Equipment and reagents

- LBX (lysis buffer *Xenopus*: 150 mM NaCl, 2 mM ethylenediamine tetraacetic acid (EDTA), 10 mM HEPES, pH 7.6, 1 mM phenylmethylsulphonyl fluoride (PMSF), 10 μg ml^{-1} leupeptin, 5 μg ml^{-1} aprotinin, 50 μg ml^{-1} benzamidine, 1 μg ml^{-1} pepstatin A, 0.1% Triton X-100 (optional))

- Freon (1,1,2-trichlorofluoroethane, Sigma T-5271)
- Pipetman
- Benchtop centrifuge

Method

1 Collect eggs/embryos and stage them. Remove jelly coat and vitelline membrane (see *Protocol 3*).

2 Collect 10–20 embryos for each stage in a 1.5-ml microtube, and remove excess liquid. Use immediately or store at $-80\,°C$ after rapid freezing in liquid nitrogen. All subsequent steps are performed on ice or at 4 °C.

Protocol 6 continued

3   Add ice-cold LBX (5 μl per embryo, typically 100 μl for 20 embryos) to each tube and homogenize by pipetting, using a Pipetman.

4   Optional: add an equal or greater volume of Freon. Vortex well. This step improves removal of yolk particles and protein solubilization.

5   When the solution is uniformly grey (without dark particles), centrifuge the sample for 5 min at 10 000 g at 4 °C.

6   Recover the upper floating phase (lipids) together with the clear yellow supernatant by careful pipetting. Discard the dark-grey pellet (yolk particles and other cellular debris).

7   Analyse the supernatant fraction by SDS–polyacrylamide gel electrophoresis (PAGE) and immunoblotting with specific antibodies to tight junction proteins.

Protocol 7

Biochemical fractionation of *Xenopus* eggs/embryos

Equipment and reagents

- Evans lysis buffer 1 (50 mM NaCl, 10 mM MgCl$_2$, 20 mM Tris–HCl, pH 7.6, 1 mM PMSF, 10 μg ml^{-1} leupeptin, 5 μg ml^{-1} aprotinin, 50 μg ml^{-1} benzamidine, 1 μg ml^{-1} pepstatin A)

- Sucrose cushion (20% sucrose in Evans lysis buffer 1)

- Evans lysis buffer 2 (110 mM NaCl, 1.9 mM KCl, 8 mM Na$_2$HPO$_4$, 2 mM KH$_2$PO$_4$, pH 7.4, 0.1% Triton X-100, 1% NP-40, 1 mM PMSF, 10 μg ml^{-1} leupeptin, 5 μg ml^{-1} aprotinin, 50 μg ml^{-1} benzamidine, 1 μg ml^{-1} pepstatin A)

- Benchtop centrifuge

Method

1   Collect and store embryos as described in *Protocol 3*. Carry out all steps at 4 °C or on ice.

2   Homogenize 20 embryos in 100 μl of ice-cold Evans lysis buffer 1 by pipetting.

3   Layer the homogenate on top of a 400 μl ice-cold sucrose cushion in a 1.5-ml microtube, making sure the solutions do not mix.

4   Centrifuge at 10 000 g for 30 min at 4 °C.

5   Collect the supernatant and floating phase (cytosolic fraction) and store on ice for immediate use or at −80 °C.

6   Re-homogenize the pellet with 100–500 μl of ice-cold Evans lysis buffer 2.

7   Centrifuge at 10 000 g for 10 min at 4 °C.

8   Collect the supernatant (membrane fraction) and discard the pellet.

9   Analyse cytosolic and membrane fractions by SDS–PAGE and immunoblotting.

2.5 Microinjection of *Xenopus* oocytes, eggs, and embryos

One of the advantages of *X.laevis* as a model system is the relative ease with which one can inject oocytes, eggs, and blastomeres with solutions containing sense cDNAs, antisense oligonucleotides, mRNAs, or proteins. Injection is usually designed to modulate the expression levels of endogenous mRNAs and proteins, or to express proteins or protein fragments exogenously, to obtain a phenotype which can help to understand the developmental role of specific genes and proteins (20–22). Microinjection of cDNAs or sense mRNAs can be used to express proteins exogenously and study their function in oocytes (23, 24). We have microinjected cDNA constructs encoding different regions of cingulin into early embryos and determined their expression and subcellular localization in advanced embryo blastomeres by immunoblotting (*Figure 1*) and immunofluorescence (25). The protocol outlined below can be used in the microinjection of oligonucleotides, mRNAs, plasmids, and proteins. mRNAs suitable for microinjection into *Xenopus* are produced by *in vitro* transcription with SP6, T7, or T3 RNA polymerase using vectors (e.g. pSP64T) containing 5′- and 3′-untranslated region globin sequences. The 'Xenopus Molecular Marker Resource' (http://vize222. zo.utexas.edu) is an excellent source of information about vectors, primers,

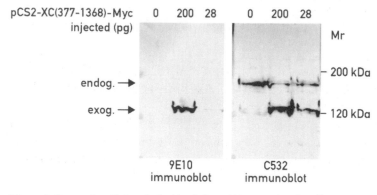

Figure 1 Expression of cingulin in microinjected *Xenopus laevis* embryos. Immunoblots with anti-Myc tag mouse monoclonal antibody (9E10) or with anti-cingulin rabbit polyclonal antiserum (C532) of *X.laevis* embryo lysates (gastrula stage, see *Protocol 6*). Embryos were microinjected (Protocol 8) with the indicated amounts (200 or 28 pg) of plasmid DNA (pCS2 vector containing the Myc-tagged cingulin C-terminal construct (25)), or with 28 ng of empty vector (0). The 9E10 immunoblot shows a polypeptide of the expected mobility only in microinjected embryos (exog.), which is clearly detectable in embryos injected with 200 ng per embryo of DNA, and barely detectable in embryos injected with a smaller amount of DNA (28 ng per embryo). The C532 immunoblot shows a polypeptide corresponding to endogeneous cingulin (endog.) in all samples, and also reacts with polypeptides corresponding to exogeneous cingulin fragments in the microinjected embryos. The C532 antiserum binds to the C-terminal region of cingulin and thus recognizes both endogeneous and exogeneous cingulin. Exogeneous cingulin in the embryo microinjected with 28 ng per blastomere of DNA is visualized better by C532 than by 9E10 antibody. This may be due to the higher titre or affinity of the C532 antibody. Immunoblots were carried out as described in ref. 12. Numbers on the right represent the mobility of pre-stained molecular size standards.

libraries, and much more. It is important to make sure that the solution has been clarified by centrifugation and to know precisely the concentration of the molecule in the injected solution. Microinjection of sufficient numbers of oocytes/eggs/embryos and control injections with solvent alone or with reporter gene mRNAs (e.g. β-galactosidase) are essential control experiments.

Protocol 8

Microinjection of *Xenopus laevis*

Equipment and reagents

- Glass microelectrode puller (Narishige PD5 or similar) and glass capillaries (1 mm external diameter, 0.75 mm internal diameter)
- Thin pipette tips (Eppendorf microloader 5242 956.003 or similar)
- Micromanipulator (Narishige NM21 or similar)
- Dissecting stereomicroscope (Olympus SZH or similar) and fibre-optic light source

- Microinjector (Pico-injector PL-100, Medical System Co., or Narishige mod. IM 300 or other suitable microinjector) equipped with compressed gas (air or nitrogen)
- Ficoll solution (6% Ficoll 400 in MMR, see *Protocol 2*)
- MMR1/4 (*Protocol 2*)

Method

1 Prepare glass needles using a microelectrode puller calibrated to obtain a tip of 5–7 mm length. Longer tips are fragile, shorter tips may damage the embryo. Bake needles at 180°C for 12 h if mRNA is injected.

2 Pipette the solution (2–4 μl) into the needle through its wide (back) opening using a thin plastic tip. Insert the needle into the microinjector holder, and fit the holder into the micromanipulator device. Gently break the tip of the needle by touching a hard metal surface.

3 To measure the diameter of the droplet produced by the tip, perform test pulses at a standard pressure and injection time (e.g. 50 p.s.i., 200 ms) under microscopic examination, using a reference standard such as the micrometric scale within the eyepiece of the microscope. Adjust pressure and pulse length to obtain a droplet of the desired volume. For a 1-cell embryo, the recommended volume is 10–20 nl, for a 4-cell embryo, 6–8 nl per cell. A 10 nl droplet has a diameter of 266 μm.

4 Dejelly eggs/embryos (*Protocol 3*) and immerse in Ficoll solution in a Petri dish. The Ficoll solution absorbs the water of the perivitelline space, and reduces the pressure exerted onto the vitelline membrane, thus reducing the possibility that injection leads to extrusion of cytoplasm.

5 Align embryos on a nylon grid (0.5 mm mesh) to keep them still during injection. Place the Petri dish on a moving stage, to speed up injection of many embryos.

Protocol 8 continued

6 Using the micromanipulator, approach the needle tip to the embryo with a tilt of about 45° in order to meet the vitelline membrane perpendicularly. When the tip of the needle touches the vitelline membrane, gently move the needle forward using the fine movement knob until the tip penetrates into the embryo.

7 Perform the injection with a pulse and gently remove the tip from the embryo.

8 Incubate the embryos in the same solution at room temperature for the time required to undergo 3–4 cell divisions. Discard embryos which do not undergo division.

9 Dilute the solution 1:1 with MMR1/4. About 1 h later, transfer the embryos into MMR1/4 and allow to develop at 15°C until examination.

3 Assays to study protein–protein interactions *in vitro*

Probably all proteins perform their functions by interacting with other proteins or with themselves (polymers). It is therefore essential to characterize such interactions in detail in order to understand biological processes at the molecular level. It is difficult to study protein–protein interactions within intact, living cells, unless specific proteins can be removed from or added to cells by genetic manipulations. *In vitro* studies can provide useful indications regarding what happens *in vivo*, especially when experiments are carried out under conditions close to the physiological situation. When carrying out *in vitro* interaction studies, ideally one would like to use purified, native, correctly folded, full-length proteins. If native proteins cannot be purified, recombinant proteins (made in bacteria, insect cells, or cultured vertebrate cells) can represent good alternatives. Fusion of the protein to a tag (GST, 6×His, Myc, or others) makes purification and/or detection easier, even though tags sometimes interfere with protein folding and function.

Often one of the proteins is a recombinant, purified fusion protein, and the other protein is present in a crude cell lysate. Working with lysates rather than purified proteins raises the concern that any interaction detected between protein A and protein B could, in principle, be indirect, e.g. mediated by protein C. Appropriate controls, competition experiments, experiments with lysates from different sources, and SDS–PAGE analysis of protein complexes should be used to make sure that interactions are direct. In any event, the physiological relevance of interactions should be verified whenever possible by quantitative analysis of interactions (see *Protocols 13* and *14*) and studies *in vivo*.

We describe here simple methods to produce recombinant proteins from bacteria, insect cells, and reticulocytes, and to prepare lysates from cultured vertebrate cells. These reagents can be used in GST pull-down assays to study protein–protein interactions. Proteins are detected by SDS–PAGE or by immunoblotting with specific antibodies (25). There are many additional techniques to study protein–protein association, one of which is immunoprecipitation (*Protocol 15*), which relies on the formation of immune complexes.

3.1 Bacterial and insect cell expression of recombinant proteins

Bacteria and cultured insect cells are used commonly to produce recombinant proteins or protein fragments. The choice of system depends mainly on the protein to be expressed and its subsequent use. Baculovirus-infected insect cells are preferred when post-translational modifications which do not occur in bacteria are required for correct protein folding and function, when proteins to express are large (>100 kDa) (the yields in bacteria are usually low), or when more than one protein must be expressed simultaneously (multisubunit protein complexes). On the other hand, when the protein to be expressed is small, and is correctly folded in bacteria, bacterial expression can offer higher yields at lower costs. The use of truncation or point mutants of specific proteins helps to identify regions and residues involved in interactions with other proteins. A number of vectors are available commercially (from Pharmacia, Stratagene, Invitrogen, Novagen, Life Technologies, Pharmingen, etc.) which are designed for bacterial or insect cell expression of proteins, with or without tags fused to the C- or N-terminal ends. We refer the reader to the website list at the end of the chapter for complete product descriptions and manuals.

Protocol 9

Preparation of bacterial lysates containing recombinant GST fusion proteins

Equipment and reagents

- DH5α and BL21(DE3) bacterial strains
- 1 M isopropyl-β-d-thiogalactopyranoside (IPTG) (stock frozen in 1 ml aliquots at −20°C)
- Triton X-100, 20% in PBS
- PBS (140 mM NaCl, 2.7 mM KCl, 10.1 mM Na_2HPO_4, 1.8 mM KH_2PO_4, pH 7.3)
- 2XTY + Amp medium
- Benchtop centrifuge

Method

1. Subclone cDNA corresponding to the protein (or protein fragment) of interest into a pGEX vector (Amersham Pharmacia) using DH5α competent cells.

2. Transform competent BL21(DE3) bacteria with the correct construct. Identify colonies which express protein after induction with 0.1 mM IPTG by running SDS–polyacrylamide gels (and/or immunoblotting) of whole bacterial lysates. The electrophoretic mobility of the recombinant polypeptide should correspond to 29 kDa (GST) plus the calculated mass of the polypeptide sequence coded by the cDNA insert.

Protocol 9 continued

3 Optimize expression levels by adjusting the temperature (30 or 37°C) and time of growth (1–6 h) after induction. Check expression by SDS–PAGE and immunoblotting, always running samples of non-induced bacteria as controls.

4 For each protein, optimize the bacterial lysis/extraction procedure, using small-scale cultures, and testing the effect of detergents, chaotropic agents, salt concentration, and pH on protein solubility. Check supernatants and pellets after bacterial lysis by SDS–PAGE and immunoblotting.

5 For large-scale preps, innoculate 1–2 l of 2XTY+Amp with freshly streaked bacteria. Grow until the $OD_{600} = 0.5$–1, induce with IPTG, and grow for a further 1–4 h.

6 Collect bacteria by centrifugation in large centrifuge tubes (Sorvall GS4 rotor), and freeze by pouring a small amount of liquid nitrogen on the pellets. Pellets can now be stored at −70°C if desired. Resuspend pellets completely on ice with PBS (50 ml for 1 l of original culture) by swirling/pipetting.

7 Add Triton X-100 to 1% final concentration. Sonicate twice for 30 s on ice at 70% power, avoiding froth. Leave on ice for 15 min. Centrifuge for 10 min at 15 000 g at 4°C. Store the lysate supernatant in aliquots at −70°C.

8 Determine the recombinant protein yield by incubating different volumes of bacterial lysate (range 50–1000 ml) with a fixed volume (20 ml) of glutathione–Sepharose resin (*Protocol 13*), and analysing the bound protein by SDS–PAGE, running a standard curve of BSA on the same gel.

Protocol 10

Preparation of baculovirus-infected insect cell lysates containing recombinant proteins

Equipment and reagents

- pFASTBac vector (Life Technologies)
- DH10BAC bacterial strain containing bacmid and helper
- LB-BAC medium (containing kanamycin 50 μg ml^{-1}, gentamicin 7 μg ml^{-1}, tetracycline 10 μg ml^{-1}) and LB-BAC plates (same composition as the medium, with the addition of X-gal 100 μg ml^{-1}, IPTG 40 μg ml^{-1}, agar 15 g l^{-1})
- Insect cell medium (TC-100 or Grace's medium (Life Technologies) with the addition of 10% fetal bovine serum, 5 U ml^{-1} penicillin, 5 μg ml^{-1} streptomycin)
- Lipid concentrate (Life Technologies cat. no. 21900-030)

- Transfection buffer A (Pharmingen cat. no. 21483: insect cell medium)
- Transfection buffer B (Pharmingen cat. no. 21483: 25 mM HEPES, pH 7.1, 125 mM CaCl$_2$, 140 mM NaCl)
- LBT (lysis buffer Triton: 150 mM NaCl, 20 mM Tris–HCl, pH 7.5, 5 mM EDTA, 1% Triton X-100, 1 mM PMSF, leupeptin 5 μg ml^{-1}, antipain 5 μg ml^{-1}, pepstatin 5 μg ml^{-1})
- Wash buffer: same composition as LBT, but with Triton X-100 at 0.5%
- Sf9 or Sf21 insect cells

Method

1 Subclone cDNA corresponding to the protein of interest into pFASTBAC, using DH5α cells.

2 Transpose the pFASTBAC construct into the viral genome by transforming competent DH10BAC bacteria with 20 ng of pFASTBAC construct and isolating white colonies growing on LB-BAC plates. The colonies require a long incubation time to be seen (usually 24–36 h at 37°C, followed by 24–48 h at 4°C). Pick 2–4 candidates and restreak to verify the white phenotype.

3 Isolate recombinant bacmid DNA by inoculating a positive colony into LB-BAC medium and preparing DNA by a modified miniprep protocol (see website manual Life Technologies). Analyse the preparation of bacmid DNA by electrophoresis on a 0.5% agarose gel.

4 Grow insect cells (*Spodoptera frugiperda*, Sf9 or Sf21) from frozen stock on tissue culture dishes/flasks (optimal density 10–80%), or as a liquid culture in Erlenmeyer flasks (0.5–2.0×10^6 cells ml^{-1}, 100 r.p.m.) at 28–30°C. Add lipid concentrate (dilute $100\times$ to $1\times$) for growth in liquid culture. Passage cells by dilution and avoid mechanical stresses, since cells are fragile. Store in liquid nitrogen in 90% insect cell medium, 10% DMSO.

5 Transfect insect cells with bacmid DNA. Seed 2×10^6 cells on a 35-mm plate and allow to attach for 1 h. Remove the medium and add 333 μl of fresh medium (Transfection buffer A). Mix 5 μl of bacmid DNA with Transfection buffer B and add dropwise to the cells. Incubate at 28–30°C for 4 h, remove transfection solution and add 3 ml of fresh medium. Check cell morphology every day. Successful infection results in morphology changes. After 4 days, remove the supernatant (initial viral stock) and analyse whole-cell lysates by SDS–PAGE or immunoblotting to look for expression of recombinant protein. If expression is not detected or is too low, repeat infection on a 6-cm plate with 100 μl of initial viral stock. After 4 days, repeat the procedure with the latest viral stock. Store phage stock in the dark at 4°C (stable for 2 months) or at −70°C after slow freezing. Virus titre can be determined by plaque assay (see Lifetech website manual).

6 Optimize protein expression levels by adjusting the temperature (27–30°C) and time of growth (12–96 h) after induction. For each protein, optimize the lysis procedure to improve yield in the soluble, supernatant fraction.

7 For large-scale production of recombinant protein and phage stock, infect a liquid culture of cells (100–500 ml per Erlenmeyer flask) at a density of 5×10^5 cells ml^{-1} with baculovirus stock at a multiplicity of infection of 1. At the end of infection, harvest cells by centrifugation (5 min at 1000 r.p.m. in 50-ml tubes). Keep the supernatant (virus stock). Wash cells twice with cold PBS, lyse with LBT (25 ml per original 250 ml cell culture) on ice, and sonicate ten times with 3-s blasts at 50% power. Centrifuge the lysate for 30 min at 100 000 g at 4°C to remove insoluble material. Store the lysate supernatant at −70°C in aliquots.

3.2 Preparation of vertebrate cell lysates containing tight junction proteins

Cultured mammalian epithelial cells (MDCK, CaCo2, etc.) or *X.laevis* epithelial cells (A6) can be used to prepare whole-cell lysates containing tight junction (and other) proteins for GST pull-down experiments (*Protocol 13*). To improve yield, cells should be lysed just before they reach confluence. The volume of lysis buffer used to lyse the cells depends on the abundance of the protein. For a 9-cm Petri dish, a convenient volume is between 1 and 3 ml. The composition of the lysis buffer can be adjusted to optimize partition of the protein(s) of interest in the soluble phase. However, high salt or ionic detergents (for instance SDS at a concentration >0.1%) should be avoided, since they affect the binding of GST fusion proteins to glutathione–Sepharose resin. This limits the applicability of this method to protein components of tight junctions which are not tightly associated with membranes. Cultures can be labelled with [35S]cysteine–methionine (Redivue PROm-MIX Amersham code AGQ0080, diluted 1:100 in cysteine-free medium) to identify polypeptides by autoradiography (26).

Protocol 11

Preparation of lysates from cultured vertebrate cells

Equipment and reagents

- PBS containing calcium and magnesium (PBS-CM: 137 mM NaCl, 0.9 mM $CaCl_2$, 2.7 mM KCl, 1.5 mM KH_2PO_4, 0.2 mM $MgCl_2$, 8.1 mM Na_2HPO_4)
- Rocker platform
- LBT (see *Protocol 10*)

Method

1 Grow cells until almost confluent in appropriate medium in a 9 or 15 cm diameter Petri dish.

2 Remove medium and wash cells twice with cold PBS-CM.

3 Add 1–3 ml (9 cm) or 5–8 ml (15 cm) of LBT and lyse for 30 min at 4 °C with rocking.

4 Collect the lysate and clarify by centrifugation at 13 000 *g* for 10 min. Keep total, supernatant, and pellet fractions to check protein solubilization by SDS–PAGE and immunoblotting.

5 Store lysates in aliquots at −70 °C.

Protocol 12

Preparation of ^{35}S-labelled proteins in TNT reticulocyte lysates

Equipment and reagents

- DNA template: cDNA subcloned in a vector with an RNA polymerase promoter (e.g. T3, T7, or SP6) positioned 5′ with respect to transcription start. The plasmid is purified by Qiagen columns and at a concentration ≥0.5 mg ml^{-1}

- [^{35}S]Methionine 1200 Ci mmol^{-1} at 10 mCi ml^{-1} (Amersham Pharmacia AG1094)

- TNT® T7 Coupled Reticulocyte Lysate System (e.g. Promega L4610 for T7 polymerase), containing:
  TNT® reticulocyte lysate (store in 50 μl aliquots at −80 °C)

  1 mM amino acid mixture minus methionine
  T7 RNA polymerase
  TNT® reaction buffer.

- Ribonuclease inhibitor (40 U μl^{-1})

- Nuclease-free water

- Amplify solution (Amersham cat. no. NAMP100)

- Gel dryer (Biorad Model 543)

- Fixing solution (40% methanol, 7% acetic acid)

Method

1 In a microtube, mix 25 μl of reticulocyte lysate (thawed at room temperature) with 2 μl of TNT® reaction buffer (thawed on ice), 1 μl of T7 RNA polymerase (from glycerol stock at −20 °C), 1 μl of amino acid mixture minus methionine, 2 μl of [^{35}S]methionine (the volume is increased if activity is decayed), 1 μl of ribonuclease inhibitor, and 2 μg of DNA template. Add nuclease-free water to a final volume of 50 μl. The reaction can be scaled down to 25 μl total volume. Smaller volumes usually do not work.

2 Incubate at 30 °C for 90 min. Lysates can be used immediately or stored at −20 °C for a few days.

3 Check expression of radiolabelled protein by running an SDS–polyacrylamide gel of a 2–5 μl aliquot of the lysate.

4 Soak the gel for 30 min in fixing solution and for 30 min in Amplify solution. Dry gel onto blotting paper for 90 min at 80 °C and expose at −70 °C overnight.

3.3 GST pull-down

The following protocol is a prototype for any protocol where a protein–protein complex is separated from a solution by taking advantage of the fact that one of the two proteins ('bait') is insoluble at high centrifugal forces (e.g. F-actin, which pellets at 100 000 g), or can be adsorbed onto a resin which is insoluble at low centrifugal forces. Affinity resins binding to a variety of sequence tags fused to recombinant proteins are commercially available (e.g. Pharmacia, Sigma, Qiagen). Tagged or untagged proteins can be produced in bacteria (*Protocol 9*), insect cells

169

(*Protocol 10*), reticulocyte lysates (*Protocol 12*), or native/transfected cultured cells (*Protocol 11*). If both target and bait proteins are tagged, it is possible to purify them before carrying out the assay. If not, the tagged protein is used as bait, and the crude lysate is used as a source of target protein. Protein tags are also useful to detect target proteins when specific antibodies are not available (*Figure 1*). Alternatively, target proteins can be detected by autoradiography if they are ^{35}S-labelled after transcription/translation in reticulocyte lysates.

Protocol 13

GST pull-down assay

Equipment and reagents

- Tube rotator (e.g. Thermolyne 400–110)
- Glutathione–Sepharose resin (Pharmacia cat. no. 17-0756-01)
- Bacterial (or insect cell) lysate containing recombinant GST fusion protein
- Cell lysate (cultured cell, insect cell, reticulocyte) containing target proteins, diluted if necessary in LBT (see *Protocol 10*)
- Wash buffer (see *Protocol 10*)

Method

1 Pipette the required amount of glutathione–Sepharose resin slurry for all pull-downs (20–30 μl per sample). Wash once with PBS in a plastic centrifuge tube (1 min at 3000 r.p.m. in a microfuge). Resuspend in PBS and aliquot in 1.5-ml microtubes. Carry out all subsequent washes/centrifugations with a microfuge (15 s, 5000 r.p.m.) and remove liquid by aspiration with a pipette tip connected to a vacuum line, being careful not to aspirate the resin.

2 Centrifuge all aliquots and remove PBS by aspiration.

3 Add an appropriate volume of bacterial lysate containing recombinant GST fusion protein (titred as described in *Protocol 9*) to the washed resin. Bring the total volume to 1.5 ml with PBS. To minimize loss of resin, it is convenient to maintain the volume of the assay within 1.5 ml.

4 Incubate for 1 h at room temperature, rotating end-over-end.

5 Wash three times with 1 ml of PBS.

6 Add cell lysate containing the target protein (0.5–1.5 ml lysate, if necessary bring the volume to 1.5 ml with LBT)

7 Incubate for 1 h at 4 °C, rotating end-over-end.

8 Wash three times with 1 ml of wash buffer.

9 Resuspend the resin in 20 μl of 2× SDS sample buffer. Vortex, boil for 3 min, centrifuge briefly, and load the supernatant onto an SDS gel.

10 Detect protein(s) bound to resin directly by Coomassie staining of the gel, or indirectly by autoradiography or immunoblotting.

3.4 Determination of the dissociation constant for protein–protein interaction (K_d)

The determination of the dissociation constant (K_d) is useful to evaluate the affinity and the potential physiological significance of protein–protein inter-actions. A low K_d value indicates high affinity and stability of interaction, meaning that the half-life of the bonds between the proteins is relatively long.

The protocol below consists essentially of a parallel series of GST pull-downs which provide a quantitative analysis of protein–protein interaction. The con-centration of one of the two proteins (bait, which can be removed from solution by centrifugation) is maintained fixed, and the target protein concentration is varied. For an accurate measurement of K_d, the protein concentration (especially target protein) must be calculated as precisely as possible. In the case of purified proteins, knowing the extinction coefficient at 280 nm is useful. When proteins are 'purified' by immunoprecipitation or pull-down from cell lysates, and they are then detectable by Coomassie staining of SDS–polyacrylamide gels, protein concentration is determined by densitometry and comparison with serial dilutions of a standard protein (BSA or immunoglobulin G). If the protein concentration is too low to be detectable by Coomassie staining, the immunoblot staining intensity of the protein in the 'unknown' samples is compared with the immunoblot stain-ing intensity of dilutions of samples with 'known' concentrations (e.g. previously established on the basis of Coomassie staining or other approaches). Whatever method is used, it is important that the curve used as a standard reference for protein concentration or immunoblot reactivity is linear. We use BSA at concen-trations ranging between 100 and 1000 ng to make the standard curve for protein concentration, and use values within the linear part of the curve.

Because of the quantitative nature of the assay, it is recommended to use purified protein solutions of target protein at known concentration whenever possible. In fact, if the target protein solution is a crude cell lysate (e.g. from epithelial cells, if the target protein is a tight junction protein), it is likely that other proteins present in the lysate interact with the target protein. Thus, the effective 'available' concentration of the target protein (for binding to the bait) is reduced, and the 'total' concentration becomes unknown. In addition, binding of the target protein to other proteins may modify its affinity for the bait. If the lysate is from a cell where the structure containing the target protein does not exist (e.g. reticulocyte lysates or insect cell lysates, which do not contain tight junctions nor specific tight junction proteins), this problem is, in principle, reduced or absent. Appropriate controls, such as SDS–PAGE analysis of protein complexes, can help to confirm that a direct interaction is being studied. Reticu-locyte lysates cannot be used for K_d determination since sufficient amounts of target protein to obtain saturation would be too expensive to produce. On the other hand, insect cell lysates containing recombinant protein have been used succesfully for measurement of K_d (25, 27), as outlined in *Protocol 14* below.

The amount of bait protein to use in the assay should be determined em-pirically in combination with the target protein concentration. If too little bait is used, the bait is saturated easily, but the target protein concentration sufficient to saturate may not be detectable or measured accurately. If too much bait is used, the volume of target protein solution required to obtain saturation may be too high, or the amount of protein bound at saturation may be too high to be measured accurately. In either case, the target protein solution must contain a sufficiently high concentration of target protein to ensure that saturation can be obtained without having to use too large a volume of the solution. A preliminary experiment to determine useful ranges of concentration can be done with 1, 2.5, 5, 10, and 25 µg of bait protein, mixed with a fixed amount of target protein solution (e.g. 1 ml of insect cell lysate).

Protocol 14

Determination of the K_d of protein–protein interaction using GST pull-down assay

Equipment and reagents

- Digital scanner
- Software for digital densitometric analysis (e.g. Biosoft Quantiscan 2.1) and linear regression analysis
- Reagents for GST pull-down assays (*Protocol 13*)

- 'Bait' protein (tagged with GST) solution and 'target' protein (lysate of insect cells infected with baculovirus) solution, at known concentrations
- BSA, 1 mg ml^{-1}

Method

1 Prepare six microtubes, each with 20 µl of washed glutathione–Sepharose resin (*Protocol 13*).

2 To each aliquot add an equal volume of bacterial lysate, corresponding to an amount of GST fusion protein previously established (1–25 µg range).

3 Attach the GST fusion protein to the resin and wash as described in steps 3–5 of *Protocol 13*.

4 Add different amounts of target protein solution to the tubes. For example, 50 µl of insect cell lysate in tube 1, 100 µl in tube 2, 250 µl in tube 3, 500 µl in tube 4, 1 ml in tube 5, 1.25 ml in tube 6. Bring the volume to 1.5 ml with LBT.

5 Incubate and wash as described in steps 6–8 of *Protocol 13*.

6 Resuspend each resin pellet in 20 µl of 2× SDS sample buffer. Vortex, boil for 3 min, centrifuge briefly, and load the supernatant onto an SDS gel. Include in the same gel 3–5 lanes with different loadings of diluted BSA (in the 100–1000 ng range, e.g. 100, 250, and 500 ng).

7 Stain the gel with Coomassie blue and destain thoroughly.

8 Digitize the image of the gel using the scanner and quantify the intensity of all bands (GST fusion protein, target protein, BSA) in each lane using Biosoft Quantiscan 2.1.

Protocol 14 continued

9 Calculate the amounts of GST fusion protein and bound target protein (expressed as nanograms or nanomoles) in each of the six assays by reference to the BSA standard curve obtained from the gel data.

10 Normalize the data to account for the loss of resin during the procedure (e.g equalize the GST fusion protein amount in all lanes).

11 Calculate the free target protein amount by subtracting the bound protein (determined as above) from the total protein added in each assay.

12 Calculate bound and free target protein concentrations (ng ml^{-1}) by dividing calculated (nanogram) amounts by 1.5 (volume of the assay in ml).

13 Plot bound target protein concentration versus free target protein concentration to obtain the saturation curve.

14 Plot the ratio of bound/free versus bound concentration of target protein and obtain a linear regression curve. This gives the Scatchard plot, where the negative reciprocal of the slope of the curve is the K_d (–1/slope). Another way to calculate the K_d is to determine the maximal binding concentration of the target protein (B_{max}). This value is given by the x intercept of the the Scatchard curve. Then evaluate, from the saturation curve, the concentration of free target protein corresponding to the half-maximal binding ($B_{max}/2$). This concentration is the K_d.

3.5 Immunoprecipitation

If two (or more) proteins form a complex which can be immunoprecipitated by antibodies against one or the other (or both) proteins, this is usually a good indication that the proteins form a complex *in vivo*. Cell lysates for immunoprecipitation can be prepared from *Xenopus* oocytes, eggs, and embryos, or from cultured cells. The composition of the lysis buffer can dramatically affect the results, since many factors, including the presence, type, and concentrations of detergents and salts, and pH affect the protein solubility and protein–protein interactions. Thus, the more stringent the assay conditions, the higher is the likelihood that high-affinity interactions are detected. Negative results are usually not very informative, since they may be due to inappropriate assay conditions, or the fact that the antibodies used for immunoprecipitation interfere with protein–protein interactions. As for GST pull-down experiments, isolation of protein complexes is achieved by adsorbing the antibody-containing complex on a Sepharose resin. For immunoprecipitations, resins conjugated to protein A, protein G, or any other molecule which interacts with high affinity with the Fc region of immunoglobulins, are used. The antibody can be adsorbed to the resin either before or after formation of the immune complex with antigen. Antibodies with high titre, specificity, and affinity/avidity are obviously preferable. Anti-tag antibodies are useful to immunoprecipitate tagged protein constructs from cell lysates. Pre-clearing of samples with pre-immune serum and *Staphylococcus aureus* (Pansorbin cells) is recommended to improve signal specificity. The following protocol was used to immunoprecipitate cingulin, ZO-1, and ZO-2 from lysates of cultured epithelial cells (25).

Protocol 15

Immunoprecipitation from Triton-soluble and SDS-soluble fractions of cultured epithelial cells

Equipment and reagents

- Tube rotator
- Pansorbin cells (Calbiochem, cat. no. 507858)
- Protein A–Sepharose CL-4B resin (Pharmacia cat. no. 17-0780-01)
- CSK (cytoskeleton) extraction buffer (50 mM NaCl, 300 mM sucrose, 10 mM piperazine-*N,N'*-bis-2-ethane sulphonic acid (PIPES), pH 6.8, 3 mM $MgCl_2$, 0.5% Triton X-100, 1 mM PMSF, 0.1 mg ml^{-1} DNase, 0.1 mg ml^{-1} RNase)
- SDS immunoprecipitation buffer (SDS-IP buffer: 1% SDS, 10 mM Tris–HCl, pH 7.5, 2 mM EDTA, 0.5 mM dithiothreitol (DTT), 0.5 mM PMSF)
- Wash buffer 1 (WB1: 150 mM NaCl, 20 mM Tris–HCl, pH 7.5, 5 mM EDTA, 0.5% Triton X-100, 0.1% SDS)
- Wash buffer 2 (WB2: 500 mM NaCl, 20 mM Tris–HCl, pH 7.5, 0.5% Triton X-100)

Method

1 Grow cells until 90% confluent.

2 Aspirate the medium and rinse cells twice with cold PBS.

3 Lyse cells in CSK buffer (7 ml per 15-cm diameter Petri dish) for 15 min at 4°C on a rocker platform.

4 Collect all lysate and debris, centrifuge for 10 min at 10 000 g at 4°C, and decant the supernatant (Triton-soluble pool) into a new tube.

5 Resuspend the pellet in 1/10 the original volume of SDS-IP buffer. Incubate for 3–5 min at 100°C. Add 9/10 the original volume of CSK buffer. This is the SDS-soluble pool. Use lysates immediately or store at −70°C in aliquots.

6 Add 100 µl of Pansorbin cell suspension and 10 µl of pre-immune serum (if available) to 1 ml of lysate. Rotate at 4°C for 1 h.

7 Centrifuge for 10 min, 10 000 g at 4°C. Transfer the supernatant into a new tube.

8 Add 5 µl of immune serum (or 5 µg of purified antibody solution) to the lysate. Rotate at 4°C for 1 h.

9 Add 50 µl of a 50% suspension of protein A–Sepharose. Rotate at 4°C overnight.

10 Wash twice with LBT (*Protocol 10*), twice with WB1, twice with WB2 at 4°C. Be careful not to aspirate beads when removing the supernatant.

11 Add 30 µl of 2× SDS sample buffer to the resin. Vortex, and incubate in boiling water for 3 min.

12 Analyse the proteins bound to the resin by SDS–PAGE and immunoblotting.

Websites for specialized suppliers

www.apbiotech.com
www.biochem.roche.com
www2.lifetech.com
www.pharmingen.com
www.narishige.co.jp
www.biosoft.com
www.nascofa.com
www.invitrogen.com
www.bio-rad.com
www.xenopus.com

Acknowledgements

We would like to thank David Shore, Stuart Edelstein, Giambruno Martinucci, and Paolo Burighel for support. Work cited from the authors' laboratories was funded by grants from MURST, CNR, EC Biomed Programme, Swiss National Fonds, and the State of Geneva.

References

1. Gumbiner, B. (1987). *Am. J. Physiol.*, **253**, C749.
2. Stevenson, B. R. and Keon, B. H. (1998). *Annu. Rev. Cell Dev. Biol.*, **14**, 89.
3. Mitic, L. L. and Anderson, J. M. (1998). *Annu. Rev. Physiol.*, **60**, 121.
4. Madara, J. L. and Hecht, G. (1989). In *Functional epithelial cells in culture* (ed. K. S. Matlin and J. D. Valentich). p. 131. Alan R. Liss, New York.
5. Citi, S. and Cordenonsi, M. (1998). *Biochim. Biophys. Acta*, **1448**, 1.
6. Muller, H. A. and Hausen, P. (1995). *Dev. Dyn.*, **202**, 405.
7. Cardellini, P. and Rasotto, M. B. (1988). *Acta Embryol. Morphol. Exp.*, **9**, 97.
8. Regen, C. M. and Steinhardt, R. A. (1986). *Dev. Biol.*, **113**, 147.
9. Kalt, M. R. (1971). *J. Embryol. Exp. Morphol.*, **26**, 51.
10. Roberts, S. J., Leaf, D. S., Moore, H.-P., and Gerhart, J. C. (1992). *J. Cell Biol.*, **118**, 1359.
11. Cardellini, P., Davanzo, G., and Citi, S. (1996). *Dev. Dyn.*, **207**, 104.
12. Cordenonsi, M., Mazzon, E., De Rigo, L., Baraldo, S., Meggio, F., and Citi, S. (1997). *J. Cell Sci.*, **110**, 3131.
13. Fesenko, I., Kurth, T., Sheth, B., Fleming, T. P., Citi, S., and Hausen, P. (2000). *Mech. Dev.* **96**, 51.
14. Chen, Y., Merzdorf, C., Paul, D. L., and Goodenough, D. A. (1997). *J. Cell Biol.*, **138**, 891.
15. Merzdorf, C. S. and Goodenough, D. A. (1997). *J. Cell Sci.*, **110**, 1005.
16. Merzdorf, C. S., Chen, Y. H., and Goodenough, D. A. (1998). *Dev. Biol.*, **195**, 187.
17. Kay, B. K. and Pweng, H. B. (ed.).(1991). *Methods in cell biology*. Vol. 36, *Xenopus laevis: practical uses in cell and molecular biology*, p. 709. Academic Press, San Diego, CA.
18. Nieuwkoop, P. D. and Faber, J. (1956). *Normal table of Xenopus laevis*. North Holland Publishing Co., Amsterdam.
19. Evans, J. P. and Kay, B. K. (1991). In *Methods in cell biology* (ed. B. K. Kay and H. B. Peng). Vol. 36, p. 133. Academic Press, San Diego, CA.
20. McCrea, P. D., Brieher, W. M., and Gumbiner, B. M. (1993). *J. Cell Biol.*, **123**, 477.

21. Heasman, J., Torpey, N., and Wylie, C. (1992). *Development* Suppl., 119.
22. Heasman, J., Ginsberg, D., Geiger, B., Goldstone, K., Pratt, T., Yoshida-Noro, C., *et al.* (1994). *Development*, **120**, 49.
23. Dahl, G. (1992). In *Cell–cell interactions: a practical approach* (ed. B. R. Stevenson, W. J. Gallin, and D. L. Paul). p. 143. Oxford Univesity Press, Oxford.
24. Paul, D. L., Ebihara, L., Takemoto, L. J., Swenson, K. I., and Goodenough, D. A. (1991). *J. Cell Biol.*, **115**, 1077.
25. Cordenonsi, M., D'Atri, F., Hammar, E., Parry, D. A., Kendrick-Jones, J., Shore, D., *et al.* (1999). *J. Cell Biol.*, **147**, 1569.
26. Cordenonsi, M., Turco, F., D'Atri, F., Hammar, E., Martinucci, G., Meggio, F., *et al.* (1999). *Eur. J. Biochem.*, **264**, 374.
27. Itoh, M., Nagafuchi, A., Moroi, S., and Tsukita, S. (1997). *J. Cell Biol.*, **138**, 181.

Chapter 8
Oocyte–granulosa cell interactions

Barbara C. Vanderhyden

Ottawa Regional Cancer Centre, Centre for Cancer Therapeutics, 503 Smyth Road, 3rd Floor, Ottawa, Ontario K1H 1C4, Canada

1 Introduction

Oocyte development occurs largely within the context of a complex and continuously changing environment, involving proliferation, differentiation, and apoptosis of the surrounding somatic or granulosa cells. Decades of investigation have resulted in protocols that enable the replication of many aspects of this process *in vitro*, some of which are able to maintain or enable to some degree the developmental capabilities of the oocyte. While the acquisition of developmental competence relies most heavily on adequate communication between the oocyte and granulosa cells, the process is complicated further by the fact that these cell–cell interactions are strongly influenced by hormones and growth factors generated by the hypothalamic–pituitary–ovarian axis. This chapter describes some of the methods currently used to investigate oocyte–granulosa cell interactions in pre-antral and antral follicles as well as the procedures that enable the study of the influence of each cell type on the other.

2 Mammalian oocyte growth and development *in vivo*

The germ cells of the ovary originate, during gestation, from a small number of primordial germ cells that migrate to the genital ridges, rapidly proliferating during transit. Their migration ends in the gonadal primordium where the germ cells intermingle with the somatic cells. The germ cells enter meiosis at this developmental stage; however, the germ cells, now called oocytes, become arrested in the first meiotic prophase, and remain in that state until fully grown and stimulated to ovulate. At the time of birth in most mammals, the ovary is populated predominantly by primordial follicles, each composed of a meiotically arrested primary oocyte surrounded by a single layer of flattened pre-granulosa cells (*Figure 1*). In mammals, follicles develop continuously from the pool of primordial follicles throughout the reproductive lifespan of the animal.

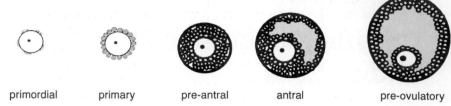

primordial   primary   pre-antral   antral   pre-ovulatory

Figure 1 Diagram illustrating the stages of follicular development.

Follicle development is characterized by an increase in the diameter of the oocytes and a synchronous proliferation of the granulosa cells, resulting in multiple layers of cells that surround each oocyte (*Figure 1*). The role of granulosa cells in supporting the growth and development of mammalian oocytes has been investigated extensively, often by removing the granulosa cells and observing the subsequent effects on specific oocyte activities. As a result, two modes of intercellular communication between oocytes and their surrounding granulosa cells have been described: gap junctions and paracrine factors. Oocyte culture systems must promote appropriate oocyte–granulosa cell interactions via both modes of communication to ensure the production of oocytes with full developmental capabilities.

Granulosa cells have clearly established roles in supporting the growth of the oocyte and maintenance of meiotic arrest (1). The last decade has yielded an increasing amount of evidence that the communication is bidirectional, i.e. that the oocyte can influence several aspects of granulosa cell development, including proliferation, differentiation, and extracellular matrix and steroid hormone production (2). While evidence for this cellular interdependence continues to accumulate, much less is known about the molecules that mediate these interactions. Some responses rely on cell–cell contact, while other responses can be elicited by co-culturing the two cell types or by culturing one cell type in the presence of conditioned medium from the other, indicating an important role for paracrine factors. Identification of these factors will be an exciting element of the study of oocyte–granulosa cell interactions in the next decade.

Although there has been long-standing interest in the mechanisms controlling oocyte maturation and those factors contributing to oocyte developmental potential, there has been a recent notable increase in the study of oocyte–granulosa cell interactions during pre-antral follicle development. With the development of culture systems that support pre-antral follicle development comes the potential for experimental manipulation of this system to identify the features that are critical for early follicular development. The larger numbers of pre-antral follicles in the ovaries of fetal or prepubertal animals have considerable potential value, with applications for transgenesis, domestic animal production, and conservation of endangered species. The ability to increase the number of oocytes that could enter the reproductive process has not gone unnoticed as a potential clinical application for infertile couples.

Infertile animals have always offered opportunities to investigate the regulatory factors that govern follicle development. Mutations causing reduced production of the oocyte-expressed Kit receptor or of its ligand by granulosa cells result in the development of ovaries in which follicles fail to grow beyond one cell layer (3, 4), strongly implicating this ligand–receptor system in the control of early follicle development. Similarly, the recent application of transgenic technology to generate mice deficient in specific genes has yielded animal models whose disrupted ovarian function can provide valuable clues to our understanding of folliculogenesis. Follicle development is arrested at the primary stage in mice carrying a null mutation at the *Gdf9* locus (5), a gene which is expressed predominantly in oocytes. Similarly, mice deficient in connexin (Cx) 37, which forms the gap junctions between oocytes and granulosa cells, show limited oocyte growth and premature luteinization of granulosa cells (6). Generation and analysis of genetically deficient mice with abnormal folliculogenesis provide novel opportunities to identify the contribution of specific genes to follicle development.

From these examples, it is clear that the continued investigation of oocyte–granulosa cell interactions promises to be an exciting field on several fronts. The availability of culture systems that can be used to determine the cellular and molecular events that control these interactions, as described in this chapter, will undoubtedly provide a valuable resource for future studies.

3 Isolation and culture of pre-antral follicles and oocyte–granulosa cell complexes

3.1 Overview

Follicle growth is characterized initially by the transition of granulosa cells from flattened to cuboidal cells (*Figure 1*). Continued growth features both an increase in oocyte diameter and proliferation of granulosa cells. This phase of pre-antral follicle growth is relatively slow, comprising about 85% of the total duration of follicle growth in some species. During pre-antral follicle growth, theca cells become associated with the growing follicle, although they remain separated by a basement membrane. As the follicle grows, a fluid-filled antrum is formed and the granulosa cells acquire differentiated characteristics, many of which are dependent upon the presence of gonadotrophins. The granulosa cells differentiate into two subpopulations: cumulus granulosa cells, which are those most closely associated with the oocyte and are ovulated with it; and mural granulosa cells, which form a multilayered wall against the basement membrane of the follicle. During oocyte growth, usually at the early stages of antrum formation, oocytes acquire meiotic competence (i.e. the ability to resume meiosis). The final stages of oocyte development are characterized by the gradual acquisition of full developmental competence, including the ability to undergo embryonic development. In all species examined to date, the duration of follicular growth *in vivo* is greater than the duration of an oestrous or a menstrual cycle.

For the past two decades, various research groups have reported the isolation and culture of pre-antral follicles. Mouse follicles have been used extensively as they are relatively easy to isolate and grow to full size within a short time. These studies have contributed significantly to our understanding of the growth requirements that contribute to the progression of pre-antral follicles to antral follicles. Although attempts to apply that knowledge to the growth of pre-antral follicles from non-rodent species is still in its infancy, successful attempts to isolate and culture bovine (7), porcine (8), and human pre-antral follicles (9) have been reported. The size and growth rates of murine, bovine, and porcine oocytes and follicles from primordial to pre-ovulatory stages have been summarized previously (10).

3.2 Isolation and culture of rodent pre-antral follicles

Oocytes isolated from pre-antral follicles and cultured in the absence of granulosa cells, or even in co-culture with granulosa cells, fail to show significant growth. All successful reports of oocyte growth have used methods that maintain or re-establish the essential gap junctional communication with granulosa cells (11, 12). Several methods have been described for the isolation of rodent pre-antral follicles, and differ primarily in whether they use enzymatic or mechanical means, or a combination of both, to isolate the follicles. The developmental capabilities of follicles isolated by the various methods are probably linked, at least in part, to the degree of theca cell investment that remains with the follicles. Pre-antral follicles isolated using digestion with collagenase lose most of their theca cell investiture (and are therefore more appropriately called oocyte–granulosa cell complexes), and yet are capable of supporting significant oocyte growth and some developmental competence in culture (13). To retain the theca cell component, follicles must be isolated by microdissection, which has as advantages the ability of those follicles to form antra (14) as well as to mimic ovulation *in vitro* (15).

Protocol 1

Isolation of rodent oocyte–granulosa cell complexes using enzymatic and mechanical means

Equipment and reagents

- Isolation medium: Waymouth MB 752/1 (Sigma) containing 3 mg ml^{-1} bovine serum albumin (BSA; ICN), 5 μg ml^{-1} insulin, 5 μg ml^{-1} iron-saturated transferrin, 5 ng ml^{-1} selenium, 50 μM 3-isobutyl methylxanthine (prepared first as a 100 mM stock solution in ethanol)

- Collagenase (Worthington; 5 mg ml^{-1} in isolation medium; sterilized by passage through a 0.22-μm syringe filter)

- Pasteur pipettes

Method

1 Isolate ovaries from 10- to 12-day-old female mice or rats into the collagenase medium in a 35-mm Petri dish. Leave to sit for 5–10 min at 37°C to allow the enzyme to begin digestion.

Protocol 1 continued

2   Using a sterile Pasteur pipette, gently pipette the ovaries 8–10 times. Repeat pipetting every 5 min until the follicles are completely dispersed (~40 min).

3   Wash the isolated oocyte–granulosa cell complexes by transferring through 4×2.5 ml of enzyme-free isolation medium. Complexes will each consist of an early growing oocyte surrounded by 2–3 tightly packed layers of granulosa cells.

4   Transfer complexes into a culture system, as shown in *Figure 2*.

   Successful cultures of pre-antral follicles or oocyte–granulosa cell complexes have been reported following isolation from young cycling golden hamsters (22), but ovaries from 8- to 12-day-old pups are most commonly used for mice (13) and rats (21). The choice of culture system is largely dependent upon the experimental objectives, as each offers unique advantages (*Figure 2*). Culture of follicles

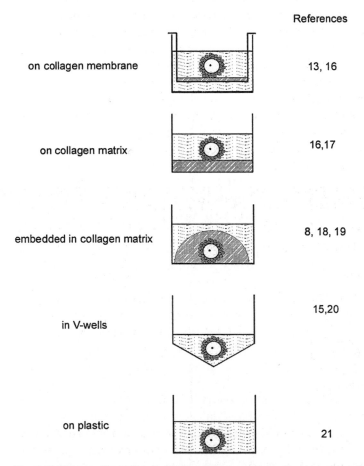

Figure 2 Diagram illustrating the types of culture systems that have been described for the culture of pre-antral follicles and oocyte–granulosa cell complexes.

is ideal for investigations concerning mechanisms of follicle development, ovulation, or oocyte production, whereas oocyte–granulosa cell interactions and oocyte development are best studied using cultures of oocyte–granulosa cell complexes.

3.3 Growth of pre-antral follicles from commercially important species

At this point in time, culture systems for growing bovine, porcine, and sheep oocytes are still under development, although there has been significant progress in the ability to isolate pre-antral follicles in these species. Isolation of pre-antral follicles from domestic species is more difficult than from rodents due to the fibrous nature of their ovaries, which renders the use of collagenase less effective. As a consequence, the majority of techniques that have been reported for the successful isolation of pre-antral follicles from both cows and sows are mechanical in nature, including grating, chopping, or mincing of the ovary followed by microdissection or filtration to obtain individual follicles. Combination treatments of mincing and trypsin digestion (23), or collagenase and microdissection (8), have yielded reasonable numbers of small and large pre-antral follicles.

In contrast to rodent follicle isolation, where the population of isolated follicles after the washing procedure is predominantly early pre-antral follicles, follicles isolated from bovine or porcine ovaries can show a wide range of sizes. The size and yield of follicles using the various techniques have been summarized previously (10), and range from 40 to 220 μm in size with a yield of 37–357 from each bovine ovary, and sizes <60 to 300 μm with a yield of more than 200 000 from each porcine ovary. Pre-antral follicles 130–250 μm in size have been isolated from sheep ovaries (24). Although it is possible to select individual follicles by size using calibrated pipettes, this method is clearly laborious and will yield fewer follicles. Graded sieves have been used with some success to separate follicles by size with some precision (10).

Initial experiments examining the ideal culture conditions for growth of bovine and porcine oocytes have examined the effects of culture substrate (notably collagen) and additives (8, 25) on the ability of the follicles to maintain their three-dimensional morphology during culture. Follicle growth, by nature of the species, is necessarily slow, but culture for up to 16 days either on (7) or within a collagen matrix (8) has supported granulosa cell proliferation, antrum formation (8), and growth of both bovine (7) and porcine oocytes (8). Mechanically isolated sheep pre-antral follicles have been grown successfully to antral stages in V-well microtitre plates (24; *Figure 2*).

An alternative to the culture of isolated pre-antral follicles is the culture of whole ovaries, pieces of ovary, or cortical slices. In addition to providing some structural support for the growth of larger follicles, this type of culture system has clear potential as a means to investigate the mechanisms that control the initiation of primordial follicle development. Once initiated, complete follicle development cannot be sustained in these organ cultures, and isolation and culture of individual pre-antral follicles is necessary to extend their developmental potential. A landmark report by Eppig and O'Brien (26) describes the complete development of mouse oocytes from primordial follicles using a series of sequential culture systems for growth, maturation, and fertilization. Initiation of primordial follicle development has been observed in cultures of bovine ovarian cortical slices (27), although the developmental competence of follicles grown under these conditions has yet to be determined.

3.4 Growth of primate pre-antral follicles

While culture of isolated pre-antral follicles from rodent and domestic species has resulted in continued oocyte development with variable levels of success, there have been few attempts to isolate follicles from primate ovaries. Roy and Treacy (28) reported the successful enzymatic isolation and long-term culture of human pre-antral follicles with gonadotrophin-inducible DNA synthesis and antrum formation after 120 h when the follicles were cultured in an agar sandwich. Despite this promising report, most investigators have preferred to culture pieces or slices of ovarian cortex, often from fetal ovaries, rather than attempt to culture individual follicles (29, 30). Primordial follicles in fetal baboon ovaries can survive and develop to the secondary stage *in vitro* when cultured within 1 mm thick ovarian slices in serum-free conditions (29).

3.5 Parameters indicative of successful follicle development

The choice of species, age, length of culture, and type of culture system, including culture medium, volume, substrate, and additives, all impact on the behaviour of pre-antral follicles grown in culture. Manipulations of the various components enable the evaluation of a wide variety of morphological and physiological observations regarding the role of oocytes and granulosa cells during this process. Parameters that can be measured to determine the success of the follicle culture include: follicle and oocyte survival; granulosa cell function, including antrum formation, cell number, and steroidogenesis; and oocyte function, including growth (size), nuclear and cytoplasmic maturation, fertilizability, cleavage, and developmental competence that is sufficient to yield viable offspring.

4 *In vitro* maturation techniques

4.1 Meiotic (nuclear) and cytoplasmic maturation of oocytes

In most mammals, meiosis is initiated in female germ cells during fetal life; however, the process is arrested at prophase I at or around the time of birth. Even after oocyte growth has been initiated, oocytes continue to be maintained in meiotic arrest, despite a significant increase in cell size. This arrest results from either an absence of essential cell cycle regulatory proteins, or the presence of meiosis-arresting substances, or both. As oocytes enter the final stages of growth, the mechanisms controlling meiosis undergo a dramatic change, such that the oocytes acquire the ability to resume meiosis spontaneously when liberated from the follicles and placed in culture. Oocytes that can achieve spontaneous oocyte maturation (i.e. have meiotic competence) clearly have acquired the molecules required for the resumption and completion of meiosis, yet they do not resume meiosis if they are retained within the follicle. The surge of luteinizing hormone (LH) triggers the resumption of meiosis, which then proceeds through completion of the first meiotic division, including first polar body formation, with subsequent arrest at metaphase II. The final phases of meiosis are completed only after sperm penetration.

Nuclear or meiotic maturation of oocytes can be achieved in chemically defined media, which helped to identify the minimal nutritional requirements for the resumption of meiosis (31). However, those conditions proved to be insufficient to promote full fertilization and developmental capabilities. Initial failures in efforts to fertilize oocytes matured *in vitro* were attributed to incomplete or aberrant cytoplasmic maturation, despite the apparent completion of nuclear maturation (31, 32). Although the most striking abnormality in these oocytes was the failure to form the male pronucleus, full cytoplasmic maturation encompasses a number of relatively ill-defined processes that prepare the oocyte for fertilization, activation, and early embryo development. Extensive investigation has resulted in the establishment of culture systems that have overcome many of these problems and that support the complete maturation of oocytes *in vitro*. The availability of these culture systems in which fully grown oocytes can be manipulated and analysed will continue to prove valuable in the elucidation of the mechanisms that control meiotic and cytoplasmic maturation.

4.2 Factors to consider when designing oocyte maturation experiments

4.2.1 Number of oocytes required

The majority of studies of oocyte–granulosa cell interactions have focused on the processes of oocyte maturation and acquisition of fertilizability and developmental competence during the later stages of oocyte development. An important consideration for this type of study is the number of oocytes required to perform each experiment. While microscopic evaluation of morphologically evident outcomes (e.g. germinal vesicle breakdown, GVBD) is quite feasible, the number of oocytes required to perform molecular analysis to determine, for example, changes in patterns of protein synthesis, render these experiments more difficult to perform. To this end, prepubertal animals are commonly treated with hormones to induce the development of multiple follicles. This procedure enables the retrieval of a larger number of oocytes (typically 40–80 per mouse) that are synchronous in their development (fully grown, meiotically competent but arrested at prophase I), as well as being more cost efficient.

4.2.2 Role of cumulus cells

Cumulus cells play an essential role in promoting normal cytoplasmic maturation of oocytes necessary for pronuclear formation and subsequent developmental capability (31, 33). The importance of the presence of cumulus cells during maturation is evident from observations that the rate of male pronucleus formation was higher in oocytes that had been matured cumulus-enclosed versus those that were matured cumulus-free (34). The specific cumulus-derived factors that promote cytoplasmic maturation of oocytes remain undefined; however, mechanisms as simple as nutrient support cannot be eliminated since male pronucleus formation following sperm penetration of porcine oocytes was reported to be much greater when the oocytes were matured in the presence of amino acids (34).

4.2.3 Control of oocyte meiosis *in vitro*

Since the follicular environment provides signals that maintain the oocyte in meiotic arrest, oocytes isolated from antral follicles will undergo meiotic maturation spontaneously. In experimental systems, it is therefore necessary to include meiotic inhibitors in the isolation and culture media if there is a need to maintain the oocytes in meiotic arrest. This has been achieved most frequently by the addition of cyclic adenosine monophosphate (cAMP) analogues such as 250 μM dibutyryl cAMP or 8-bromo-cAMP or 100 μM of the phosphodiesterase inhibitor 3-isobutyl-1-methylxanthine to the media or, in bovine oocytes, the application of general protein synthesis inhibitors (e.g. 2–10 μg ml^{-1} cycloheximide). However, recent studies have examined the effectiveness of novel oocyte maturation inhibitors that have greater specificity for the mechanisms controlling progression of meiosis in the oocyte. Inhibitors of phosphodiesterase 3 (milrinone; 35) and M-phase-promoting factor kinase activity (roscovitine; 36) effectively and specifically block GVBD in rodent and bovine oocytes and should prove to be valuable tools in the manipulation of oocyte maturation in future experiments.

4.3 *In vitro* maturation of rodent oocytes

Mouse and rat oocyte–cumulus cell complexes (OCCs) can be isolated from antral follicles by puncturing the follicles with 25-gauge needles and aspirating the complexes released into the medium. The yield can be greatly increased by the intraperitoneal injection of pregnant mares' serum gonadotrophin (5 IU of PMSG for mice; 5–20 IU for rats) 40–44 h before isolation of the OCCs. The effectiveness of the hormonal stimulation in promoting follicular development is greatest when the animals are still sexually immature (20- to 26-day-old mice or 22- to 28-day-old rats) and is reduced as the animals age. Isolation and maturation of the OCCs is best performed in 2.5 ml of Waymouth MB752/1 with supplements (37, 38) in 35-mm Petri dishes (e.g. Falcon cat. no. 1008), not tissue culture dishes, to minimize cell adhesion to the dish. After follicular puncture, OCCs can be aspirated using a flame-pulled glass pipette and washed by serial transfer through four dishes containing 2.5 ml of medium.

Although a number of types of media have been used for isolation and maturation of oocytes, Waymouth medium has been shown in a previous study comparing maturation media to yield the highest percentage of mouse oocytes capable of development to the blastocyst stage (37). If fertilization of the *in vitro* matured oocyte is part of the proposed experiment, then the medium used for both collection and maturation should be supplemented with 5% fetal bovine serum (FBS) or 1 mg ml^{-1} fetuin (39) to prevent hardening of the zona pellucida. Oocytes can be matured in the presence of follicle-stimulating hormone (FSH), either on a Petri dish in drops of medium under washed oil placed in a 5% CO_2 incubator overnight, or in 1 ml of medium in snap-cap tubes placed in a 37°C waterbath overnight. Some investigators have found the use of a modular incubator chamber (Billups-Rothenburg) flushed with a gas mixture of 5% CO_2, 5% O_2, 90% N_2 (13) to be an advantageous component of *in vitro* maturation regimens.

4.4 *In vitro* maturation of oocytes from domestic species

Bovine oocytes are most often collected from ovaries obtained from abattoirs and transported at 30°C in 0.9% saline containing antibiotics (100 000 IU l^{-1} penicillin, 100 mg l^{-1} streptomycin, and 250 μg l^{-1} amphotericin B). The larger size of the follicles of domestic species allows the OCC to be collected by follicular aspiration. Quality of the OCC is determined by microscopic evaluation of the number and continuity of the layers of cumulus cells and the homogeneity of the oocyte cytoplasm (40). One of the methods routinely used for the collection and *in vitro* maturation of bovine oocytes (41) is described in *Protocol 2*.

Protocol 2

Collection and *in vitro* maturation of bovine oocytes

Equipment and reagents

- CO_2 incubator
- *N*-(2-hydroxyethyl) piperazine-*N'*-(2-ethanesulphonic acid) (HEPES)-buffered Tyrode's medium[a] supplemented with 10% heat-treated FBS, 0.2 mM pyruvic acid (Sigma), and 50 μg ml^{-1} gentamicin (Sigma).

- TCM-199 medium with Earle's salts (Gibco) and bicarbonate, supplemented with 5–10% heat-treated FBS, 200–500 ng ml^{-1} FSH, 5 μg ml^{-1} LH, 1 mg ml^{-1} oestradiol-17β (Sigma), 0.2 mM pyruvic acid, 50 μg ml^{-1} gentamicin.
- Mineral oil

Method

1 Aspirate selected antral follicles using an 18-gauge needle fitted to a 5-ml syringe.

2 Wash OCCs three times in HEPES-buffered Tyrode's medium.

3 For *in vitro* maturation, prepare 50 μl droplets of TCM-199 medium.

4 Overlay the droplets of medium with mineral oil and incubate for a minimum of 2 h before adding the OCCs.

5 Transfer the OCCs in groups of 10 to each droplet and culture for 24 h at 38.5–39°C in an atmosphere of 5% CO_2 in humidified air.

[a] See ref. 42 for details.

For *in vitro* maturation of porcine oocytes, OCCs can be collected from ovaries of prepubertal gilts after slaughter and transported in saline with antibiotics at 30–37°C. At the laboratory, ovaries are washed three times with saline and OCCs can be aspirated from 2–6 mm antral follicles with an 18-gauge needle. OCCs are then washed twice in 10 ml of medium. Only oocytes with many layers of compact cumulus and homogenous cytoplasm should be selected for *in vitro* maturation. Groups of 50 OCCs can be cultured in each well of a 4-well multidish (eg. Nunc) containing 500 μl of maturation medium, which is TCM-199 supplemented with 0.1 mg ml^{-1} sodium pyruvate, 10 IU ml^{-1} human chorionic gonadotrophin (hCG), 10 IU ml^{-1} PMSG, 1 mg ml^{-1} oestradiol, and 10% porcine follicular fluid (43). The medium is covered with mineral oil and oocytes are incubated at 38.5°C with 5% CO_2 in humidified air for 40–45 h. After 42 h of culture, more than 90% of the oocytes should have undergone GVBD and polar body extrusion.

4.5 *In vitro* maturation of primate oocytes

Although the ability of human oocytes to establish pregnancies after *in vitro* oocyte maturation was first reported in 1983 (44), and has been reported sporadically since that time, the culture systems clearly remain less than optimal. *In vitro* maturation conditions for oocytes from non-human primates (e.g. marmoset and rhesus monkey) that support subsequent pre-implantation embryo development have been described (45), and conditions for macaque (46) and baboon (47) oocyte maturation have been reported recently. The culture conditions generally include the presence of gonadotrophins, and most studies have found the retention of cumulus cells to be advantageous. Gilchrist *et al.* (45) reported the successful maturation of marmoset monkey oocytes cultured in 100 μl droplets of Waymouth MB 752/1 (Gibco) supplemented with 6 IU ml^{-1} hFSH, 116 IU ml^{-1} hLH, 1 μg ml^{-1} oestradiol, 20% FBS, 0.5 mM sodium pyruvate, 1 mM glutamine, 10 mM sodium lactate, 4 mM hypotaurine, and antibiotics under washed silicon oil in 35-mm Petri dishes. For most species, GVBD should occur within 24 h of culture; however, longer periods may be necessary depending on the timing of oocyte removal from the follicle.

4.6 Evaluation of *in vitro* maturation

Although the nucleus or germinal vesicle (GV) can sometimes be seen through the cumulus mass, assessment of the nuclear status is confirmed most reliably by first removing the cumulus cells (see Section 5). At the end of culture, denuded oocytes can simply be examined microscopically for the presence or absence of the GV and a polar body. In domestic animals, visualization of the nucleus is facilitated by staining the oocytes with aceto-orcein or by incubating them in Hoechst solution (50 μg ml^{-1}) for 1 h at 39°C, transferring the oocytes onto a slide spotted with a paraffin wax–vaseline mixture, and placing a coverslip directly over the oocytes.

For most species, the fertilization capacity and developmental competence of *in vitro* matured oocytes are influenced by the age and hormonal status of the donor animal, and the presence or absence of hormones or other supplements in the maturation medium. With current methodologies, PMSG-primed mice show the highest developmental capabilities after *in vitro* maturation, with 80–100% of GV-stage oocytes matured in the presence of FBS undergoing subsequent fertilization and cleavage.

5 Evaluating granulosa cell effects on oocyte development

Oocytes cannot grow in isolation from their granulosa cells, although some development is possible when in co-culture with soluble factors from granulosa cells. This dependence of the oocyte upon the granulosa cells includes not only oocyte growth, but also regulation of meiosis. Previous studies have demonstrated this dependence by simply stripping the oocytes of their adherent granulosa cells and evaluating oocyte behaviour in the denuded versus granulosa cell-enclosed oocytes.

For all species, isolation of OCCs prior to stripping of the granulosa cells is achieved using the methods described for *in vitro* maturation (Section 4). The methods that work effectively to remove the granulosa cells vary slightly between species, but all entail the gentle trituration of the OCCs through a small-bore pipette. Oocytes that have become denuded of their granulosa cells should be transferred immediately to another dish, as repeated pipetting of cumulus-free oocytes will cause their death. Cumulus cells can be removed from porcine and bovine oocytes by repeated passage through a small-bore pipette or by simply vortexing a group of complexes in medium (e.g. TCM-199 with HEPES).

If the oocytes to be denuded are in expanded OCCs (e.g. after ovulation or *in vitro* maturation), cumulus cells can be removed with 0.1% hyaluronidase in medium, allowing time for the enzyme to act before pipetting. The removal of cumulus cells from primate OCCs also fares better when a combination of mechanical and enzymatic methods is used. Cumulus cells can be stripped using a fine-bore micropipette in medium with 500 μg ml^{-1} hyaluronidase (Sigma) and 10 μg ml^{-1} soybean trypsin inhibitor (ICN; 45).

Isolation of denuded oocytes from pre-antral follicles is best achieved using methods similar to those described in *Protocol 1*, with the exception that the ovaries are placed in 3 ml of calcium-free phosphate-buffered saline (PBS) with 0.5% collagenase and 0.01% bovine pancreas DNase I (Sigma; 16). Once stripped of their surrounding granulosa cells, denuded oocytes can then be used in various culture systems to evaluate growth or maturation. The same methods described here can be used for the preparation of denuded oocytes for use in conditioning medium.

6 Evaluating oocyte effects on granulosa cell development

Ovarian follicles do not develop without oocytes in them, indicating that oocytes participate in follicular development from the earliest stages. Recent studies investigating the importance of the oocyte in regulating granulosa cell activity have found multiple roles for the oocyte in modulating granulosa proliferation, differentiation, and extracellular matrix and steroid hormone production (48). The specific effects of oocytes on granulosa cell function may change throughout follicle development and, thus, studies on granulosa cell development must consider the stage of follicle development. Antral follicle development involves the differentiation of granulosa cells into two subpopulations: cumulus granulosa cells and mural granulosa cells. The differentiation of these two subpopulations involves the acquisition of unique characteristics, such as the expression of LH receptors by mural granulosa cells, and the ability of cumulus cells to secrete hyaluronic acid and undergo expansion. As the methods used to study cumulus and mural granulosa cells differ, they will be presented in turn.

6.1 Isolation and culture of cumulus granulosa cells

Although cumulus granulosa cells can be cultured as a monolayer, as described for mural granulosa cells below, methods that maintain the normal three-dimensional organization of cumulus cells are also likely to retain more of the normal physiological interactions between the cells. To this end, the procedure of oocytectomy, which microsurgically removes the oocyte from an OCC, has been used very effectively to investigate the oocyte effects on granulosa cells. For example, this procedure has been used to demonstrate that cumulus cells lose the ability to produce hyaluronic acid in response to FSH when separated from oocytes (49, 50). The equipment and procedures used to oocytectomize OCCs are the same for any mammalian species. Those with limited experience with the preparation of micropipettes and the set-up of a micromanipulation station are referred to previous publications (51, 52) where these aspects of micromanipulation are described in greater detail than can be provided here.

Protocol 3

Procedure for oocytectomy

Equipment and reagents

- Inverted microscope, equipped with phase contrast optics (preferably Nomarski differential interference contrast optics)

- Two micromanipulators (e.g. Narishige, Leica)

- Hamilton syringe or syringe micrometer connected by plastic tubing to one of the manipulators, the entire system filled with light mineral oil (Fisher), without bubbles

- Pipette puller (e.g. Sutter Instruments), to pull the pipettes

- Lancing pipette: borosilicate glass pipettes (outer diameter 1.0 mm; Leica or World Precision Instruments), pulled at one end to a fine tip

- Microforge (e.g. Narishige), to heat-polish and bend the tip of the holding pipette

- Holding pipette: borosilicate glass pipettes (as described above), pulled at one end. Break the tip where the outer diameter is 80–100 μm to create a blunt end and, using a microforge, bend and heat-polish the tip.

Protocol 3 continued

- Dissecting microscope with transmitted light illumination
- Plexiglass slide, prepared as described in *Protocol 4*
- Light mineral oil, equilibrated with medium by stirring for 4 days and then filter-sterilized

Method (see *Figure 3*)

1. Backfill the holding pipette with medium and attach to the oil-filled tubing connected to a Hamilton syringe. Attach the lancing pipette to the tubing connected to the other syringe.

2. Place a small drop (~25 μl) of medium in the centre of the coverslip bottom of the plexiglass slide and fill the well with enough washed mineral oil to just cover the drop.

3. Transfer 50 OCCs into a small area within the drop of medium and gently place the slide on the stage of the microscope.

4. Lower the lancing and holding pipette slowly into the drop of medium, slightly away from the complexes, such that the lancing pipette can be inserted into the opening of the holding pipette at the same plane of focus as the complexes.

5. Shift the stage so that the complexes are at the bottom of the field of view. Using the Hamilton syringe, develop a small amount of negative pressure in the holding pipette and move the pipette to grasp one of the complexes and return it to the centre of the field of view.

6. With the holding pipette grasping the complex, push the lancing pipette through the full diameter of the complex such that the tip of the lancing pipette enters the holding pipette. Increase the negative pressure within the holding pipette very slightly and withdraw the lancing pipette. The complex should collapse as the oocyte is aspirated into the holding pipette.

7. Release the oocytectomized complex at the top of the field of view, and repeat the process for the remaining OCCs.

8. Wash all oocytectomized complexes at least twice before proceeding with the experiment.

Protocol 4

Preparation of plexiglass slides for micromanipulation

Equipment and reagents

- Plexiglass (plastic) slides
- Beeswax (Fisher)
- Heating plate

Method

1. Cut plexiglass to the size of a microscope slide (~75 × 25 mm).

Protocol 4 continued

2   Drill a hole with dimensions of 25 × 15 mm in the centre of the slide.

3   Slowly melt the beeswax in a beaker on a warm heating plate.

4   Apply melted beeswax completely around the hole on one side of the slide and quickly affix a 24 × 50 mm coverslip so that it completely covers the hole and is sealed to the slide when the beeswax has hardened.

5   The glass surface can be cleaned and sterilized with 70% ethanol immediately before use.

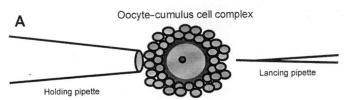

The complex is held in position with the holding pipette.

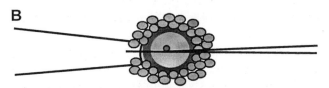

The lancing pipette is pushed through the complex and into the holding pipette.

Withdrawal of the lancing pipette allows the negative pressure in the holding pipette to aspirate the oocyte, and the zona pellucida collapses.

Figure 3 Microsurgical procedure to oocytectomize oocyte–cumulus cell complexes.

6.2 Isolation and culture of mural granulosa cells

The vast majority of studies with objectives to examine the factors that regulate granulosa cell proliferation and differentiation have been performed using monolayers of mural granulosa cells. The effect of oocytes on these cells can be investigated simply by placing oocytes on the granulosa cell monolayer. Placement of oocytes on a monolayer of granulosa cells has shown that gap junctional communication can be restored (11), and that the presence of oocytes can alter steroid hormone production (53).

Rat granulosa cells are obtained most frequently from animals that have been treated for 3 days with diethylstilbestrol (DES) or oestradiol, which augments the number of growing pre-antral follicles. DES administration begins when rats are 18–21 days old and can be given by daily subcutaneous injections of 1 mg of DES/ 0.1 ml sesame oil or by subcutaneous silastic implants containing 5 mg of DES. Ovaries are then isolated into Dulbecco's modified Eagle's medium (DMEM; with glucose at 4.5 g l^{-1}; Ham's F12 (F12; 1:1, Gibco), and the granulosa cells retrieved by puncturing the follicles with 30-gauge needles and then expressing the granulosa cells from the follicles by gentle pressure with the heat-sealed end of a Pasteur pipette. Granulosa cells obtained from DES-primed ovaries are considered to be undifferentiated, pre-antral-stage granulosa cells, although it should be noted that the hormonal treatment may have fostered the development of granulosa cells whose physiological state may not completely resemble that of untreated animals. As an alternative, granulosa cells can be isolated from the ovaries of untreated, immature rats by incubating the ovaries in 8.8 mM ethylene glycol-bis(2-aminoethyl) tetraacetic acid (EGTA) and 0.2% BSA in M199, followed by incubation in 0.5 M sucrose, 1.8 mM EGTA, and 0.2% BSA in M199 (54, 55).

Preparations of granulosa cells isolated by expression from follicles may have contaminating theca/interstitial cells. If these are a concern, they can be eliminated by layering the crude granulosa cell suspension over a 40% Percoll solution in saline and centrifuging at 400 g for 20 min. The purified granulosa cells can be aspirated from the top of the Percoll solution and resuspended in DMEM:F12. Cells can be seeded at 3×10^5 cells in 96-well tissue culture plates and incubated at 37°C with 5% CO_2.

Bovine granulosa cells can be aspirated from follicles of different sizes in much the same way as OCCs described earlier. Clearly, less differentiated granulosa cells will be obtained from smaller follicles. The granulosa cells can then be washed and seeded onto tissue culture wells or dishes; for example, 5×10^5 cells can be plated in a 35-mm dish in 2 ml of DMEM:F12 (1:1) supplemented with 2.2 g l^{-1} sodium bicarbonate in the presence of 10% FBS and antibiotics. When plated at this density, they can be cultured for at least 2 days.

6.3 Granulosa cell differentiation in culture

Cumulus and mural granulosa cells differ in several aspects of gene expression and function. For example, expression of LH receptors is a marker of the mural granulosa cell phenotype in pre-ovulatory follicles, as it is expressed at low or undetectable levels in the granulosa cells of pre-antral or small antral follicles and in cumulus cells. However, in addition to differences based on granulosa cell subtype, investigators wishing to study oocyte effects on granulosa cells must pay particular attention to the ability of these cells to undergo spontaneous luteinization once placed in culture, such that analysis of cellular and molecular events in these cells must consider their state of differentiation. Recent studies have been carried out to determine the conditions required for maintenance of aromatase expression and activity in long-term cultures of pig granulosa cells (56), a feature characteristic of granulosa cells before luteinization. Culture of these cells in the presence of serum or high concentrations of FSH tended to encourage their luteinization, while culture in serum-free conditions in the presence of low concentrations of FSH was optimal for the maintenance of high oestradiol production by granulosa cells after 144 h of culture (56). Identification of the conditions that maintain aromatase activity, and other features of the non-luteinized follicle, is critical if one wishes to use long-term cultures of granulosa cells to make physiologically relevant observations regarding follicular function.

7 Analysis of oocyte–granulosa cell gap junctional communication

Gap junctions play a most important role in providing essential components to the oocyte during growth. In addition, the presence of this type of intercellular communication among the granulosa cells permits coordinated cellular activity. The structural proteins comprising these channels, called connexins, are members of a multigene family that consists of at least 13 members (57). A recently developed mouse model with Cx37 deficiency clearly demonstrates that disruption of gap junction-mediated communication between oocytes and granulosa cells profoundly affects the development of both the oocyte and the granulosa cells (6). Determination of expression of specific connexins during follicle development in cows (58) and rats (59) will provide opportunities to investigate the importance of these molecules in granulosa–granulosa and oocyte–granulosa cell communication in these species as well.

Gap junctions mediate the transfer of metabolites from cumulus cells to oocytes, as has been described using a variety of metabolic cooperativity studies. In this type of study, uptake of radiolabelled compounds (e.g. uridine, choline, or 2-deoxyglucose) by granulosa cells is monitored for the transfer of radioactivity to unlabelled oocytes (60). Since gap junction channels can be selective for different tracer molecules, it is advisable to try several different dyes before conclusions are drawn concerning the presence or absence of coupling. As an alternative, if micromanipulation equipment (as described in *Protocol 3*) is available, injection of Lucifer yellow into oocytes can be performed using the methods described in *Protocol 5*, which will enable the evaluation of transfer in the opposite direction, i.e. from oocyte to cumulus cell (61). In these experiments, disruptors of gap junctional communication can prove effective in inhibiting the transfer of dyes or labelled compounds between the cells, thereby interfering with oocyte–granulosa cell communication (62). While these methods are sufficient to demonstrate cell coupling, quantification of the communication is more difficult, but can be achieved (60).

8 Antisense and other strategies for interfering with oocyte–granulosa cell interactions

While oocyte–granulosa cell interactions can be investigated using methods to ablate one or the other cell type or to interfere with cellular communication, at some point it becomes desirable to focus on specific genes that can be identified as key regulators of the cellular interactions. To this end, manipulation of gene expression to yield loss- or gain-of-function mutations in either oocytes or granulosa cells can prove highly effective. These manipulations can be used in short-term studies using some of the culture systems described above, or can be used to generate long-term alterations in gene expression that allow evaluation of the consequences of the altered expression *in vivo* (transgenesis). Identification of the promoters of genes expressed in a cell-specific manner (e.g. ZP3 in oocytes (63); inhibin in granulosa cells (64)) has enabled the generation of transgenic mice with cell-specific gene expression, some of which have altered follicular development (65). Production of transgenic mice is a lengthy process and has been described in detail elsewhere (51, 52); therefore, methods to manipulate gene expression in short-term studies will be described here.

8.1 Manipulation of gene expression in oocytes

Loss of function of a particular gene in oocytes can be achieved by microinjecting antisense oligonucleotides into the cytoplasm of oocytes (66–68). The antisense sequences are thought to pair with their complementary (targeted) mRNA, and double-stranded structures are targeted for destruction by enzymes such as RNase H. The following sections describe the design and synthesis of the oligonucleotides (Section 8.1.1) and the procedures used for oocyte microinjection (Section 8.1.2).

8.1.1 Oligonucleotides

Oligonucleotides for microinjection into oocytes are usually 20–30 nucleotides in length and are designed to be complementary to a region of the targeted mRNA close to the start codon to prevent translation of the transcripts into protein. Missense oligonucleotides, which include all the nucleotides of the antisense oligonucleotide, but in random order, and/or sense oligonucleotides should be used as controls; however, the sequences should be checked (e.g. with GenBank) to ensure no significant homology to any known sequences other than the targeted transcripts. The oligonucleotides should be purified by high-performance liquid chromatography (HPLC), dried under vacuum, and resuspended in distilled water or calcium- and magnesium-free Dulbecco's PBS at a concentration of 100 ng ml^{-1}.

8.1.2 Microinjections into oocyte cytoplasm

Although microinjection into cumulus-free oocytes is much easier than injection into cumulus-enclosed oocytes, oocytes injected in the presence of their surrounding cumulus cells have enhanced survival rates (67). The equipment required for microinjection into oocytes is essentially the same as that described previously for oocytectomy (see *Protocol 3*), with three exceptions:

(1) The lancing pipette is now an injecting pipette, which can be backloaded with the oligonucleotide solution more easily if the pipette has a filament fused to the inner wall (e.g. Kwik-Fil, World Precision).

(2) Once the injection pipette has been lowered into the drop of medium in the plexiglass slide, special care must be taken in applying some positive pressure through the injection pipette to prevent the medium being drawn in by capillary action.

(3) The final modification is optional: an automated injector (e.g. Eppendorf injector 5246) can prove helpful if large numbers of oocytes are being injected.

Protocol 5

Microinjection of oligonucleotides into oocytes

Equipment and reagents

- Microscopy and micromanipulation equipment (see *Protocol 3*)
- Culture medium (e.g. Waymouth MB 752/1, *Protocol 1*) supplemented with 5% FBS
- Mineral oil
- Plexiglass slide (see *Protocol 4*)
- Oligonucleotide solution (see Section 8.1.1)

Method

1. Collect OCCs and wash twice in culture medium with 5% FBS[a]. The washing and presence of serum reduce the stickiness of handling the complexes during micromanipulation. If the state of meiotic maturation is a concern, meiotic inhibitors should be included in the medium (see Section 4.2.3).

2. Backload the oligonucleotide solution into the injecting pipette.

3. Place 40–50 OCCs in 100 µl of medium under mineral oil on a plexiglass slide (see *Protocol 4*) and place the slide on the microscope stage.

4. Lower the injecting and holding pipette slowly into the drop of medium, and focus in the same plane as the complexes.

5. Using a small amount of negative pressure in the holding pipette, grasp one of the complexes. If the GV can be seen, orient the complex such that the GV is at 90° to the holding pipetting.

6. Quickly push the injecting pipette through the zona pellucida and into the cytoplasm of the oocyte. Deliver 2–10 pl of the oligonucleotide solution by applying

Protocol 5 continued

positive pressure in the injecting pipette, either manually or with an automatic injector. The success of this will be evident by changes in the consistency of the cytoplasm near the tip of the injecting pipette. If a bubble appears, the injecting pipette has not pierced the cytoplasmic membrane and so must be pushed further into the oocyte.

7   Withdraw the pipette slowly and release the complex from the holding pipette. Repeat the process for the remaining OCCs. The drop of medium in the plexiglass slide should be changed whenever a different oligonucleotide solution is being used.

8   Wash all complexes at least twice before proceeding with the experiment.

[a] If denuded oocytes are being injected, higher concentrations of serum (i.e. 10–20%) can be advantageous in promoting better oocyte survival following microinjection.

During their growth phase, oocytes synthesize a large amount of mRNA which is then stored, rather than translated immediately. Since application of antisense technology to these stored, untranslated transcripts would fail to yield a response, it is essential to determine that oocytes are producing the appropriate protein at the stage of interest, and that application of antisense technology has reduced the expression of this protein (Plate 3). This can be achieved using immunodetection of the protein, applying standard immunocytochemical procedures and recommendations provided by the source of the antibodies. In some cases, immunodetection of oocyte proteins may require some special consideration (as described in *Protocol 6*) due to the presence of the zona pellucida and the size of the oocyte.

Protocol 6

Immunofluorescent detection of proteins expressed in oocytes

Equipment and reagents

- Acid Tyrode's (NaCl, 8.0 g l^{-1}; KCl, 0.2 g l^{-1}; CaCl$_2$, 0.24 g l^{-1}; glucose, 1.0 g l^{-1}, pH 2.5)
- Culture medium (e.g. Waymouth MB 752/1 with BSA, *Protocol 1*)
- Microtest Terasaki flat-bottom tissue culture plates
- TBS/BSA (100 mM Tris–HCl, 150 mM NaCl, pH 7.5, supplemented with 5% BSA)
- 4% paraformaldehyde in PBS

- Primary antibodies diluted in TBS/BSA with 0.02% Triton X-100
- Secondary antibodies (e.g. conjugated to a fluorescent tag) diluted in TBS/BSA
- Humidified light-protected chamber
- 15% glycerol
- Fluorescent compound or confocal microscope
- Nail varnish

Method

1 Strip oocytes of cumulus cells by repeated pipetting or as described in Section 5.

2 Briefly submerge oocytes (in groups of 10–15) in acid Tyrode's to dissolve the zona pellucida.

3 Incubate zone-free (ZF) oocytes for 45 min in culture medium to allow for recovery.

4 Rinse ZF oocytes in microtest Terasaki flat-bottom tissue culture plates three times over 10 min with TBS/BSA.

5 Fix ZF oocytes in 4% paraformaldehyde in PBS for 15 min at room temperature.

6 Briefly rinse oocytes twice, and then wash three times for 5 min in TBS/BSA.

7 Incubate oocytes with the appropriate concentration of primary antibodies for 1 h at room temperature (or overnight at 4°C) in a humidified light-protected chamber.

8 Rinse oocytes twice, wash three times for 5 min in TBS/BSA, and expose to the appropriate secondary antibodies for 45 min at room temperature in a humidified light-protected chamber. Control experiments in which one or both antibodies are omitted will help to reveal any non-specific staining (e.g. autofluorescence inherent to the oocyte).

9 Construct a glass chamber on the glass slides by cutting pieces from a coverslip and using permount to affix them to the glass slide to form a frame

10 Mount oocytes in 15% glycerol on glass slides within the constructed glass chamber, apply the coverslips, and seal with nail polish to prevent drying. The glass chamber will preserve the three-dimensional structure of the stained oocytes.

11 Fluorescent staining can be visualized under a fluorescent compound or confocal microscope using an appropriate filter (see Plate 3).

While these methods enable the reduction or ablation of expression of a particular gene in oocytes, the same procedures can be applied to examine the consequences of gene overexpression or expression of genes that are not normally expressed. In this case, *in vitro* transcribed RNA can be injected with the aim that these transcripts will be translated in the oocyte after microinjection. As with the injection of oligonucleotides, it is crucial to confirm that the appropriate modifications in protein expression have been achieved, i.e. that the microinjected mRNA has been translated. Such gain-of-function manipulations have enabled the expression in pig oocytes of receptors that are not normally expressed (69).

It should be noted that manipulation of gene expression in oocytes is not limited to techniques that affect the translation of mRNA. Microinjection of antibodies into the cytoplasm of mouse (70) and bovine (71) oocytes has been shown to be effective in interfering with the normal function of the targeted protein.

8.2 Manipulation of gene expression in granulosa cells

Due to the size of granulosa cells, microinjection of oligonucleotides is not a practical approach to manipulating gene expression in these cells; however, standard techniques for the introduction of cDNAs or oligonucleotides into somatic cells have been applied to granulosa cells with some success. Like many somatic cells, simple culture of rat granulosa cells in the presence of antisense oligonucleotides has proven sufficient to interfere with expression of targeted transcripts (72). Transient expression in primary cultures of rat (55) and porcine (73) granulosa cells has been observed following introduction of vectors using transfection mediated by lipofectamine or calcium phosphate precipitation.

9 *In vivo* experimental systems for studying oocyte–granulosa interactions

While culture systems are extremely valuable for identifying individual components and determining their potential contribution to follicle development, there is always the need to confirm these observations using *in vivo* models. Early investigators attempting to examine oocyte–granulosa cell interactions *in vivo* simply removed the oocyte from Graafian follicles or induced oocyte degeneration *in situ*, and monitored the effects on subsequent luteinization and progesterone production (74, 75). More recent approaches include the generation of transgenic animals that express transgenes or show gene deficiency in a cell-specific manner (5). Application of antisense technology to *in vivo* systems allows the physiological observations normally provided by knockout animals to be obtained under more rapid and more easily reversible conditions. For example, inhibition of follicular gene expression in rat (76) and mouse ovaries (77) *in vivo* has been achieved by administration of antisense oligonucleotides directly into the ovarian bursa, the sac-like structure surrounding the ovary of these species.

In future, the refinement and application of these approaches should enhance our understanding of the importance of specific genes in oocyte–granulosa cell communication.

Acknowledgements

The author wishes to thank Drs John Eppig, Jerry Kidder, Marc-André Sirard, and Evelyn Telfer for their helpful suggestions and comments during the preparation of this chapter.

References

1. Eppig, J. J. (1991). *BioEssays*, **13**, 569.
2. Eppig, J. J., Chesnel, F., Hirao, Y., O'Brien, M. J., Pendola, F. L., Watanabe, S., *et al.* (1997). *Hum. Reprod.*, **12**, 127.
3. Kuroda, H., Terada, N., Nakayama, H., Matsumoto, K., and Kitamura, Y. (1988). *Dev. Biol.*, **126**, 71.

4 Huang, E., Manova, K., Packer, A. I., Sanchez, S., Bachvarova, R. F., and Besmer, P. (1993). *Dev. Biol.*, **157**, 100.

5. Dong, J., Albertini, D. F., Nishimori, K., Kumar, T. R., Lu, N., and Matzuk, M. M. (1996). *Nature*, **383**, 531.

6. Simon, A. M., Goodenough, D. A., Li, E., and Paul, D. L. (1997). *Nature*, **385**, 525.

7. Telfer, E. E. (1998). *Theriogenology*, **49**, 451.

8. Hirao, Y., Nagai, T., Kubo, M., Miyano, T., Miyake, M., and Kato, S. (1994). *J. Reprod. Fertil.*, **100**, 333.

9. Roy, S. K. and Treacy, B. J. (1993). *Fertil. Steril.*, **59**, 783.

10. Telfer, E. E. (1996). *Theriogenology*, **45**, 101.

11. Herlands, R. L. and Schultz, R. M. (1984). *J. Exp. Zool.*, **229**, 317.

12. Buccione, R., Cecconi, S., Tatone, C., Mangia, F., and Colonna, R. (1987). *J. Exp. Zool.*, **242**, 351.

13. Eppig, J. J. and Schroeder, A. C. (1989). *Biol. Reprod.*, **41**, 268.

14. Nayudu, P. L. and Osborn, S. M. (1992). *J. Reprod. Fertil.*, **95**, 349.

15. Boland, N. I., Humpherson, P. G., Leese, H. J., and Gosden, R. G. (1993). *Biol. Reprod.*, **48**, 798.

16. Vanderhyden, B. C., Caron, P. J., Buccione, R., and Eppig, J. J. (1990). *Dev. Biol.*, **140**, 307.

17. Eppig, J. J. and Telfer, E. E. (1993). In *Methods in enzymology* (ed. P. M. Wassarman and M. L. DePamphilis). Vol. 225, p. 78. Academic Press, London.

18. Torrance, C., Telfer, E., and Gosden, R. G. (1989). *J. Reprod. Fertil.*, **87**, 367.

19. Hulshof, S. C. J., Figueiredo, J. R., Beckers, J. F., Beyers, M. M., Van der Donk J. A., and Van den Hurk, R. (1995). *Theriogenology*, **44**, 217.

20. Spears, N., Boland, N. I., Murray, A. A., and Gosden, R. G. (1994). *Hum. Reprod.*, **9**, 527.

21. Daniel, S. A. J., Armstrong, D. T., and Gore-Langton, R. E. (1989). *Gamete Res.*, **24**, 109.

22. Roy, S. K. and Greenwald, G. S. (1989). *J. Reprod. Fertil.*, **87**, 103.

23. Morbeck, D. E., Flowers, W. L., and Britt, J. H. (1993). *J. Reprod. Fertil.*, **99**, 577.

24. Cecconi, S., Barboni, B., Coccia, M., and Mattioli, M. (1999). *Biol. Reprod.*, **60**, 694.

25. Hulshof, S. C. J., Figueiredo, J. R., Beckers, J. F., Beyers, M. M., Van der Donk, J. A., and Van den Hurk, R. (1995). *Theriogenology*, **44**, 217.

26. Eppig, J. J. and O'Brien, M. J. (1996). *Biol. Reprod.*, **54**, 197.

27. Wandji, S.-A., Sršen, V., Voss, A. K., Eppig, J. J., and Fortune, J. E. (1996). *Biol. Reprod.*, **55**, 942.

28. Roy, S. K. and Treacy, B. J. (1993). *Fertil. Steril.*, **59**, 783.

29. Wandji, S.-A., Sršen, V., Nathanielsz, P. W., Eppig, J. J., and Fortune, J. E. (1997). *Hum. Reprod.*, **12**, 1993.

30. Hovatta, O., Wright, C., Krausz T., Hardy, K., and Winston, R. M. (1999). *Hum. Reprod.*, **14**, 2519.

31. Vanderhyden, B. C. and Armstrong, D. T. (1989). *Biol. Reprod.*, **40**, 720.

32. Thibault, C. (1977). *J. Reprod. Fertil.*, **51**, 1.

33. Armstrong, D. T., Zhang, X., Vanderhyden, B. C., and Khamsi, F. (1991). *Ann. NY Acad. Sci.*, **626**, 137.

34. Ka, H. H., Sawai, K., Wang, W. H., Im, K. S., and Niwa, K. (1997). *Biol. Reprod.*, **57**, 1478.

35. Wiersma, A., Hirsch, B., Tsafriri, A., Hanssen, R. G., Van de Kant, M., Kloosterboer, H. J., *et al.* (1998). *J. Clin. Invest.*, **102**, 532.

36. Mermillod, P., Tomanek, M., Marchal, R., and Meijer, L. (2000). *Mol. Reprod. Dev.* **55**, 89.

37. van de Sandt, J. J., Schroeder, A. C., and Eppig, J. J. (1990). *Mol. Reprod. Dev.*, **25**, 164.

38. Eppig, J. J. and Telfer, E. E. (1993). In *Methods in enzymology* (ed. P. M. Wassarman and M. L. DePamphilis). Vol. 225, p. 78. Academic Press, London.

39. Schroeder, A. C., Schultz, R. M., Kopf, G. S., Taylor, F. R., Becker, R. B., and Eppig, J. J. (1990). *Biol. Reprod.*, **43**, 891.

40. Leibfried-Rutledge, M. L., Critser, E. S., and First, N. L. (1986). *Biol. Reprod.*, **35**, 850.

41. Sirard, M. A., Parrish, J. J., Ware, C. B., Leibfried-Rutledge, M. L., and First, N. L. (1988). *Biol. Reprod.*, **39**, 546.

42. Bavister, B. D., Leibfried, M. L., and Lieberman, G. (1983). *Biol. Reprod.*, **28**, 235.

43. Funahashi, H. and Day, B. N. (1993). *Theriogenology*, **39**, 965.

44. Veeck, L. L., Wortham, J. E. W., Witmyer, J., Sandow, B. A., Acosta, A. A., Garcia, J. E., *et al.* (1983). *Fertil. Steril.*, **39**, 594.

45. Gilchrist, R. B., Nayudu, P. L., and Hodges, J. K. (1997). *Biol. Reprod.*, **56**, 238.

46. Schramm, R. D. and Bavister, B. D. (1999). *Hum. Reprod.*, **14**, 2544.

47. Brzyski, R. G., Leland, M. M., and Eddy, C. A. (1999). *Fertil. Steril.*, **71**, 1153.

48. Eppig, J. J., Chesnel, F., Hirao, Y., O'Brien, M. J., Pendola, F. L., Watanabe, S., *et al.* (1997). *Hum. Reprod.*, **12**, 127.

49. Buccione, R., Vanderhyden, B. C., Caron, P. J., and Eppig, J. J. (1990). *Dev. Biol.*, **138**, 16.

50. Vanderhyden, B. C., Caron, P. J., Buccione, R., and Eppig, J. J. (1990). *Dev. Biol.*, **140**, 307.

51. Hogan, B., Constantini, F., and Lacy, E. (ed.) (1986). *Manipulating the mouse embryo: a laboratory manual.* p. 166. Cold Spring Harbor Laboratory Press, Cold Spring Harbor, NY.

52. Monk, M. (ed.) (1987). *Mammalian development: a practical approach.* p. 220. IRL Press, Oxford.

53. Vanderhyden, B. C. and Tonary, A. M. (1995). *Biol. Reprod.*, **53**, 1243.

54. Campbell, K. L. (1979). *Biol. Reprod.* **21**, 773.

55. Sharma, S. C., Clemens, J. W., Pisarska, M. D., and Richards, J. S. (1999). *Endocrinology*, **140**, 4320.

56. Picton, H. M., Campbell, B. K., and Hunter, M. G. (1999). *J. Reprod. Fertil.*, **115**, 67.

57. Bruzzone, R., White, T. W., and Paul, D. L. (1996). *Eur. J. Biochem.*, **238**, 1.

58. Johnson, M. L., Redmer, D. A., Reynolds, L. P., and Grazul-Bilska, A. T. (1999). *Endocrine*, **10**, 43.

59. Granot, I. and Dekel, N. (1994). *J. Biol. Chem.*, **269**, 30502.

60. Heller, D. T., Cahill, D. M., and Schultz, R. M. (1991). *Dev. Biol.*, **84**, 455.

61. Isobe, N., Maeda, T., and Terada, T. (1998). *J. Reprod. Fertil.*, **113**, 167.

62. Li, R. and Mather, J. P. (1997). *Endocrinology*, **138**, 4477.

63. Lira, S. A., Kinloch, R. A., Mortillo, S., and Wassarman, P. M. (1990). *Proc. Natl Acad. Sci. USA*, **87**, 7215.

64. Hsu, S. Y., Lai, R. J., Nanuel, D., and Hsueh, A. J. (1995). *Endocrinology*, **136**, 5577.

65. Morita, Y., Perez, G. I., Maravei, D. V., Tilly, K. I., and Tilly, J. L. (1999). *Mol. Endocrinol.*, **13**, 841.

66. Ismail, R. S., Dubé, M., and Vanderhyden, B. C. (1997). *Dev. Biol.*, **184**, 333.

67. Paules, R. S., Buccione, R., Moschel, R. C., Vande Woude, G. F., and Eppig, J. J. (1989). *Proc. Natl Acad. Sci. USA*, **86**, 5395.

68. Sallés, F. J., Richards, W. G., Huarte, J., Vassalli, J.-D., and Strickland, S. (1993). In *Methods in enzymology* (ed. P. M. Wassarman and M. L. DePamphilis). Vol. 225, p. 351. Academic Press, London.

69. Kim, J. H., Machaty, Z., Cabot, R. A., Han, Y. M., Do, H. J., and Prather, R.S. (1998). *Biol. Reprod.*, **59**, 655.

70. Simerly, C., Nowak, G., de Lanerolle, P., and Schatten, G. (1998). *Mol. Biol. Cell*, **9**, 2509.

71. Liu, L. and Yang, X. (1999). *Biol. Reprod.*, **61**, 1.

72. Shapiro, D. B., Pappalardo, A., White, B. A., and Peluso, J. J. (1996). *Endocrinology*, **137**, 1187.

73. LaVoie, H. A., Garmey, J. C., Day, R. N., and Veldhuis, J. D. (1999). *Endocrinology*, **140**, 178.

74. El-Fouly, M. A., Cook, B., Nekola, M., and Nalbandov, A. V. (1970). *Endocrinology*, **87**, 288.

75. Hubbard, G. M. and Erickson, G. F. (1988). *Biol. Reprod.*, **39**, 183.
76. Piontkewitz, Y., Enerback, S., and Hedin, L. (1996). *Dev. Biol.*, **179**, 288.
77. Matsumoto, K., Nakayama, T., Sakai, H., Tanemura, K., Osuga, H., Sato, E., *et al.* (1999). *Mol. Reprod. Dev.*, **54**, 103.

Chapter 9

Cell–cell interactions in early mammalian development

Tom P. Fleming, Judith J. Eckert, Wing Yee Kwong, Fay C. Thomas, Daniel J. Miller, Irina Fesenko, Andrew Mears and Bhavwanti Sheth

School of Biological Sciences, University of Southampton, Bassett Crescent East, Southampton SO16 7PX, UK

1 Introduction

Cell–cell interactions during early development prior to implantation play a key role in the morphogenesis of the blastocyst. The essential processes in blastocyst formation that are dependent upon intercellular contact and signalling and which occur concurrantly are the differentiation of the trophectoderm epithelium and the segregation of the inner cell mass (ICM) as a separate cell lineage from the epithelium. The trophectoderm forms the wall of the blastocyst, is responsible for generating the blastocoel cavity by transport processes, and, later in development, gives rise to trophoblast and most placental cell lineages. The ICM is a non-epithelial cell cluster located on the inner surface of the trophectoderm at one pole of the blastocyst from which the entire fetus forms after implantation (*Figure 1*).

1.1 Cell–cell interactions in blastocyst morphogenesis

Studies on cell interactions during trophectoderm differentiation have concentrated on either the molecular maturation of the epithelial adhesion systems of the adherens, tight, and desmosomal junction types as cells progress through stages of cleavage of the egg (reviewed in refs 1, 2) or on the expression and assembly of key components of the vectorial transport systems, such as the Na^+,K^+-ATPase pump (3, 4). A critical advantage in employing the early embryo model for analysing mechanisms of epithelial differentiation rather than using the more conventional approach of cultured epithelial cell lines (e.g. Madin–Darby canine kidney (MDCK) cells, ref. 5) is that the trophectoderm is a truly differentiating cell layer with recognizable steps in its development from non-epithelial precursors. These steps can correlate with particular cell cycles during cleavage and represent the unfolding of a temporal programme of gene

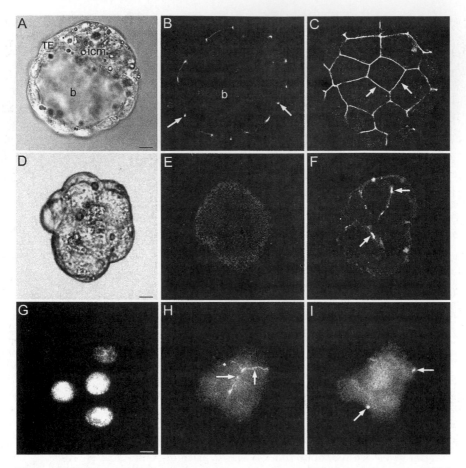

Figure 1 (A–C) Immunolabelled whole-mount mouse embryos, **(D–F)** inner cell masses (ICMs) isolated following immunosurgery, and **(G–I)** a cell cluster derived from a 2/16 synchronized cluster stained for tight junction-associated proteins and viewed by confocal microscopy. (A–C) Blastocyst shown in brightfield (A) to indicate the position of trophectoderm (TE), inner cell mass (icm), and blastocoel cavity (b), and for localization of zonula occludens 1 (ZO-1) in the same mid-plane view (B) showing ZO-1 located at contact sites between TE cells (arrows) and absent from the ICM. In a tangential view of the embryo surface (C), the circumferential staining pattern for ZO-1 around each TE cell (arrows) is evident. (D–F) ICM shown in brightfield (D) and following occludin staining (E and F); occludin is undetectable immediately after immunosurgery (E) but by 2 h of culture is evident at contact sites between ICM cells (F, arrows). (G–I) Cluster of four cells (4/32 cluster) stained with Hoechst dye to assess cell number (G) and for localization of cingulin at cell contact sites (arrows) viewed tangentially (H) and in mid-plane (I). Bar (A–C) = 10 μm, (D–I) = 5 μm.

expression that determines the timing of development. Thus, for example, genes contributing towards the formation of the epithelial tight junction may initiate transcription at defined times during cleavage such that only by the 32-cell stage in the mouse embryo will tight junction assembly be complete. The timing of gene expression can therefore act as a regulator of development since the permeability seal of the trophectoderm provided by the tight junction is essential for vectorial transport to generate a blastocoel cavity. In contrast, most epithelial cell lines comprise fully mature cells that are commonly induced to undergo 'differentiation' by artificial means, such as modulation of extracellular calcium concentration (calcium switching). Significantly, the temporal control of differentiation exerted by an intrinsic gene expression programme is not contributing to this process, which therefore may not be fully representative of actual development.

Cell interactions in the early embryo model have been shown to act upon and influence the temporal programme of gene expression so as to coordinate lineage segregation during blastocyst formation. Thus, the asymmetry of cell contact experienced by cells of the trophectoderm lineage (i.e. each cell has an outer, contact-free face) acts as a positive regulator of translation of tight junction gene mRNAs. Conversely, the symmetry of contacts experienced by cells within the ICM lineage (i.e. total enclosure by surrounding cells) acts as a negative regulator of tight junctional protein translation, thereby restricting tight junction assembly to the outer epithelial layer (1, 6–10).

Here, we will review some of the technical methodologies that we and other laboratoroes use to study cell–cell interactions during this critical period in early development. For methods used in the collection and culture of mouse early embryos, the primary species model used for such studies, the reader should consult Chapter 10. The inherent asynchrony in the mammalian embryo cell cycle can disguise the extent of temporal control of gene and protein expression that occurs. Manipulations that can be used to generate cell cycle synchrony in clusters of cells from early embryos will therefore be included here. One major disadvantage of the early embryo as a model for differentiation compared with cell lines is, of course, the limitation in the amount of material available for analysis. We will therefore include our protocols for analysis of gene expression by semi-quantitative means, suited for small embryo samples. In addition, methods for examining intact pre-implantation embryos by confocal microscopy for localization of proteins and mRNAs are provided. Since a major significance of cell interactions during trophectoderm and ICM diversification is concerned with the allocation of appropriate numbers of cells towards these two lineages, we also include protocols that will help to distinguish between them These include methods for isolation of ICMs, for differential counting of trophectoderm and ICM cell numbers, and for determining the extent of apoptosis within these lineages. Lastly, the rate of proliferation and differentiation during pre-implantation development has been shown to be gender-specific in many mammalian species, with male embryos reaching the blastocyst stage in advance of female embryos (11). The distinction between the sexes is likely to derive from differences in their gene expression programme coupled with cell interactions, and can have long-term consequences in terms of susceptibility to the maternal environment (12). Lastly, we provide a method to couple sexing of individual pre-implantation embryos with their gene expression profile.

2 Generation of synchronous embryos and cell clusters

As stated above, one important problem with the use of intact mouse embryos at different stages of cleavage is that cell cycle length may vary within a single embryo. Consequently, blastomeres from different cycles may be present alongside each other. This asynchrony is well borne out in studies of cell numbers within embryos (13). In some circumstances, this may not be too critical, and staging of intact embryos in culture can be achieved by, for example, hourly inspection of the stock for the presence of a particular developmental transition, such as entry into the 8-cell stage or initiation of compaction. Embryos reaching

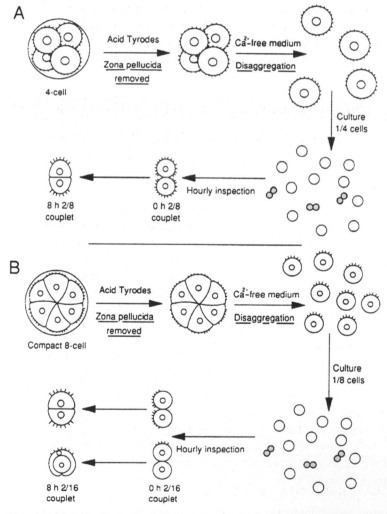

Figure 2 Strategies for production of synchronized cell clusters derived from 1/4 or 1/8 mouse blastomeres isolated from disaggregated 4-cell and 8-cell embryos, respectively; see *Protocol 1* for details.

this transition at successive inspection times can be collected and cultured separately from stock embryos and will be crudely synchronized for subsequent experimental analysis. This simple approach has been useful for determining the timing of lineage segregation in the embryo (e.g. 14). Moreover, where embryos are to be used subsequently in microscopic examination (e.g. immuno-fluorescence confocal microscopy), these synchronized embryos can be co-labelled with a nuclear dye for cell number determination, thereby providing a quality control for the extent of synchrony maintained (8, 9).

In other circumstances, however, it is critical to know the precise cell cycle in which a particular gene expression event or a protein membrane assembly event occurs. Here, the generation of synchronized cell clusters derived from single blastomeres has been a valuable approach. Crucially, manipulation of mouse embryos in this way does not appear to influence their developmental capacity, and the method has been used successfully in many circumstances, originally by Johnson and Ziomek (e.g. 15, 16) and subsequently by other workers, including ourselves (e.g. 8, 9). The method is described in *Protocol 1* and illustrated in *Figure 2*. Images of immunolabelled cell clusters are shown in *Figure 1*.

Protocol 1

Generation of synchronized cell clusters from mouse pre-implantation embryos

Equipment and reagents

- Stereomicroscope
- Microforge (e.g. Fonbrune, Narishige)
- Alcohol wick lamp
- Glass capillaries (Clark Electromedical, type GC100-15)
- Culture medium (e.g. T6, H6 media with 4 mg ml^{-1} bovine serum albumin (BSA); T6+BSA, H6+BSA; see reference 17)
- Mineral oil (Sigma, cat. no. M8410)
- Microdrop culture dishes (Sterilin type above) of T6+BSA overlain with mineral oil containing, for each dish, seven rows of eight approximately 5 μl microdrops in a grid-like pattern plus an additional larger (~20 μl) drop at the top of the grid for washing cells and orientation of the grid

- Culture dishes (Bibby Sterilin: cat. no. 122; Falcon: Becton Dickinson cat. no. BS3004)
- Cavity blocks with lids (square watch glasses; sterile)
- Ca^{2+}-free H6+BSA (H6+BSA without CaCl$_2$ and with 6 mg ml^{-1} BSA); prepare drops in two Falcon dishes overlain with mineral oil
- Acid Tyrode's medium (140 mM NaCl, 2.7 mM KCl, 1.8 mM CaCl$_2$, 0.5 mM MgCl$_2$·6 H$_2$O, 5.5 mM D-glucose, 0.4% polyvinyl pyrrolidone (PVP) in sterile pure water, pH adjusted to 2.5; stored in 1 ml aliquots at $-20\,°C$)
- Flame-pulled Pasteur pipettes and mouth pipette for manipulating embryos

A. Preparation of flame-polished micropipettes

1   Pull glass capillaries over the flame of the alcohol lamp and break off in the centre to form two micropipettes.

Protocol 1 continued

2 Place each micropipette in the holder of the microforge with the pulled region in contact with a pre-formed glass bead on the filament. Heat the filament to fuse the micropipette with the bead. Turn off the filament and retract the micropipette to form a clean break at the tip.[a]

3 Flame-polish the tip by placing it close to the heated filament.

B. Generation of synchronized 8-cell stage clusters

1 Flush embryos from oviducts at the late 2-cell or early 4-cell stage (~46 h post-human chorionic gonadotrophin (hCG) injection) and culture overnight so that early the next morning most stock embryos (aim for around 200) will be at the late 4-cell stage.[b]

2 The following morning, collect all 4-cell embryos together and incubate in acid Tyrode's for a brief period (15–30 s) in a cavity block to remove the zona pellucida from each embryo and transfer immediately to H6+BSA in a cavity block for embryos to recover at 37 °C in the incubator.

3 Collect about 20 embryos from the stock zona-free embryos maintained in H6+BSA and place them in Ca^{2+}-free H6+BSA dish 1. Wash in one drop and transfer into another to ensure removal of calcium, placing the embryos apart in the drop. Return to culture for 15 min.

4 After about 10 min, repeat step 3 for a second group of embryos but use Ca^{2+}-free H6+BSA dish 2.

5 Take Ca^{2+}-free H6+BSA dish 1 and disaggregate the embryos to single cells using an appropriate flame-polished micropipette attached to the mouth pipette. Gently suck each embryo into the micropipette and eject it several times to separate the blastomeres from each other.

6 Transfer about eight single 4-cell blastomeres (called 1/4 cells) first into the larger top drop of one of the T6+BSA microdrop dishes and from there into the first microdrop on row 1. Place the single cells around the periphery of the microdrop to keep them separate from each other.

7 Repeat step 6 until all the 1/4 cells have been transferred to succeeding microdrops in the T6+BSA dish, limiting the number of cells within each drop to about eight.

8 Transfer another batch of zona-free 4-cell embryos from H6+BSA into Ca^{2+}-free H6+BSA dish 1, using a fresh drop for the next 15 min culture period, and place in the incubator.

9 Disaggregate 4-cell embryos in Ca^{2+}-free H6+BSA dish 2 as above and transfer to T6+BSA microdrops.

10 Repeat until all 1/4 cells are in microdrop culture. With experience, steps 3–10 should take around an hour to complete.

Protocol 1 continued

11 Examine the microdrop culture every hour for evidence of division of single blasto-meres to sister couplets (called 2/8 pairs).

12 At each hourly inspection, collect all newly divided 2/8 pairs and transfer them individually into microdrops of T6+BSA in another dish. Use the rows to track the location of 2/8 pairs collected from that time point and mark the underside of the dish with a fine felt-tip pen to show the location of each hourly group.

13 Collect groups of synchronized 2/8 couplets over several hours, thereby filling up a number of microdrop dishes.

14 Use the 2/8 pairs at the required hours post division for experimental analysis (e.g. gene expression assay, immunocytochemistry, and confocal microscopy).

C. Generation of synchronized 16-cell stage clusters

1 Flush embryos (aim for around 100) from oviducts either at the early/mid 4-cell stage (evening before experiment) or at the early/mid 8-cell stage (morning of experiment).[c]

2 Collect together compacted embryos[d] from the stock culture on the morning of the experiment.

3 Remove the zona pellucida using acid Tyrode's and store in H6+BSA (step B2 above).

4 Treat groups of about 20 embryos with Ca^{2+}-free H6+BSA in dishes 1 and 2, dis-aggregate to single cells (called 1/8 cells), and transfer to a microdrop culture using the methods described above for 1/4 cells (steps B3–10).[e]

5 Inspect hourly for cell division of 1/8 cells to 2/16 couplets and transfer each newly formed couplet to individual microdrops for continued culture as described for 2/8 couplets (steps B11–14).

[a] The internal bore of the micropipette should be just smaller than the diameter of an embryo, around 50 μm.

[b] The optimal time for inducing superovulation in mice to achieve this staging would be for pregnant mares' serum and hCG injections to be at 6 p.m.

[c] The optimal time for inducing superovulation in mice to achieve this staging would be for pregnant mares' serum and hCG injections to be at 2 p.m.

[d] These embryos are likely to be 8-cell stage but some may be advanced, with a proportion of cells already at the 16-cell stage.

[e] Discard those cells of any embryo which are at the 16-cell stage, clearly recognized by their smaller size and the presence of more that eight cells within an embryo.

These methods for preparaing synchronized cell clusters are also appropriate for generating reconstituted synchronized embryos by, for example, aggregating four 2/8 couplets or eight 2/16 couplets, all of the same age post-division. Manipulation of reconstituted embryos to include fluorescent-tagged cells of differing ages has been used in cell lineage studies (18, 19).

3 Analyses of gene and protein expression

3.1 Gene expression analysis

The use of reverse transcriptase–polymerase chain reaction (RT–PCR) protocols is now commonplace in pre-implantation research. We developed a method for single blastomere RT–PCR analysis at the blastocyst stage following positional marking specifically of trophectoderm cells and embryo disaggregation. This method permitted analysis of individual trophectoderm or ICM cells for at least two transcripts of genes involved in cell interactions (desmocollin (DSC), E-cadherin) using a two-stage PCR amplification, and has been fully published elsewhere (20, 21).

More recently, we have developed methods suited to semi-quantitative analysis of up to six transcripts per single embryo based on protocols developed for analysis of bovine embryos (22, 23). The methods for storage of embryos for this technique, extraction of mRNA from individual embryos, and RT–PCR amplification of transcripts are given in *Protocols 2*, *3*, and *4* respectively. This approach has been used successfully for analysis of gene expression of cell adhesion and junction constituents in bovine pre-implantation embryos and also works in rat and mouse embryos (papers in preparation).

Protocol 2

Long-term storage of embryos for mRNA extraction and RT–PCR

Equipment and reagents

- 0.5-ml siliconized microfuge tubes
- Plastic Petri dishes
- UV cross-linker (Spectrolinker XL1000, Spectronics Corporation)
- Pulled Pasteur pipettes and mouth pipette
- Dry ice
- PBS+PVP (PBS supplemented with 0.1% PVP (Sigma, P0903; prepared with DNase/RNase-free water, Anachem)

Method

1 Prepare 50 μl droplets of PBS+PVP in a plastic Petri dish and irradiate in the UV cross-linker at 400 000 mJ cm^{-2}.

2 Collect the embryos from the culture medium and place them in the PBS+PVP droplets.

3 Wash the embryos three times in separate PBS+PVP droplets.

4 Pipette the embryos individually in a minimal volume of PBS+PVP (~2–5 μl) in the bottom of a siliconized tube and snap-freeze on dry ice.

5 Store the tubes with the embryos at −70°C. Do not allow them to thaw before their use for mRNA extraction and RT–PCR (*Protocol 3*).

Protocol 3

Isolation of mRNA for multiple semi-quantitative RT–PCR analysis of embryos

Equipment and reagents

- Thermocycler (DNA Engine, PTC-225, MJ Research) or heating block
- Magnetic separator (Dynal)
- UV cross-linker (Spectrolinker XL1000, Spectronics Corporation)
- Vortex
- 0.5-ml microfuge tubes (sterile and UV irradiated at 400 000 mJ cm^{-2})

- Dynabead mRNA Direct kit (Dynal; contains lysis buffer, bead solution, wash buffer 1 and 2)
- DNase/RNase-free water (e.g. Anachem)
- Standard mRNA (1 μg μl^{-1} luciferase; Promega)
- Ice

Method

1 Remove frozen microfuge tubes containing individual embryos from the freezer (see *Protocol 2*) and immediately add 150 μl of lysis buffer to each. Vortex and incubate the mixture for 10 min at room temperature. During this time, perform step 2.

2 Prepare the Dynabeads by pipetting sufficient bead solution for each embryo tube to receive 10 μl into a 0.5-ml microfuge tube and placing the tube on the magnetic separator to isolate the beads from the storage buffer. Remove the supernatant and discard. Add the same amount of lysis buffer and repeat this procedure.

3 Add 10 μl of resuspended beads in lysis buffer and 1 pg (1 μl) of luciferase mRNA into each tube containing embryos and allow the endogenous and standard mRNA to hybridize to the beads by gentle mixing for 10 min at room temperature in a horizontal position.

4 Place each tube on the magnet and separate the beads from the solution for 2 min at room temperature.

5 Remove the supernatant carefully with a thin pipette tip and discard. Add 100 μl of wash buffer 1 per tube. Remove the tube from the magnet, mix gently by flicking, and replace the tube on the magnet for another separation at room temperature.

6 Repeat step 5.

7 Remove the supernatant carefully with a thin pipette tip and discard. Add 100 μl of wash buffer 2 per tube. Remove the tube from the magnet, mix gently by flicking, and replace the tube on the magnet for another separation at room temperature.

8 Repeat step 7 twice.

9 After removal of the supernatant, add 10 μl of DNase/RNase-free water[a] per tube, mix gently by flicking, and incubate the tube at 65 °C for 2 min to elute the mRNA from the beads.

10 Locate the magnet on ice, place the tube on it, and separate the beads from the mRNA solution.

Protocol 3 continued

11 Remove the supernatant containing the mRNA, subdivide into 8- and 2-μl portions[b] and add to the RT mix (see *Protocol 4*, step 1) for RT–PCR experimental and control (minus reverse transcriptase) samples, respectively.[c] Add 10 μl of storage buffer to the beads and store at 4°C.

[a] The volume will depend on the size required for the subsequent reverse transcription and could be from 4 to 30 μl.

[b] The ratio of these volumes may be varied depending upon the desired allocation of template between experimental and control samples.

[c] Do not freeze these samples as RNA may degrade, and use as rapidly as possible for RT–PCR analysis.

Protocol 4

Semi-quantitative RT–PCR of pre-implantation embryos[a]

Equipment and reagents

- Thermocycler (DNA Engine, PTC-225, MJ Research)
- UV cross-linker (Spectrolinker XL1000, Spectronics Corporation)
- Gel electrophoresis apparatus
- Image analyser with digitized camera system and analytic software (Alpha Imager, Alpha Innotech Corporation)
- Pipette tips for PCR (with filter)
- 0.5-ml and 0.2-ml microfuge tubes (PCR-grade, thin-walled, sterile, and UV irradiated at 400 000 mJ cm^{-2})
- Random hexamer primers (Perkin Elmer)
- Reverse transcriptase (SuperscriptII RNase H-, 200 U μl^{-1}, Gibco-BRL)
- Reverse transcriptase (RT) buffer (for SuperscriptII, Gibco-BRL; 50 mM Tris–HCl, pH 8.3, 75 mM KCl, 3 mM MgCl$_2$, 500 μM dithiothreitol (DTT))
- RNase inhibitor (Perkin Elmer)
- dNTPs (Gibco-BRL)

- Standard mRNA (luciferase mRNA, Promega)
- Nuclease-free water (Anachem)
- Mineral oil (Sigma)
- *Taq* DNA polymerase (Gibco-BRL)
- PCR buffer (Gibco-BRL; 10× stock: 20 mM Tris–HCl, pH 8.4, 50 mM KCl, 1.5 mM MgCl$_2$)
- Gene-specific primers
- 5× TBE buffer (0.45 M Tris base, 0.44 M boric acid, 0.01 M ethylenediamine tetraacetic acid (EDTA), pH 8 in water)
- Ethidium bromide
- Agarose gel (Sigma; 2% agarose in 0.5× TBE buffer containing 1 μg ml^{-1} ethidium bromide)
- Loading buffer (15% Ficoll, 0.02% bromophenol blue, Sigma)
- Molecular weight marker (100-bp ladder, Gibco-BRL)

Protocol 4 continued

Method

1 Prepare the RT mix for experimental and control (minus-RT) samples which, after addition of mRNA (step 2) and RT (step 3), will comprise 500 μM of each dNTP, 1$\times$ RT buffer, 5 μM random hexamer primers, RNase inhibitor (20 U) and water up to a total volume of 20 μl.[b] Vortex and centrifuge briefly before use.

2 Add the extracted mRNA samples (8 μl experimental and 2 μl control, *Protocol 3*, step 11). Overlay the tube with 50 μl of mineral oil and place in the thermocycler programmed for 10 min at 25 °C.

3 Set the thermocycler temperature to 37 °C for 1 h, and add the reverse transcriptase (200 U; 1 μl) to the experimental tube after 2 min incubation through the oil layer. Add only water to the control tube. After 1 h, inactivate the reverse transcriptase by heating to 75 °C for 15 min.

4 Prepare the PCR mix which, after addition of primers and cDNA (step 5), will comprise 200 μM of each dNTP, 1$\times$ PCR buffer, and water to a final volume of 50 μl for each PCR. Vortex and centrifuge.

5 Aliquot the PCR mix into the required number of 0.2-ml tubes,[c] add 400 μM of the respective primer pairs and the respective amount of cDNA from the RT reaction.[d] Mix well before overlaying with 50 μl of oil.

6 Place the tubes into the thermocycler programmed to 95 °C for 3 min, followed by 72 °C for 30 s. During the 72 °C step, add 2.5 IU of *Taq* polymerase to each tube underneath the oil and mix well by carefully pipetting up and down.

7 Run the appropriate number of cycles[d] for each transcript of interest using denaturation for 30 s at 95 °C, annealing at the appropriate annealing temperature for 30 s, and elongation for 30 s at 72 °C. Follow the cycling protocol by a final extension step for 10 min at 72 °C and a holding temperature of 10 °C.

8 Prepare the agarose gel.

9 For loading of the product on the agarose gel, add 10 μl of the PCR product to 2 μl of 6$\times$ loading buffer, mix well, and place into wells. Run for 30 min at 80 mA.

11 Visualize the PCR products under UV light using a digitized camera system and quantify the band intensities using adequate computer software. Measure the band intensities as integrated density values which are calculated as the sum of all pixel values after background correction. Subtract a control value from a gel region of no product from the density data.

12 Semi-quantify mRNA expression by dividing the band intensity of the transcript of interest by the band intensity of the standard mRNA (luciferase) which was run alongside test samples throughout the whole procedure (mRNA extraction, RT–PCR, agarose gel electrophoresis).

[a] Adapted from refs 22 and 23.

[b] This volume could be altered according to experimental circumstances.

Protocol 4 continued

^c The number of PCRs should include all the endogenous transcripts under investigation, a luciferase standard, the RT negative control, a negative control without cDNA, and a positive control using a suitable reverse-transcribed RNA.

^d For each primer pair, PCR conditions may vary slightly and have to be optimized, for example, at different $MgCl_2$ concentrations and annealing temperatures. For quantification, ensure for each transcript of interest separately that the PCR amplification is within the linear range (test different cycle numbers). To examine several transcripts from a single embryo, adjust the percentage of the embryo needed to amplify each transcript of interest reproducibly in the PCR. This depends on the abundance of each transcript. Together with adjusting cycle numbers (between 35 and 45), up to six different transcripts have been amplified so far from one single embryo. Finally, verify your products by direct sequencing at least once for each type of tissue examined.

3.2 Protein localization at cell contact sites in pre-implantation embryos

Cell–cell interactions in the pre-implantation embryo have been studied extensively by protein immunolocalization. In view of the importance of spatial organization and the interrelationship between trophectoderm and ICM lineages in blastocyst morphogenesis, confocal microscopy of intact embryos is the most suitable approach for protein localization, permitting digital reconstruction of collapsed *z*-series from optical section analysis (8–10; *Figure 1*). In our laboratory,

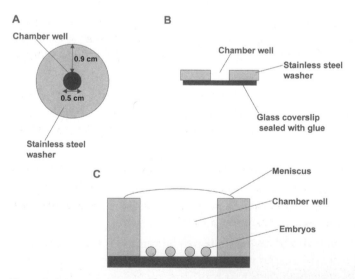

Figure 3 Schematic representation of embryo chambers used for immunocytochemisty processing and viewing of specimens on the confocal microscope. Chambers were first used in ref. 24 and are described here in *Protocol 5*. (**A** and **B**) Chambers in top and side view following sealing of the coverslip to a stainless steel washer. (**C**) Higher magnification view of the well of the chamber with embryos attached to the coverslip and the meniscus distant from the embryos, thereby minimizing the chance of their detachment during processing.

we use for routine work the Bio-Rad MRC-600 series confocal imaging system with Bio-Rad Confocal Assistant software. The small size of pre-implantation embryos (~80 μm diameter) imposes some technical difficulties in processing them reliably for analysis of cell–cell interactions. In our laboratory, we routinely use small 'chambers' composed of a steel washer sealed to a coverslip to which embryos are attached to protect them during processing (*Figure 3*). This method was first developed by Maro *et al* (24) and ensures that during processing, embryos sitting within the 'well' of the chamber do not pass through a meniscus and so are more likely to be retained attached to the coverslip. Here, we describe first our method for chamber preparation (*Protocol 5*) before giving details of fixation and attachment of embryos to chambers (*Protocol 6*) and immunolabelling using the chambers (*Protocol 7*).

Protocol 5

Preparation of immunolabelling chambers for embryos

Equipment and reagents

- Stainless steel washers (25 mm overall diameter, 5 mm hole diameter, 1 mm thick, smooth both sides; Glenwoods Ltd, or other hardware retailers)
- Teepol detergent
- Coverslips (round, 22 mm diameter, no. 0; Chance Propper Ltd).
- Cyanoacrylate adhesive (2 g dispenser, RS cat. no.159-3979)

Method

1 Wash the steel washers in hot water and detergent to remove any grease and then in distilled water twice to remove detergent before placing and drying on lint-free tissue paper.

2 Wearing gloves, clean the coverslips with lens tissue, removing all dust and fingerprints.

3 Dot small drops of glue around the hole in the washer and about 5 mm distance from it.

4 Place the coverslip onto the glued washer surface; the glue should fill the undersurface of the coverslip to avoid leaking and yet should not seep into the hole.

5 Leave to dry and seal for at least 30 min. Store in a suitable dust-free container (e.g. 50-ml plastic syringe housing with plunger removed).

Western blotting and immunoprecipitation methods have been used to determine levels of expression of proteins involved in cell–cell adhesion and interactions in embryos (8–10, 25). Due to the small size of embryos, and depending on antibody antigenicity, a large number of staged embryos may be required for analysis of changing patterns of expression at different periods of cleavage. Broadly, our methods otherwise are similar to those used in cell biology, and the reader is referred to Chapter 1. An alternative approach, developed by Warner and colleagues, is the immuno-PCR technique to quantify levels of protein expression in small groups of embryos (26, 27).

Protocol 6

Fixation and attachment of embryos to immunolabelling chambers

Equipment and reagents

- Stereomicroscope
- Bench-top centrifuge
- Immunolocalization chambers (*Protocol 5*)
- Filter paper circles (25 mm; Schleicher and Schuell, cat. no. GB004)
- Centrifuge tubes (made in-house of the exact internal diameter to hold chambers in a stack and also to fit the centrifuge well. The tubes also need a removable bottom platform (solid plastic) upon which chambers are placed and a hole in the base to insert a plunger (stiff wire) to lift the platform for loading or removing chambers.)
- Acid Tyrode's solution (see *Protocol 1*)
- H6+BSA culture medium (see *Protocol 1*)

- H6+PVP medium (H6 supplemented with 0.6% PVP, Sigma P0903, instead of BSA)
- PBS
- 1-ml syringe with needle
- Poly-L-lysine hydrobromide (PLL, Sigma, 1.5 mg ml^{-1} in PBS)
- Concanavalin A (0.1 mg ml^{-1}; Sigma) in PBS
- Falcon tissue culture dishes (Becton Dickinson, Falcon cat. no. BS3004)
- Fixative (e.g. 1 or 4% formaldehyde or paraformaldehyde in PBS; methanol or acetone at $-20\,°C$)
- Coverslips (round 19 mm diameter, no. 1; Chance Propper Ltd)

A. Aldehyde fixation

1 Remove the zona pellucida from embryos with acid Tyrode's solution at 37 °C in a cavity block, transfer to another cavity block containing H6+BSA as described in *Protocol 1* step B2, and return to the incubator to recover for about 20 min.[a]

2 While embryos are recovering, check chambers for any leaks by pipetting a small drop of PBS into the well only and leaving for a few minutes.

3 Remove the PBS from each chamber well using the syringe and needle. Replace with PLL solution and leave for 15 min.

4 Fix zona-free embryos in drops of aldehyde fixative at room temperature in a Falcon dish for 10–20 min.

5 While embryos are fixing, draw off the PLL solution from each chamber well and wash three times with PBS, leaving each well filled with PBS such that there is a positive meniscus but none spread over the remainder of the steel surface of the chamber (see *Figure 3*).

6 Transfer the embryos from the fixative to a PBS wash drop in another Falcon dish

and from there to the chambers using the stereomicroscope. Place embryos apart at the bottom of the well and ensure that they adhere to the PLL-coated coverslip.

7 Carefully slide a coverslip over each well without introducing any air bubbles and dry the top of the covered chamber with tissue paper.

8 When all chambers are completed, place them carefully into the centrifuge tubes to a maximum of four per tube, with each separated by a filter circle. Ensure that equal numbers of chambers are present in each tube for balancing.

9 Centrifuge at 15 000–17 000 r.p.m. for 10 min.

10 Remove the chambers from the tube and place on absorbent tissue paper on the bench top.

11 Remove the the top coverslip carefully by sliding off using either gloved fingers or the prongs of a pair of forceps.

12 Wash the chambers by flooding with PBS including the steel upper surface surrounding the well. Add the PBS slowly to the edge of the chamber so as not to dislodge the embryos.

13 Store the chambers if necessary in a PBS bath[b] or proceed directly to immuno-labelling (*Protocol 7*).

B. Methanol and acetone fixation

1 Remove the zona pellucida from embryos and allow to recover as described in step A1 above.

2 Whilst embryos are recovering, fill the well of chambers with concanavalin A in PBS to make the coverslip base adhesive for living embryos, and leave for 15 min. Wash with PBS twice and finally with H6+PVP.

3 Wash zona-free embryos in three drops of H6+PVP in a Falcon dish to remove BSA and, using the stereomicroscope, transfer to the wells of the chambers, allowing the embryos to adhere to the surface of the coverslip.

4 Cover each chamber with a top coverslip and centrifuge as described in steps A7–11 above.

5 After removing the top coverslip, gently lower the chambers into a methanol or acetone bath at −20 °C and leave in the freezer for 10 min.

6 Remove the chambers from the fixative, place on absorbent tissue paper on the bench top, and carefully flood the chambers with PBS from the edge. Repeat this PBS wash three times over 15 min. Store the chambers in a PBS bath[b] or proceed directly with immunolabelling (*Protocol 7*).

[a] All solution transfers of embryos are achieved using drawn Pasteur pipettes attached to a mouth pipette.

[b] For convenience, chambers can be slotted into an in-house plastic rack within the PBS bath.

Protocol 7

Immunolabelling of embryos within chambers

Equipment and reagents

- 0.25% Triton X-100 in PBS
- PBS
- NH_4Cl solution (2.6 mg ml^{-1} in PBS)
- PBS-Tween (0.1% Tween-20 in PBS)
- Citifluor mounting medium (Citifluor Ltd; PBS solution type)
- Antibody solutions (primary or secondary at the appropriate dilution in PBS-Tween and, optionally, 1% BSA to block non-specific binding)
- Coverslips (19 mm; Chance Propper Ltd.)
- Clear nail varnish

A. Aldehyde-fixed embryos

1 Remove chambers from PBS storage, place on absorbent tissue paper on the bench top, and flood the chamber surface with Triton X-100 in PBS for 15 min to permeabilize embryos.

2 Wash the chambers three times with PBS.

3 Treat the chambers with NH_4Cl solution for 10 min to neutralize the fixative and wash three more times in PBS.

4 Carefully dry the periphery of the upper surface of each chamber with tissue, avoiding drawing liquid from the well.

5 Carefully add 20–35 µl of diluted primary antibody solution to the centre well and leave for 1 h at room temperature or overnight at 4°C. Leave chambers on moist tissue paper and cover to prevent evaporation.

6 Wash three times with PBS-Tween over 30 min.

7 Add secondary antibody solution in the same way as the primary and leave covered for 1 h at room temperature.

8 Wash three times with PBS-Tween over 30 min.

9 Carefully dry the periphery of the chambers and add 40 µl of Citifluor (PBS) to the centre well. Cover the chamber with a coverslip and seal the edges with nail varnish.

B. Methanol- or acetone-fixed embryos

1 Proceed as above in A from step 4 onwards.

4 Analysis of cell lineage segregation

Cell–cell interactions have been studied in relation to the segregation of trophectoderm and ICM cell lineages during blastocyst formation. The lineage-specific pattern of mRNA expression of tight junction and desmosome constituents has been studied in different ways. First, immunosurgery has been used to isolate ICMs followed by RT–PCR as discussed above (see *Protocols 2–4*) to compare ICM and whole blastocyst profiles for specific mRNAs (8, 9, 20). The protocol for immunosurgery is described below (*Protocol 8*) and confocal microscope images of isolated ICMs shown in *Figure 1*. Secondly, *in situ* hybridization of whole-mount embryos has been used in conjunction with confocal microscopy for comparing the distribution of zonula occludens-1 (ZO-1) isoforms between trophectoderm and ICM (8). This method is provided in *Protocol 9*. Thirdly, fluorescent labelling of trophectoderm prior to blastocyst disaggregation and separation of trophectoderm (fluorescent-positive) and ICM (fluorescent-negative) cells was used prior to single-cell RT–PCR analysis for lineage-specific comparison of DSC-2 mRNA expression (20).

Protocol 8

Immunosurgery of mouse blastocysts

Equipment and reagents

- Plastic Petri dishes
- Pulled Pasteur pipettes for embryo handling
- Pulled and polished micropipettes for ICM isolation (see *Protocol 1*)
- Cavity blocks
- Mineral oil (Sigma; M8410)
- Acid Tyrode's (see *Protocol 1*)
- H6+BSA (see *Protocol 1*)
- H6+PVP (see *Protocol 6*)

- Trinitrobenzene sulphonic acid (TNBS; 1:10 dilution in PBS with 0.1% PVP, adjusted to pH 7.4 with NaOH; store in a dark bottle at 4°C)
- Goat anti-dinitrophenyl-BSA antibody solution (ICN Biochemicals; 3 mg ml^{-1} in PBS with 0.1% PVP)
- Guinea pig complement (Cedarlane; reconstituted in 1 ml of ice-cold water and diluted 1:10 in cold H6+BSA, stored in aliquots at −70°C)

Method

1. Remove the zona pellucida of blastocysts using acid Tyrode's in cavity blocks as described in *Protocol 1* and transfer immediately to H6+BSA to recover for 20 min at 37°C.

2. Place the embryos in a large drop (300 μl) of TNBS solution under oil for 10 min at room temperature.

3. Transfer the blastocysts to three sequential washes in H6+PVP in 100 μl droplets under oil until TNBS coloration is no longer visible.

4. Incubate the blastocysts in a 25 μl drop under oil of goat anti-dinitrophenyl-BSA solution for 10 min at room temperature.[a]

5. Transfer the embryos to three sequential washes in H6+PVP in 100 μl drops under oil.

6. Incubate the embryos in a 25 μl drop under oil of guinea pig complement for 10 min at 37°C.[b]

7. Place the embryos in 50 μl drops of H6+BSA under oil and incubate for 10–20 min at 37°C.

Protocol 8 continued

8   Remove the lysed trophectoderm cells mechanically by repeated pipetting through a pulled and polished micropipette with a bore diameter just greater than the ICM and place the isolated ICMs in a fresh drop of H6+BSA before further use.

[a] An alternative protocol for immunosurgery would be to omit TNBS and goat anti-dinitrophenyl-BSA treatments (steps 2–4) and replace with rabbit anti-mouse spleen antibody (diluted 1:10 in H6+BSA and stored at −20°C) for 10 min at room temperature. If isolated ICMs are to be used subsequently in an immunolocalization protocol, the host species of antibodies used in immunosurgery should be compatible to ensure no cross-reactivity.

[b] The guinea pig complement loses its activity very quickly when thawed, therefore prepare fresh just before use.

Protocol 9

Whole-mount fluorescence *in situ* hybridization of pre-implantation embryos

Equipment and reagents[a]

- Digoxigenin RNA labelling kit (Roche DIG RNA, cat. no. 1 175 025, following the manufacturer's protocol)
- Cavity blocks
- Acid Tyrode's (see *Protocol 1*)
- PBS
- Formaldehyde (3.7% in PBS)
- PBS-Tween: 0.1% Tween-20 in PBS
- Fixative in PBS-Tween: 3.7% formaldehyde, 0.2% glutaraldehyde in PBS-Tween
- Embryo holders: cut out the middle region of a yellow pipette tip to form a small tapered funnel and cover the narrow end with a piece of polycarbonate filter (pore size 2.0 or 5.0 μm) held in place by a small rubber ring, cut from tubing. Autoclave before use.
- Methanol
- PBS-Tween/methanol at 3:1, 1:1, and 1:3 ratios
- Proteinase K (PK; 10 mg ml^{-1} in PBS-Tween)
- Glycine solution (2 mg ml^{-1} in PBS-Tween; freshly prepared from stock at 10 mg ml^{-1} stored at −20°C.

- 20× standard saline citrate (SSC) stock (1× SSC: 150 mM NaCl, 15 mM sodium citrate), pH 5.0, adjusted with citric acid
- 50× Denhardt's reagent (1× Denhardt's: 0.02% Ficoll, 0.02% PVP, 10 mg ml^{-1} RNase-free BSA)
- Total yeast RNA stock (160 mg ml^{-1} total yeast RNA (Boehringer Mannheim) in water, extracted twice with RNase-free phenol/chloroform, precipitated with ethanol, and then dissolved in water)
- Heparin (Sigma cat. no. H9399; 50 mg ml^{-1} stock in water)
- Hybridization buffer (50% formamide, 5× SSC, 1× Denhardt's, 1% sodium dodecyl sulphate (SDS), 160 μg ml^{-1} total yeast RNA (10 μl of stock per 12 ml of hybridization buffer), 100 mg ml^{-1} heparin, 2% Roche blocking reagent cat. no. 1 096 176). Sterilize by filtering (0.22 μm) and store at −20°C
- Solution 1 (50% formamide, 5× SSC, 1% SDS)
- Solution 2 (50% formamide, 2× SSC, 1% SDS)

Protocol 9 continued

- Solution 3 (50% formamide, 2× SSC)
- 10× Tris-buffered saline (TBS; 1.4 M NaCl, 27 mM KCl, 0.25 M Tris–HCl, pH 7.6)
- TBST (0.1% Tween-20 in 1× TBS)
- Levamisole stock solution, 100 mM in water, stored at −20°C
- Antibody solution (anti-DIG-alkaline phosphatase Fab fragment, Roche, cat. no. 1 093 274, diluted 1:2000, 1% sheep serum,[b] 0.1% Roche blocking reagent, in TBST)
- Blocking solution (15% sheep serum,[b] 2 mM levamisole, 1% Roche blocking reagent, in TBST)
- Levamisole/TBST (2 mM levamisole in TBST, freshly prepared)
- 2-Hydroxy-3-naphthoic acid-2-phenylanilide phosphate (HNPP) fluorescence detection set (Roche, cat. no. 1 758 888; see manufacturer's protocol for preparation)

A. Preparation of digoxigenin-labelled RNA probes

1 Clone the required cDNA insert into an RNA expression vector according to standard procedures.[c]

2 Linearize the RNA expression vector with the appropriate restriction enzyme to allow *in vitro* synthesis of both sense and antisense RNA probes.[d]

3 Purify the linearized plasmid with phenol/chloroform extraction and ethanol precipitation.

4 Generate digoxigenin-labelled RNA probes in both sense and antisense directions by *in vitro* transcription with the DIG RNA labelling kit following the manufacturer's protocol.

B. Fixation and pre-hybridization treatment of embryos

1 Remove the zona pellucida of embryos using acid Tyrode's as described in *Protocol 1*, step B2.

2 Wash the embryos in ice-cold PBS.

3 Fix embryos in 3.7% formaldehyde in PBS for 20 min at room temperature using a cavity block.

4 Wash the embryos three times for 2 min each in ice-cold PBS-Tween.

5 Place the embryos into the holders in PBS-Tween and, on ice, sequentially dehydrate the embryos in PBS-Tween/methanol at 3:1, 1:1, and 1:3 ratios, 10 min each, followed by two washes in 100% methanol for 10 min each.

6 Store the embryos in methanol at −20°C if required.

7 Rehydrate the embryos in PBS-Tween/methanol on ice in reverse order to that in step 2.

8 Treat with PK solution for 20–30 min at room temperature.

9 Stop the reaction by washing twice in glycine solution, 5 min each at room temperature.

10 Wash three times in PBS-Tween, 5 min each on ice.

11 Re-fix with freshly made fixative in PBS/Tween for 10 min on ice.

Protocol 9 continued

12 Wash four times in PBS-Tween, 5 min each, on ice.

13 Wash in pre-warmed hybridization buffer twice, 10 min each, at 65 °C.

14 Replace with a further change of pre-warmed hybridization buffer for storage of embryos, if necessary, at −20 °C.

15 Replace with hybridization buffer at 70 °C and pre-hybridize for 2 h at 70 °C and then for 2 h at 64 °C. The embryos can be stored after this step at −20 °C.

C. Hybridization and immunochemical processing of embryos

1 Add RNA probes to hybridization buffer at 0.5–1 µg ml⁻¹, pre-warm buffer to 64 °C, and incubate embryos in this mix overnight at 64 °C.

2 Wash embryos at 64 °C in hybridization buffer for 20 min.

3 Wash embryos at 64 °C in solution 1, three times, 20 min each.

4 Wash embryos at 64 °C in solution 2, twice for 15 min each, and then twice for 20 min each.

5 Wash embryos at 64 °C in solution 3, three times, 20 min each.

6 Wash the embryos at room temperature three times in TBST, 10 min each.

7 Heat embryos at 70 °C for 20 min to inhibit endogenous alkaline phosphatases.

8 Open the holders in TBST, remove the embryos, and place them in baked cavity blocks in TBST.

9 Incubate embryos in blocking solution in the cavity blocks for 40–60 min at room temperature.

10 Transfer embryos into antibody solution and incubate overnight at 4 °C.

11 Wash embryos with levamisole/TBST three times at 1 min each, three times at 30–40 min each, and overnight at 4 °C.

12 Stain embryos with the HNPP fluorescence set.

13 Transfer embryos to immunolocalization chambers (see *Protocol 5*) and analyse using confocal microscopy.

[a] Treat water for all solutions with 1% diethylpyrocarbonate (DEPC) to avoid RNase contamination, wear gloves at all stages, and bake glassware at 200 °C for 5–6 h before use.

[b] Sheep serum is heat-inactivated for 30–45 min at 56–60 °C before use.

[c] See ref. 28.

[d] An alternative to linearized plasmids would be PCR-generated templates containing an RNA polymerase promoter sequence for *in vitro* transcription (see ref. 29).

Cell–cell interactions in the early embryo mediate differential rates of proliferation between trophectoderm and ICM lineages (13). These changes can be brought about by lineage-specific differences in the activity of growth factor ligand–receptor systems and/or rates of apoptosis (30–34). The cell number present within trophectoderm and ICM lineages of the blastocyst is determined ideally by differential nuclear labelling using the technique first reported by Handyside and Hunter (35). Our method based on this report is given in *Protocol*

10. The morphological detection of apoptosis in blastocysts by TUNEL labelling, based on the method by Brison and Schultz (34), is given in *Protocol 11*. Finally, it is now recognized that gender differences may also explain asynchrony in rates of early development, with male embryos proliferating at a faster rate that female embryos in a number of species (11). Thus, cell interactions between blastomeres combine with endogenous sex differences to determine the rate of early development. We also provide our method (*Protocol 12*) for sexing pre-implantation embryos which is suitable for combining with RT–PCR to relate gene expression profile with gender.

Protocol 10

Differential nuclear labelling of mouse or rat blastocysts[a]

Equipment and reagents

- H6+BSA (see *Protocol 1*)
- Acid Tyrode's (see *Protocol 1*)
- Rabbit anti-rat or mouse spleen cell antiserum[b] (RASA)
- Guinea pig complement (VH Bio)[c]
- Antibody/complement/PI solution: RASA-diluted 1:10, complement diluted 1:10, propidium iodide (0.5 mg ml^{-1}, Sigma), in H6+BSA)

- H6+PVP (see *Protocol 6*) with 0.1 mg ml^{-1} propidium iodide
- Ethanol/Hoechst: 25 μg ml^{-1} Hoechst 33258 dye (Sigma) in absolute ethanol
- Plastic culture dishes
- Mineral oil (Sigma, cat. no. M8410)
- Glycerol
- Microscope slides and coverslips

Method

1   Remove the zona pellucida from blastocysts by brief incubation in warmed acid Tyrode's as described in *Protocol 1*, step B2, and return to H6+BSA to recover for 10–15 min at 37 °C.

2   Incubate embryos in a 20 μl drop of antibody/complement/PI solution under oil at 37 °C for 30 min to cause partial immunosurgery and labelling of trophectoderm nuclei.

3   Wash embryos in H6+PVP containing propidium iodide.

4   Complete differential labelling by transferring embryos to ethanol/Hoechst for a minimum of 1.5 h at 4 °C or for up to 2 weeks for storage.[d]

5   Mount embryos on a slide in a drop of glycerol and spread cells under a coverslip. Count nuclei by fluorescence microscopy under UV excitation.[e]

[a] Adapted from ref. 35.

[b] See ref. 36 for description of preparation.

[c] Guinea pig complement becomes inactive very rapidly even at 4 °C, hence, once reconstituted in water, it should be aliquoted and stored at −70 °C. It is also important to avoid freeze and thaw repeatedly as this can lead to reduction in potency.

[d] Storage can also be in absolute ethanol after ethanol/Hoechst staining.

[e] Nuclei of propidium iodide-labelled trophectoderm cells will appear pink or red, whereas nuclei of Hoechst-labelled ICM cells will appear blue.

Protocol 11

Detection of apoptotic cells in blastocysts by terminal deoxynucleotidyl transferase (TdT)-mediated dUTP nick end labelling (TUNEL)[a]

Equipment and reagents

- Bench-top centrifuge
- PBS+PVP (3 mg ml^{-1} PVP in PBS)
- Paraformaldehyde (Agar Scientific Ltd, cat. no. R1018; 3.7% in PBS)
- Immunolocalization chambers, coated with PLL (see *Protocols* 5 and 6A)
- TUNEL reagents (Boehringer Mannheim, cat. no. 1684795)
- TUNEL labelling mix (5 μl TdT + 45 μl dUTP-FITC)
- DNase (Promega, cat. no. M6101; 50 mU ml^{-1})

- Triton/PBS (0.5% Triton X-100 in PBS)
- RNase A (Boehringer Mannheim, cat. no. 109142)
- RNase A solution (50 μg ml^{-1} RNase A in 40 mM Tris–HCl, pH 8, 10 mM NaCl, 6 mM MgCl$_2$)
- PI/RNase A solution (50 μg ml^{-1} propidium iodide in RNase A solution)
- Citifluor/PBS (Citifluor Ltd, cat. no. AF3)
- Immunolabelling chambers (see *Protocol* 5)
- Nail varnish

Method

1 Wash embryos four times in PBS+PVP.

2 Fix in paraformaldehyde for 1 h at room temperature.

3 Wash four times in PBS+PVP.

4 Transfer embryos to PLL-coated chambers and spin at 1000 g for 10 min.

5 Permeabilize embryos with Triton/PBS for 1 h at room temperature.

6 Wash twice in PBS+PVP.

7 Treat the positive control embryo chamber with DNase solution at 37°C for 20 min.

8 Incubate embryos in TUNEL labelling mix (~40 μl per chamber) at 37°C for 1 h in the dark.

9 Wash twice in Triton/PBS and once in PBS+PVP.

10 Wash embryos three times in RNase A solution and incubate in this solution at 37°C for 20 min in the dark.[b]

11 Remove RNase A solution and incubate embryos in PI/RNase A solution for 40 min at 37°C in the dark.

12 Wash twice in Triton/PBS and once in PBS+PVP.

13 Replace PBS+PVP in the chamber with Citifluor/PBS, place the top coverslip on the chamber and seal with nail varnish (see *Protocol* 7), wrap in foil and store at 4°C.

14 Analyse samples by fluorescence microscopy within 1 week, but they can be readable for 3–4 weeks.

[a] Adapted from ref. 34.

[b] It is important to treat embryos with RNase A before labelling with PI/RNase A since this fluorochrome can bind to RNA and make it difficult to count the number of nuclei.

Protocol 12

Determination of pre-implantation embryo gender by PCR[a]

Equipment and reagents

- Shaking water bath
- Bench-top centrifuge
- Rotating wheel
- Thermocycler (Hybaid Thermal Reactor)
- Gel electrophoresis equipment
- Transilluminator
- Rat liver
- Lysis buffer (50 mM Tris–HCl, pH 8, 100 mM EDTA, 100 mM NaCl, 1% SDS in water)
- PK solution (Sigma, cat. no. P2308; 1 mg ml^{-1} in water)
- PK/lysis buffer mix (35 μl of PK + 550 μl of lysis buffer, use immediately)
- Saturated NaCl solution
- Absolute ethanol[b]
- 70% ethanol
- Microfuge tubes (autoclaved before use)
- TE buffer (10 mM Tris, 1 mM EDTA, pH 8; prepare in DNase-free water, Anachem, cat. no. E476)
- RNase A (Boehringer Mannheim, cat. no. 109142)

- RNase A solution (0.2 mg ml^{-1} in TE buffer)
- *Taq* DNA polymerase (Gibco-BRL, cat. no. 18038-026)
- DNA ladder 100 bp (Gibco-BRL, cat no. 15628-019)
- Mineral oil (Sigma, cat. no. M5904)
- Agarose (Sigma, cat. no. A9539)
- dNTPs (Amersham Pharmacia Biotech., cat. no. US77100)
- Thermotubes (0.5 ml, Advanced Biotechnologies, cat. no. AB0533)
- Oligonucleotide primers for gender-specifying genes (see *Table 1*)
- Master mix (20 mM Tris–HCl, pH 8.3, 50 mM KCl, 1.5 mM MgCl$_2$, 200 μM each dNTPs, 40 pmol each outer primer (see Table 1), prepared in thermotubes
- Ethidium bromide solution (1 μg ml^{-1} aqueous)
- Tris-borate buffer (TBE buffer; μ10 from Sigma, cat. no. T4415 comprising 0.89 M Tris-borate, pH ~8.5, 0.02 M EDTA)

A. Extraction of rat genomic DNA by salting-out procedure[c]

1 Homogenize 0.2 g of rat liver in PK/lysis buffer mix to hydrolyse cellular and nucleosome proteins.

2 Incubate at 50°C overnight in a shaking water bath.

3 Centrifuge at 10 000 g for 10 min.

4 Add 560 μl of saturated NaCl solution to 400 μl of supernatant and shake vigorously to precipitate protein, the solution turning cloudy.

5 Centrifuge at 10 000 g for 10 min.

6 Add 750 μl of supernatant to 750 μl of absolute ethanol, inverting the tube several times, to precipitate DNA.[b]

7 Transfer the DNA to a microfuge tube containing 1 ml of 70% ethanol, and rinse briefly.

8 Transfer the DNA to another microfuge tube containing 1 ml of 70% ethanol, and place on a rotating wheel for several hours at room temperature to wash.

Protocol 12 continued

9 Repeat step 8 twice more.

10 Remove the ethanol and air dry the DNA for 10 min at room temperature.

11 Add 20 μl of TE buffer to the DNA and incubate at 50°C for 10 min to dissolve.

12 Add 20 μl of RNase A solution and measure the absorbance at 230–330 nm.[d]

B. Sexing rat blastocysts by PCR

1 Add 47.5 μl of master mix to 2 μl of genomic rat DNA (0.1 ng, see A above) or a single embryo (2 μl in PBS, see Protocol 2).

2 Overlay the samples with 50 μl of mineral oil.

3 Heat at 94°C for 8 min, cool to 60°C for 2 min, and repeat this cycle twice more before adding 2.5 U of *Taq* DNA polymerase.[e]

4 Carry out 30 PCR cycles of amplification with a denaturing step at 94°C for 1 min, an annealing step at 60°C for 2.5 min, and an extension step at 72°C for 2.5 min. After the last cycle, keep the samples at 72°C for a further 10 min.

5 Transfer 2 μl of the first-stage product into two thermotubes each containing 47.5 μl of master mix either with 40 pmol each of HPRT and ZFY inner primers or with SRY inner primers (see *Table 1*). Add 2.5 U of *Taq* DNA polymerase and overlay with 50 μl of mineral oil.

6 Perform PCR using the same condition as described in step B4 above.

7 Electrophorese 15 μl of the second-stage product in a 1.8% agarose gel (prepared in 0.5× TBE buffer) for 1 h at 120 V with a 100-bp DNA ladder as marker.

8 Stain the gel with ethidium bromide solution for 3–5 min and destain in 0.5× TBE buffer for 1 h.

9 Visualize the stained DNA bands under UV illumination.

[a] It is important to perform all steps carefully to avoid contamination. This includes the wearing of gloves at all times, the use of RNase, DNase-free water and pipette tips, treatment of tubes with UV cross-linker before use (see *Protocol 2*), and the preparation and dispensing of master mix in a room separate from DNA amplification and separation procedures. In addition, negative controls such as master mix without DNA template and PBS used to collect embryos should be included in each PCR run.

[b] It is important to ensure that the ethanol is at room temperature since ice-cold ethanol can also lead to RNA precipitation.

[c] Adapted from ref. 37. Before sex determination of the embryos by PCR, it is necessary to establish the PCR conditions that are suitable for primers, hence the need to extract male and female genomic DNA from liver.

[d] This extraction method gives an A_{260}/A_{280} ratio of 1.8–2, indicative of good deproteinization. By using 0.2 g of liver, 30–50 μg of DNA can be obtained and can be stored for extended period at 4°C in this buffer.

[e] This heating–cooling cycle can enhance the efficiency of the amplification; see ref. 37.

Table 1 Primers used for sexing rat blastocysts

Gene product	Primer	Sequence	Product size (bp)
Outer primers			
HPRT	5′	5′-GTTCTCTTCAATTGCTGGTCCA-3′	618
	3′	5′-TGACAACGATTCACACTGCTGA-3′	
ZFY[5]	5′	5′-AAGATAAGCTTGCATAATCACATGGA-3′	611
	3′	5′-CCTATGAAACCCTTTGCTGCACATGT-3′	
SRY	5′	5′-CACAAGTTGGCTCAACAGAATC-3′	300
	3′	5′-AGCTCTACTCCAGTCTTGTCCG-3′	
Inner primers			
HPRT	5′	5′-ATGCTGGTGTTGTCTCTTCAGA-3′	318
	3′	5′-ATCTGTCTGTCTCACAAGGGAA-3′	
ZFY[5]	5′	5′-GGAAGCATCTTTCTCATGCTGG-3′	207
	3′	5′-TTTTGAGCTCTGATGGGTGACGG-3′	
SRY	5′	5′-AGCATGCAGAATTCAGAGAT-3′	248
	3′	5′-ATAGTGTGTAGGTTGTTGTCC-3′	

Reference 38 also has the primer sequences for sexing mouse pre-implantation embryos.

Acknowledgements

We are grateful to the Medical Research Council, The Wellcome Trust, European Union, Wessex Medical Trust, and University of Southampton for financial support of our research projects.

References

1. Fleming, T. P., Papenbrock, T., Fesenko, I., Hausen, P., and Sheth, B. (2000). *Semin. Cell Dev. Biol.*, **11**, 291.

2. Collins, J. E. and Fleming, T. P. (1995). *Trends Biochem. Sci.*, **20**, 307.

3. MacPhee, D. J., Jones, D. H., Barr, K. J., Betts, D. H., Watson, A. J., and Kidder, G. M. (2000). *Dev. Biol.*, **222**, 486.

4. Jones, D. H., Davies, T. C., and Kidder, G. M. (1997). *J. Cell Biol.*, **139**, 1545.

5. Yeaman, C., Grindstaff, K. K., and Nelson, W. J. (1999). *Physiol. Rev.*, **79**, 73–98.

6. Fleming, T. P. and Hay, M. J. (1991). *Development*, **113**, 295.

7. Fleming, T. P., Hay, M., Javed, Q., and Citi, S. (1993). *Development*, **117**, 1135.

8. Sheth, B., Fesenko, I., Collins, J. E., Moran, B., Wild, A. E., Anderson, J. M., *et al.* (1997). *Development*, **124**, 2027.

9. Sheth, B., Moran, B., Anderson, J. M., and Fleming, T. P. (2000). *Development*, **127**, 831.

10. Sheth, B., Fontaine, J.-J., Ponza, E., McCallum, A., Page, A., Citi, S., *et al.* (2000). *Mech. Dev.*, **97**, 107.

11. Erickson, R. P. (1997). *BioEssays*, **19**, 1027.

12. Kwong, W. Y., Wild, A. E., Roberts, P., Willis, A. C., and Fleming, T. P. (2000). *Development*, **127**, 4195.

13. Chisholm, J. C., Johnson, M H., Warren, P. D., Fleming, T. P., and Pickering, S. J. (1985). *J. Embryol. Exp. Morphol.*, **86**, 311.

14. Fleming, T. P. (1987). *Dev. Biol.*, **119**, 520.

15. Johnson, M. H. and Ziomek, C. A. (1981). *J. Cell Biol.*, **91**, 303.

16. Johnson, M. H. and Ziomek, C. A. (1983). *Dev. Biol.*, **95**, 211.

17. Quinn, P., Barros, C., and Whittingham, D. G. (1982). *J. Reprod. Fertil.* **66**, 161.

18. Garbutt, C. L., Chisholm, J. C., and Johnson, M. H. (1987). *Development*, **100**, 125.

19. Pickering, S. J., Maro, B., Johnson, M. H., and Skepper, J. N. (1988). *Development*, **103**, 353.

20. Collins, J. E., Lorimer, J. E., Garrod, D. R., Pidsley, S. C., Buxton, R. S., and Fleming, T. P. (1995). *Development*, **121**, 743.

21. Collins, J. E. and Fleming, T. P. (1995). *Trends Genet.*, **11**, 5.

22. Wrenzycki, C., Herrmann, D., Carnwath, J. W., and Niemann, H. (1998). *J. Reprod. Fertil.*, **112**, 387.

23. Eckert, J. and Niemann, H. (1998). *Mol. Hum. Reprod.*, **4**, 957.

24. Maro, B., Johnson, M. H., Pickering, S. J., and Flach, G. (1984). *J. Embryol. Exp. Morphol.*, **81**, 211.

25. Javed, Q., Fleming, T. P., Hay, M., and Citi, S. (1993). *Development*, **117**, 1145.

26. McElhinny, A. S. and Warner, C. M. (1997). *Biotechniques*, **23**, 660.

27. McElhinny, A. S., Kadow, N. and Warner, C. M. (1998). *Mol. Hum. Reprod.*, **4**, 966.

28. Sambrook, J., Fritsch, E., and Maniatis, T. (1989). *Molecular cloning: a laboratory manual*. Cold Spring Harbor Laboratory Press, Cold Spring Harbor, NY.

29. Young, I. D., Ailles, L., Deugau, K., and Kisilevsky, R. (1991). *Lab. Invest.*, **64**, 709.

30. Harvey, M. B. and Kaye, P. L. (1990). *Development*, **110**, 963.

31. Harvey, M. B. and Kaye, P. L. (1992). *Mol. Reprod. Dev.*, **31**, 195.

32. De Hertogh, R., Vanderheyden, I., Pampfer, S., Robin, D., and Delcourt, J. *Diabetologia*, **35**, 406.

33. Lea, R. G., McCracken, J. E., McIntyre, S. S., Smith, W., and Baird, J. D. (1996). *Diabetes*, **45**, 1463.

34. Brison, D. R. and Schultz, R. M. (1997). *Biol. Reprod.*, **56**, 1088.

35. Handyside, A. H. and Hunter, S. (1984). *J. Exp. Zool.*, **31**, 429–434.

36. Solter, D. and Knowles, B. B. (1975). *Proc. Natl Acad. Sci. USA*, **72**, 5099.

37. Mercier, B., Gaucher, C., Geugeas, O., and Mazurier, C. (1990). *Nucleic Acids Res.*, **18**, 5908.

38. Kunieda, T., Xian, M., Kobayashi, E., Imamichi, T., Moriwaki, K., and Toyoda, Y. (1992). *Biol. Reprod.*, **46**, 692.

Chapter 10

Model systems for the investigation of implantation

John D. Aplin, Helen Lacey, Chie-Pein Chen, Carolyn J. P. Jones

School of Medicine, University of Manchester, Research Floor, St Mary's Hospital, Manchester M13 0JH, UK

Susan J. Kimber, Melanie J. Blissett

School of Biological Sciences, 3.239 Stopford Building, University of Manchester, Oxford Road, Manchester M13 9PT, UK

Carlos Simón

Instituto Valenciano de Infertilidad, University of Valencia, Spain

1 Embryo–endometrial interactions

1.1 Co-culture of human embryos with human endometrial epithelial cells as a model for implantation studies

1.1.1 Introduction

Embryo implantation is fundamental to reproduction, yet we know little of its cellular and molecular mechanisms, especially in humans (1). The initial interaction of the human embryo is with luminal endometrial epithelial cells at the lining of the uterine cavity. These maternal cells appear to switch from a non-receptive to a receptive state about 6 days after ovulation (2). Following attachment of trophectoderm to the epithelial apical surface, intrusion is thought to occur, with transient adhesion to the lateral epithelial borders. This is followed by direct contact with, and then penetration of, the basement membrane. Within 1–2 days, the embryo becomes embedded fully in the stroma (3). Given the obvious ethical and practical difficulties in carrying out investigations *in vivo*, progress will depend on advances in methodology for investigation of cellular interactions *in vitro*.

1.1.2 Endometrial epithelial monolayer cultures

Endometrial cells are isolated as described (4–6).

Protocol 1

Isolation and culture of human non-polarized and polarized endometrial epithelial cells (EECs)

Equipment and reagents

- Collagenase type IA (Sigma): the digestion is performed with 0.1% collagenase in Dulbecco's modified Eagle's medium (DMEM). Add 10 ml of embryo transfer water (Sigma) to 100 mg of collagenase. Stock concentration is 10 mg ml^{-1}. Store 1 ml aliquots at $-20\,°C$. Digestion volume is 10 ml, therefore add 1 ml of collagenase stock to 9 ml of DMEM to obtain 0.1% collagenase in DMEM

- DMEM (Sigma)

- MCDB-105 (Sigma): reconstitute with embryo transfer water (Sigma). Measure out 90% of the final required volume of water (culture tested) at 15–20 °C. Add the powdered medium. Stir until dissolved but do not heat. Medium is yellow coloured. Rinse original package with a small amount of water to remove all traces of powder. Add to solution with additional water to bring the solution to final volume. Adjust the pH of the medium. pH at room temperature must be 5.1 $\pm$ 0.3. Add 1 M NaOH until pH reaches 7.2 because at 37 °C pH increases 0.1–0.3. Final pH must be 7.4. Sterilize immediately by filtration using a membrane with 0.22 μm pore (Millipore Ibérica, Spain). Store at 4 °C in the dark

- Fungizone (Gibco): the vial contains 250 μg ml^{-1} of amphotericin B. Reconstitute with 20 ml of embryo transfer water. Recommended concentration is between 0.25 and 2.5 μg ml^{-1}. Our working dilution is 0.5 μg ml^{-1}

- Gentamycin (Gibco): the activity is unaffected by the presence of serum. In cell culture, it is normally used at 50 μg ml^{-1}. Our working concentration is 100 μg ml^{-1}

- Insulin (Sigma): stable at 2–8 °C for 1 year. Soluble insulin is available in acidified water. For a vial of 100 mg, to prepare a 10 mg ml^{-1} stock solution, add 10 ml of acidified water (pH $<$2.0) prepared by the addition of 100 μl of glacial acetic acid. Working concentration is 5 μg ml^{-1}

- Fetal bovine serum (FBS; Gibco) is available as a heat-inactivated form, if not it should be heat inactivated by heating the sera at 56 °C for 30 min. Appropriate aliquots should be prepared using sterile containers and then stored at $-20\,°C$

- Endometrial epithelial culture medium: the medium is composed of three parts of DMEM to one of MCDB-105 supplemented with 10% FBS (heat inactivated), gentamycin (100 μg ml^{-1}), fungizone (0.5 μg ml^{-1}), insulin (5 μg ml^{-1}). It is sterilized by passage throughout a 0.22 μm filter (Millipore) and stored at 4 °C

- Extracellular matrix (ECM, Sigma)

- 30 μm nylon sieves (Spectra/mesh, Spectrum Microgon)

- DNase (Sigma)

- Water bath

- Culture plate inserts (Millipore)

- 3 μm nylon sieves

A. Cell isolation

Endometrial biopsies are obtained in the luteal phase with a plastic probe (Gynetics, Amsterdam).

1 Mince tissue into small pieces of less than 1 mm. Put into a conical tube with 10 ml of 0.1% collagenase type IA and subject to mild collagenase digestion with agitation in a 37 °C water bath for 1 h.

Protocol 1 continued

2   Stand the tube in a vertical position for 10 min in a horizontal laminar flow. Remove the supernatant (with stromal cells) and wash by resuspending the pellet (glandular and luminal epithelial cells) in 3–5 ml of DMEM three times for 5 min each.

3   Finally, resuspend the pellet in 4–5 ml of 1% FBS in DMEM. Recover all and put into a 25-ml culture flask (Falcon, Beckton Dickinson).

4   Incubate the flask for 15 min. Then recover the supernatant and put it into a fresh flask and add 3 ml of 1% FBS in DMEM.

5   Return this second flask to the incubator for 15 min. Recover all supernatant composed mainly of EECs since stromal cells attach easily to the plastic.

B. Non-polarized EEC cultures

1   Place 800 µl of culture medium with 200 µl of recovered supernatant with EECs into culture wells.

2   Culture EECs for approximately 4–6 days until confluent (*Figure 1*).[a]

C. Polarized EEC cultures with stromal cells

1   The day before endometrial biopsy is expected, coat 4 µm culture plate inserts with ECM at 1/4 dilution, and sterilize overnight under a UV light.

2   Mince endometrial samples into pieces smaller than 1 mm and digest with collagenase at 37 °C for 1 h.

3   After sedimentation of EECs, filter the supernatant through 30 µm nylon sieves and wash three times in DMEM, filtering the supernatant each time to remove mainly stromal and blood cells.

4   Centrifuge this filtrate and treat the pellet with DNase for 1 min, and then stop the reaction with 1.5 ml of culture medium. Count the stromal cells and plate out at 200 000 cells per well in a 24-well dish. Incubate the EECs obtained in the first pellet twice for 20 min each with 1% FBS as described above, and then plate into inserts coated with ECM. After the stromal cells have reached confluence (1–2 days), place the inserts containing EECs in the wells (*Figure 2*).

[a] The homogeneity and purity of cultures may be determined by morphological characteristics and verified by immunohistochemical localization of cytokeratin, vimentin, and CD68 (4–6). Under these culture conditions, EEC monolayers contain less than 3% of endometrial stromal cells (ESCs) and less than 0.1% of macrophages. Functionality of EEC monolayers may be demonstrated by the production of prostaglandin E2 in response to interleukin-1 (4) and scanning electron microscopy (SEM) features displayed in the presence of a human blastocyst (6, 7).

1.1.3 Isolation and handling of human embryos

The ovarian stimulation protocol using gonadotrophin-releasing hormone analogues and gonadotrophins, and the *in vitro* fertilization (IVF) protocols have been described (8).

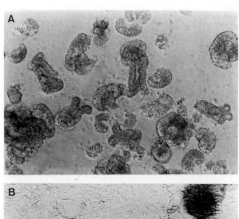

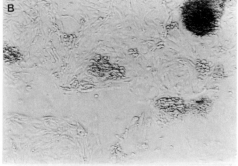

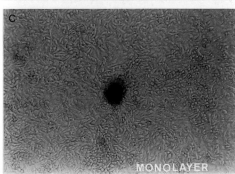

Figure 1 Non-polarized endometrial epithelial cell (EEC) culture. (**A**) Day 0: pieces of glandular epithelial cells were plated. (**B**) Day 2: EECs start to grow and proliferate. (**C**) Day 4–6: a confluent monolayer is formed.

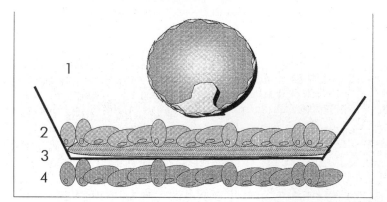

Figure 2 Diagram of the bidimensional model of polarized EECs with stromal cells:1 = blastocyst; 2 = endometrial epithelial cells; 3 = extracellular matrix; 4= endometrial stromal cells.

1.1.3.1 *Development* in vitro *to the blastocyst stage*

(i) Clinical

A long protocol is used for pituitary desensitization with administration of leuprolide acetate (LA), 1 mg per day subcutaneously (Procrin, Abbot S.A.), starting in the luteal phase of the previous cycle. Serum oestradiol (E2) levels less than 60 pg ml^{-1} (conversion factor to SI unit, 3.671) and a negative vaginal sonographic scan are used to define ovarian quiescence. Recombinant follicle-stimulating hormone (FSH; Gonal-F, Serono Laboratories; Puregon, Organon) is administered for ovarian stimulation, and routine criteria used for human chorionic gonadotrophin (hCG) administration (10 000 IU, Profasi, Serono Laboratory). Oocyte retrieval is performed 36–38 h after hCG administration.

(ii) Laboratory stages

In the IVF laboratory, embryo development from oocyte to blastocyst occurs as follows:

Day 0: oocyte retrieval day.

Day 1: fertilization day. Analyse the polar body and pro-nucleus number. Only fertilized eggs showing two polar bodies and two pro-nuclei can be considered to be fertilized correctly.

Day 2: first cleavage. Evaluate the fertilized embryos for blastomere number and fragmentation rate. The embryos generally have 2–4 cells at this stage.

Day 3: the embryos have approximately eight cells.

Day 4: subsequent divisions form the morula stage, a 16- to 32-cell embryo. Individual blastomeres become indistinct as they become more adherent to each other (*Figure 3*). This phenomenon is called compaction. On day 4, classify the type of morula as uncompacted or compacted.

Day 5: spaces appear between the compacting cells and result in the formation of an outer layer of trophoblast and a group of centrally located cells, the inner cell mass, which gives rise to the embryo. At this stage of development, the embryo is called a blastocyst. Two types of blastocysts principally can be observed on day 5: early blastocysts, in which spaces appear between the blastomeres, and cavitated blastocysts, in which the cavity is more than 50% of the total volume (*Figure 4*).

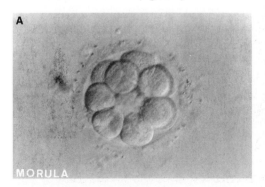 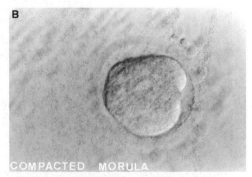

Figure 3 Human morula stage. (**A**) Uncompacted morula and (**B**) compacted morula.

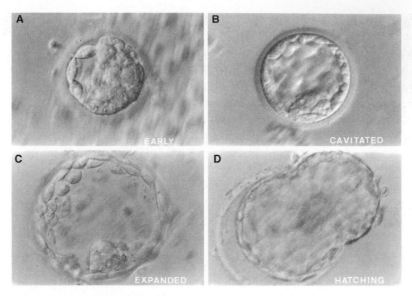

Figure 4 Human blastocyst stages. (**A**) Early blastocyst, (**B**) cavitated blastocyst, (**C**) expanded blastocyst, and (**D**) hatching blastocyst.

Day 6: when the blastocoelic cavity enlarges in size, the embryo is referred to as an expanded blastocyst, and when the embryo begins to hatch out from the zona pellucida, it is called a hatching blastocyst (*Figure 4*). This ideal development does not always occur. Sometimes embryo development is delayed or blocked. Furthermore, there is considerable variability in morphology at the same developmental stage.

1.1.4 Embryo–epithelial interaction

The initial impetus for co-culture was that paracrine interaction between autologous endometrial cells and embryos developing from the cleavage stages might create conditions that were favourable for embryo development to the blastocyst stage, thus increasing the success rates in embryo replacement after IVF. In this case, blastocysts are removed from culture and transferred back to the recipient prior to hatching. In a series of 2119 human embryos co-cultured, at the end of this procedure at day 6, we observed a 60% rate of blastocyst formation (7). Blastocysts (mean 2) were transferred back to the mother, resulting in a 51% pregnancy rate and a 30.7% implantation rate (7). After embryo transfer, EEC monolayers can be divided according to the embryonic status into EECs with embryos that reached the blastocyst stage, EECs with arrested embryos, and EECs without embryos, and investigations can be performed to determine embryonic regulation of EEC characteristics (6, 9). The method has been extended to allow hatching to occur *in vitro* and the hatched blastocysts shown to attach to endometrial monolayers (9).

Protocol 2

Co-culture of human epithelium with embryos

Equipment and reagents

- Endometrial epithelial monolayer (see *Protocol 1*)
- 24-well tissue culture plates (Falcon, Becton Dickinson)
- IVF/CCM medium (1/1; Scandinavian IVF Science)
- Frydman catheter
- CO_2 incubator

Method

1 On day 2, evaluate fertilized embryos for blastomere number and fragmentation.

2 Transfer individual human embryos to an endometrial epithelial monolayer from first cleavage (day 2), when the embryos consist of 2–4 cells. Culture each embryo individually on the epithelial monolayer in 24-multiwell tissue culture plates at 37°C in an atmosphere of 5% CO_2 in air.

3 Grow embryos in 1 ml of IVF/CCM (1/1) until they reach the 8-cell stage.

4 Change the medium every 24 h and evaluate each embryo morphologically.

5 Continue co-culture until transfer day (day 6), by which time the embryos have reached the blastocyst stage.

6 Transfer to the mother with a Frydman catheter.

1.2 Murine uterine epithelial cell cultures

1.2.1 Introduction

The uterus is not readily accessible and it is difficult to examine the physiology and other properties of its cells. One approach to circumvent this problem is to develop model *in vitro* systems for these cells. This has been attempted by a number of groups. In rodents, some of the pioneers have been Glasser and co-workers (10, 11) for the luminal epithelium (LE) and Kennedy and colleagues (12, 13) for stromal cells. *In utero*, the luminal and glandular epithelium form a simple epithelial layer which provides a homeostatic barrier between the connective tissue of the stroma and the uterine and glandular lumen. It also forms the initial interface between the implanting embryo and the uterus. In our laboratory, we use a method for culture of LE on suspended membranes which results in a semi-polarized phenotype that retains many of the features of LE *in vivo*. This technique is modified from one originally developed for use in the rat (10), based on a similar method used for Madin–Darby canine kidney (MDCK) cells. In contrast to culture on plastic, cells grown on suspended membranes retain a polarized phenotype. In this system, differential secretion of molecules into basal and apical media can be measured and the effect of reagents on the apical (equivalent to luminal surface) and basal (equivalent to stromal surface) compartments can be examined.

Protocol 3

Establishment of mouse uterine luminal epithelium (LE) cell cultures

Equipment and reagents[a]

- Immature female MF-1 mice and sexually mature stud MF-1 males
- 50-ml Falcon tubes
- Haemocytometer
- Cellagen™ discs (CD-24 14-mm diameter discs, ICN cat. no. 152316)
- Millicell®-ERS transepithelial resistance meter (Millipore)
- 24-well tissue culture plates (Nunc)
- Humidified gassed incubator
- Rotamixer (Hook and Tucker Instruments)
- Sigma Howe 3K10 centrifuge with swing-out rotor (rotor number 11133)
- Sigma 202MC centrifuge with fixed-angle rotor (rotor number 12043)
- Leitz Fluovert inverted phase microscope
- Pregnant mare's serum gonadotrophin (PMSG; Folligon®, Intervet)
- hCG (Chorulon®, Intervet)
- Ham's F10 medium (Gibco-BRL) supplemented with 4 mg ml^{-1} bovine serum albumin (BSA, crystallized powder, ICN cat no. 81-0012)
- Hank's balanced salts solution without magnesium and calcium (HBSS, Gibco-BRL)
- Trypsin dissociation medium: 20 ml of HBSS containing 0.1 g of trypsin (type III bovine Sigma cat. no. T8253, 0.5% w/v*) and 0.135 ml of pancreatin (10× liquid at 25g l^{-1}, Gibco-BRL cat. no. 45720-018, 0.675% v/v*). Filter sterilize as above and store at 4°C

- Culture medium: 10 ml of Ham's F12 (with sodium bicarbonate, without L-glutamine, Sigma cat. no. N4888) and 10 ml of DMEM (with 4500 g l^{-1} glucose and sodium bicarbonate, without L-glutamine, Sigma cat. no. D5671) containing (* indicates final concentration) 0.02 g of BSA (0.1%*); 0.53 ml of pen/strep (10000 U of penicillin and 10 mg of streptomycin ml^{-1} in normal saline, Gibco-BRL, cat. no. 151-40-031; 100 mg ml^{-1}); 0.55 ml of NuSerum (Collaborative Biomedicals supplied by Becton Dickinson, 2.5%*); 0.55 ml of heat-inactivated fetal calf serum (HIFCS, low in endotoxin and haemoglobin, Sigma cat. no. F2442, 2.5%*); 0.3 ml of N-(2-hydroxyethyl) piperazine-N'-(2-ethanesulphonic acid) (HEPES) buffer (1 M, Sigma cat. no. H0887, 15 mM*); 0.25 ml of L-glutamine (200 mM 100× liquid, Gibco-BRL cat. no. 25030-032; 29.2 mg ml^{-1}). Filter sterilize using a 0.2 mm pore size low protein binding syringe filter (Gelman Sciences) and store overnight at 4°C
- DNase medium: 20 ml of HBSS containing 20 µl of DNase solution made up at 1 µg ml^{-1} in normal saline (type II from bovine pancreas, Sigma cat. no. D4527), 100 µl of 10 mM MgCl$_2$ and 200 µl of HIFCS (0.1%*). Filter sterilize as above and store at 4°C
- Trypsin-ethylenediamine tetraacetic acid (EDTA): 10 ml of HBSS containing 0.025 g of trypsin (0.25% w/v*) and 0.050 g of EDTA (0.5% w/v*); 150 ml aliquots can be stored at −20°C

Method

1 Superovulate 20-day-old female MF-1 mice which have not yet entered the first oestrous cycle by an intraperitoneal injection of 5 IU 100 µl^{-1} PMSG, followed 46–48 h later by an intraperitoneal injection of 5 IU 100 µl^{-1} hCG. Following the hCG injection, cage the females overnight with stud MF-1 males and check for plugs the

Protocol 3 continued

following morning to confirm mating. Pregnancy is indicated by the presence of vaginal plugs. The day on which the plug is observed is designated as day 1 (NB: some others designate this day 0.5).

2   On day 2 of pregnancy, kill the mice by cervical dislocation and cut open the abdominal skin and body wall to reveal the bilateral uterine horns. During dissection, hold the uteri by the attached mesentery using forceps. Cut individual uterine horns just proximal to the cervix and gently trim along the length of the tissue to remove any fat and most mesentery. Care must be taken not to stretch the uteri. Release the oviducts from the ovaries by cutting between the two, taking care not to damage the oviductal tube. Then trim them from the uteri at the level of the utero-tubal junction and place them in a dish of Ham's F10 medium. Place the dissected uterine horns in HBSS.

3   The following steps are a modified version of the technique described by McCormack and Glasser (14) performed using aseptic technique. Using a dissecting microscope, wash the uteri through HBSS and trim away any remaining fat and mesentery. After a second wash through HBSS, cut open the uteri longitudinally and place in trypsin dissociation medium for 1 h at 4°C followed by 1 h at room temperature.

4   Remove the dissociation medium from the uteri, discard, and replace with 5 ml of cold DNase medium before vortexing at medium speed for 10 s on a Rotamixer. Transfer the resulting epithelial cell suspension to a sterile 50-ml Falcon tube and place on ice. Repeat the whole process, pool the supernatants, and chill on ice for approximately 1 min.

5   Centrifuge the epithelial cell suspension at 200 g for 5 min at 4°C in a Sigma Howe 3K10 centrifuge using the 11133 swing-out rotor. Remove the supernatant, discard, and replace with 10 ml of cold DNase medium. Gently resuspend the cell pellet by mixing and leave on ice for approximately 1 min. Repeat the centrifugation step, remove and discard the supernatant, and add 10 ml of cold DNase medium to resuspend the cells. Repeat this procedure for a total of three washes.

6   After the third wash, resuspend the cell pellet in 10 ml of cold HBSS and place the Falcon tube on ice at an angle of approximately 45° for 15 min. This allows uterine epithelial cell plaques to separate under gravity from contaminating stromal cells. The supernatant is carefully removed from the loose cell pellet and discarded. Resuspend the enriched epithelial cell population in 10 ml of fresh cold HBSS and repeat the process for a total of four gravitational separations.

7   After the third separation, remove 100 μl of the 10 ml epithelial cell suspension and use it to perform a cell count. Place the 100 μl in a 0.5-ml microfuge tube and centrifuge at 1000 g for 5 min in the Sigma 202MC centrifuge with fixed-angle rotor. Remove the supernatant, discard and replace with 100 μl of fresh HBSS. Repeat the centrifugation step and remove and discard the supernatant before adding 100 μl of trypsin-EDTA. Resuspend the cells by vigorous vortexing and incubate at 37°C for 10

min. Vortex the cell solution again to obtain a single cell suspension and estimate the cell density using a haemocytometer.

8 After the final gravitational separation, carefully remove the supernatant, discard and resuspend the cells in 1 ml of culture medium, pre-equilibrated in a humidified atmosphere with 5% CO_2 in air at 37 °C. Based on the result of the cell count, adjust the volume of the cell suspension to 8.0×10^5 cells ml^{-1} (2.0×10^5 cells per 250 µl).

9 Place Cellagen™ discs in the centre of eight wells of a 24-well plate[b] and pre-incubate with culture medium by adding 250 µl to the apical compartment of the disc and 450 µl into the culture well. The Cellagen™ discs[c] should be left to equilibrate for approximately 20 min in a humidified atmosphere with 5% CO_2 at 37 °C.

10 After pre-incubation of the discs, remove the medium in the basal compartment and replace with 450 µl of fresh equilibrated culture medium. Remove medium in the apical compartment of the disc and replace with 250 µl of pre-warmed equilibrated uterine epithelial cell suspension. The cells are then incubated in a humidified atmosphere with 5% CO_2 in air at 37 °C.

11 Change the culture medium in both apical and basal compartments every day for the first 3 days of culture and then every 48 h until the end of the experiment. After every medium change, assess the progress of the uterine epithelial cell monolayers by phase microscopy, such as with a Leitz Fluovert inverted phase microscope (*Figure 5*).

[a] Normally solutions should be prepared on the morning of culture, but they have been prepared up to 18 h in advance, with no detrimental effects being observed. Culture media should be prepared fresh every 48 h as required.

[b] Use of only the centre wells avoids edge effects on cells, but in a well-humidified incubator these should not be seen. We have only rarely seen such effects. Therefore, it is likely that all wells can be used in all but the most sensitive experiments.

[c] Other types of culture well insert are available made from a number of materials and with a variety of coatings from different manufacturers. Some of these may also be used successfully, but the Cellagen™ membranes are most successful in our hands for murine LE (see refs 10, 11, for an alternative).

1.2.2 Transepithelial resistance in uterine epithelial cell cultures

Transepithelial resistance (TR) is a measure of the development and sealing of tight junctions in the cell culture. TR is maximal between about 4 and 9–11 days, after which it falls. Our methodology of TR measurement is described in *Protocol 4*. The cells exhibit a cuboidal rather than columnar appearance reminiscent of LE cells on day 4 of pregnancy *in vivo*, rather than on day 1 when the cells are much taller. The cells exhibit moderate apical microvilli and apical protrusions, again similar to those seen at day 4 of pregnancy. By immunocytochemistry, they show the expected apico-lateral junctional complex stained for tight junction proteins, and desmosomal components. E-cadherin is expressed at lateral membranes and

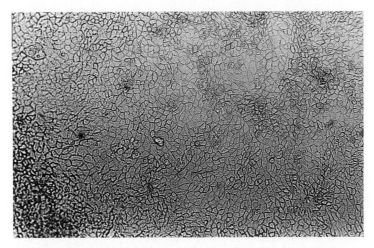

Figure 5 Primary murine epithelial cells in monolayer culture.

there is a well-developed cytokeratin network. Many but not all cells also express carbohydrate antigens shown previously to be present at the cell surface of murine LE cells (Le-x, sialyl-Le-x, Le-y H-type-1). The cells deposit laminin at the basal surface on the insert membrane and to a lesser extent on their lateral surfaces.

Protocol 4

Measurement of transepithelial resistance in uterine epithelial cell cultures

Equipment and reagents

- Transepithelial resistance meter and apparatus (e.g. Millicell®-ERS, P236 or Evom from World Precision Instruments Ltd)
- Laminar flow hood
- Mouse uterine epithelium cell culture (see *Protocol 3*)

Method

1. Transepithelial resistance of cultured uterine epithelial cell monolayers can be measured on alternate days from day 3 of culture using a transepithelial resistance meter. The procedure must be carried out in a laminar flow hood using aseptic techniques.

2. Place a sterile Cellagen™ disc in an empty well of the 24-well plate being used and add 250 µl of equilibrated culture medium to the apical compartment and 450 µl of the same medium to the basal compartment. Return the 24-well plate to the in-cubator and allow the medium to equilibrate for a minimum of 1 h before making measurements. This gives the resistance across the insert disc without cells.

3 Sterilize the transepithelial resistance meter electrodes by soaking in 70% ethanol for 15 min and then equilibrate in sterile culture medium for 15 min prior to use. The transepithelial resistance meter can then be used according to the manufacturer's instructions to measure the resistance across cultured monolayers and the control Cellagen™ insert.

Transepithelial resistance (R) of the cultured epithelial cell monolayers can be determined using the following equation:

$$\text{Rcell monolayer} = \text{Rsample} - \text{Rblank}$$

R_{sample} = resistance measurement obtained from cells cultured on a Cellagen™ insert (Ω) and R_{blank} = resistance measurement obtained from a control Cellagen™ insert without cultured cells (Ω). Effective membrane area of Cellagen™ inserts = 0.64 cm². Therefore, transepithelial resistance of cultures (Ω cm²) = $R_{cell\ monolayer} \times 0.64$ cm².

Cells grown in this way reach confluence by about 36–48 h and can be cultured for up to 12–14 days. Discard cell layers with a TR of less than 400 Ω cm².

1.2.3 Uterine epithelial cell culture with embryos

The culture of blastocysts on the apical surface of uterine epithelial cell cultures is used to examine the interaction between embryo and epithelial layer. Our methods are described in *Protocols* 5 and 6. In *Protocol* 7, the method for assessing the effect of test reagents in attachment assays is described.

Protocol 5

In vitro culture of mouse embryos

Equipment and reagents

- Laminar flow hood
- CO_2 incubator
- Dissecting microscope (Leica)
- 35 × 10 mm tissue culture dishes (Nunc)
- Light paraffin oil (Analar, BDH)
- A hypodermic needle (34 gauge × 10 mm; Coopers Needle Works)
- 1-ml syringe
- Sterile Pasteur pipettes
- Mouth pipette

- 20 ml of Ham's F10 medium (Sigma) containing 4 mg ml⁻¹ BSA where required and filter sterilized.
- M16/BSA medium (Sigma; or may be more successful if made in-house: see 15)
- KSOM medium: more recently, an alternative medium, KSOM (Speciality Media), has been produced (16–18) which allows improved development to blastocyst stage for some strains of mice (such as MF1).

Protocol 5 continued

Method[a]

1   Place 10 μl drops of M16/BSA or KSOM medium into a 35 × 10 mm tissue culture dish, cover with pre-equilibrated light paraffin oil, and incubate in a humidified atmosphere with 5% CO_2 in air at 37 °C for at least 2 h until required.

2   Using a Leica dissecting microscope, place dissected oviducts from superovulated female mice 48 h after hCG injection (see Protocol 3) in a drop of Ham's F10 medium at room temperature. Attach the hypodermic needle to a 1-ml syringe containing Ham's F10/BSA medium, insert into the fimbriated end of the oviduct, and flush out 2-cell embryos by gently expelling a small amount of Ham's F10/BSA medium through the oviduct.

3   Wash the embryos through several drops of F10/BSA medium by transfer between drops using the mouth pipette connected to a drawn out Pasteur pipette (bore just bigger than the embryos). Remove all debris, oviduct cells, and unfertilized eggs using a finely drawn out Pasteur pipette attached to a mouth pipette.

4   Wash embryos quickly through three drops of pre-equilibrated M16/BSA or KSOM medium. When the washing steps are complete, pipette the embryos into a single drop of equilibrated M16/BSA or KSOM under oil and culture for 2–3 days to the blastocyst stage. By the blastocyst stage, cultured embryos tend to be about 12 h behind their *in vivo* counterparts in development. Alternatively, blastocysts can be flushed from the uterus on the morning of day 4 of pregnancy using a 5–10-ml syringe with a ≤30 gauge hypodermic needle.

[a] Freshly prepare all culture media on the morning of embryo culture. Carry out the preparation of embryo cultures in a laminar flow hood using aseptic techniques.

Protocol 6

Assay of the rate of attachment of hatched blastocysts to cultured uterine epithelial cell monolayers

Equipment and reagents

- Uterine epithelial cell culture (see *Protocol 3*)

- *In vitro* culture of mouse embryos (see *Protocol 5*)

Method

1   Remove cultured day 5 blastocysts from the M16/BSA culture drop and wash through several drops of fresh equilibrated F12:DMEM to remove all residual traces of M16/BSA. Alternatively, directly flushed day 4 blastocysts can be used. Epithelial monolayers are generally used between day 6 and day 9 of culture.

2   Add approximately 15 hatching blastocysts to each suspended monolayer used.

Protocol 6 continued

Return monolayers with attaching blastocysts to the incubator and leave overnight in a humidified atmosphere with 5% CO_2 in air at 37 °C.

3 Blastocysts can be scored for hatching and attachment. Score hatching by counting the number of blastocysts free from their zonae pellucidae on each monolayer. Score attachment by gently tapping the side of the 24-well plate: attachment is recognised by tandem motion of uterine epithelial cell monolayers and blastocysts. Score hatching and attachment in this manner, every hour. By maintaining two sets of monolayers and staggering the scoring so that an individual well is scored only every other hour, the disruption caused by the assessment can be minimized.

Protocol 7

Assessment of the effect of test reagents in attachment assays

Equipment and reagents

- 60 × 15 mm tissue culture dishes (Nunc)
- Day 5 cultured blastocysts (see *Protocol 5*)
- Day 6–9 cultured epithelial cell monolayers (see *Protocol 3*)
- Complete F12:DMEM (see *Protocol 1*)
- Reagent of interest, diluted appropriately in F12:DMEM

Method

1 Dilute the reagent of interest appropriately in complete F12:DMEM. Then place 10 µl drops in a 60 × 15 mm tissue culture dish under oil and pre-equilibrate in the incubator.

2 Add approximately 15–20 washed blastocysts to each drop and allow to equilibrate in the reagent solutions for at least 30 min. The medium in the apical compartment of the cultured monolayers must be replaced with medium containing the reagent and allowed to equilibrate for at least 30 min.

3 Add the blastocysts to the epithelial monolayer with the corresponding reagent dilution and leave overnight for 12–14 h[a], after which time start to score.

4 Carry out scoring for attachment every 2 h to approximately 22 h as described in *Protocol 6*. Leave the attachment assays overnight and take a final score about 36 h after addition of blastocysts.

5 Fix and stain monolayers for immunocytochemistry, as required.

[a] We have found that scoring embryo attachment during the first 12 h after addition to the epithelial monolayers can seriously disrupt the ability of blastocysts to attach. This is probably due to the breaking of tenuous developing adhesions. However, this may be improved by staggered scoring and possibly by use of a heated stage.

2 Human placental development

2.1 Introduction: placental cell lineages

The placenta develops from a combination of trophoblast and extraembryonic mesenchyme, each of which gives rise to a series of different cell lineages (19, 20). The principle differentiative pathways are from stem cytotrophoblast into villous syncytium, the major transporting epithelium; from stem cytotrophoblast into migratory extravillous cytotrophoblast; and from primitive mesenchymal cells into endothelium, smooth muscle cells, and myofibroblasts. Surprisingly little is known about the intercellular interactions that control placental morphogenesis.

2.2 Cell culture methods

Tissue is generally available either at term or in late first trimester after termination of pregnancy. Most cytotrophoblast primary cell culture protocols are derived from the method originally reported by Kliman *et al.* (21) for term placenta. The villous surface of the placenta is covered by a trophoblast bilayer with stem cytotrophoblast beneath a multinucleated syncytium (22). In this protocol, trophoblast is released enzymatically from the villous surface, then purified using Percoll gradient centrifugation. There are problems with contamination of the resulting preparations with leukocytes and fibroblasts, and a need for characterization of the cells with a careful selection of markers (23). Cytokeratin-7 is a useful and moderately specific marker for trophoblast. However, there is also evidence that mononuclear syncytial fragments are formed during the preparation procedure and that these are capable of adhesion to culture dishes (24) and may be misidentified as undifferentiated cytotrophoblast. At any rate, in preparations from term placenta, trophoblast proceeds to migrate across the culture surface, aggregate, and then fuse to form differentiated hCG-producing multinuclear giant cells within about 96 h of plating (21). No reliable current method exists for maintaining these cells in a mononucleated or undifferentiated condition, though trophoblast stem cells have been isolated from early mouse placenta (25). In preparations from human first trimester tissue, some trophoblast cells also differentiate into giant cells, but other subpopulations may remain mononuclear and acquire certain characteristics of the extravillous lineage such as human leukocyte antigen (HLA) G and β_1-integrins (19, 26–28). The lifespan of cells prepared in this way has been prolonged by SV40 large T-mediated transformation (29). Choriocarcinoma cell lines such as BeWo, JAr, and JEG3 have been widely used as models but, while they are undisputedly trophoblastic, these cells do not clearly resemble any one normal lineage (19, 30, 31).

In light of these difficulties, we describe here an explant culture method that produces a pure population of mononuclear extravillous trophoblasts from first trimester tissue (*Protocols* 8 and 9). The method has been used to show the role of the interaction between fibronectin and integrin $\alpha_5\beta_1$ in placental anchorage (32, 33). We also describe protocols for the preparation and characterization of placental fibroblasts that may be useful in studying paracrine interactions in placental development (23).

2.3 Extravillous trophoblast

From early post-implantation stages to approximately 18 weeks gestation, cytotrophoblast at the tips of anchoring villi at the placental periphery proliferates to form columns. From the distal edges of these, single cells break away and infiltrate the adjacent maternal decidual stroma, progressing as far as the inner one-third of the myometrium (19, 28, 34, 35). These cells invade maternal spiral arteries and transform their walls, with loss of smooth muscle and the associated elastic and collagenous ECM, which is replaced by a fibrinoid substance.

This method enables the researcher to explant terminal placental villi onto a three-dimensional matrix of either collagen or Matrigel and monitor the formation of anchoring cytotrophoblast columns. The resulting cells can be stimulated to assume a migratory phenotype by growth factors (36, 37). Transforming growth factor-β (TGF-β) inhibits outgrowth, and antisense oligonucleotide-mediated knockdown of its receptor counteracts this effect (38). A similar method has been used to show that extravillous cytotrophoblast proliferation is enhanced and differentiation inhibited when oxygen partial pressure is reduced to values that are estimated to resemble those in the first trimester placenta (39).

2.3.1 Placental tissue

Tissue should be obtained in accordance with local ethical guidelines.

Protocol 8

Preparation of villous tissue from first trimester villous placenta[a]

Equipment and reagents

- Dissecting microscope
- Class II laminar flow hood
- Sterile dissecting kit including scissors, fine forceps
- Dissecting tray
- Sterile PBS (autoclaved)

- 100× antibiotic–antimycotic mix (AAM; Sigma)
- 500-ml disposable collection pots
- Universal bottles
- Bleach tablets
- 10% formalin

Method

1 Sterilize the dissecting kit and stand it in a beaker of 70% ethanol in the class II laminar flow hood. Keep a waste beaker containing a bleach tablet for discarded

Protocol 8 continued

medium and a second container where unwanted tissue can be transferred prior to formalin fixation.[b]

2   Collect fresh tissue into pre-washed 500-ml plastic buckets with lids, containing approximately 50 ml of sterile PBS and 100 μl of 100× AAM.

3   Remove tissue from the stockinette with scissors. Wash the tissue carefully to remove excess blood, clots, and other contaminants; this is particularly important for suction termination material.

4   Place placental and decidual tissue into approximately 20 ml of PBS containing AAM for washing and examination under the dissecting microscope.[c]

[a] Normal first trimester placental tissue is obtained at elective termination of pregnancy by suction or medical intervention (antiprogestin RU486/prostaglandin). The amount of tissue obtained varies with gestational age: at 6 weeks, the placenta is a disc of approximately 4 cm diameter, while at 12 weeks it is as much as three times this size. Suction termination material is usually obtained in a stockinette which contains trophoblast, fetus, and decidual tissue. The placenta is likely to be near one end and several pieces of tissue are usually found, especially if the gestational age is 12 weeks or more. For easier tissue recognition and improved viability, the tissue should be washed and separated as soon as possible. It may be stored at 4°C overnight in serum-free culture medium prior to explanting. In medical termination tissue, unlike in suction termination material, the placenta is normally intact, with less contamination from blood, and minimal decidual tissue. Thus placentas obtained at medical termination require less washing.

[b] Formalin-fixed tissue must be disposed of according to your institution guidelines.

[c] The villous placental tree opens into fronds when suspended in PBS, with a resemblance to seaweed. Decidual tissue is found in sheets with a shiny surface and prominent wide capillary loops running parallel to the surface. The trophoblast may be found associated with membranous material, still attached to the umbilical cord. Alternatively, sponge-like pieces may have to be picked out individually using forceps.

2.3.2 Explant culture of first trimester placental tissue for *de novo* generation of extravillous trophoblast

The protocols described include explant on either collagen I or Matrigel. The results are similar, but trophoblast outgrowth on collagen occurs radially at the surface while on Matrigel it occurs predominantly in a downward direction into the gel. The former experiment is therefore monitored more conveniently by whole-mount microscopy (*Figure 6*). Under control conditions of 20% oxygen and serum-free medium, cell proliferation occurs during the first 24 h, after which it ceases, and further outgrowth on collagen, which continues for up to about 5 days, represents the reorganization of an existing pool of cells from a thick column at the villous tip to a sheet only a few cells thick at the collagen gel surface.

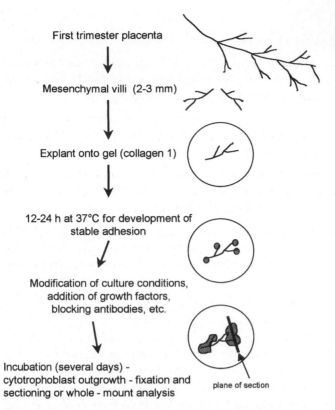

First trimester placenta

Mesenchymal villi (2-3 mm)

Explant onto gel (collagen 1)

12-24 h at 37°C for development of
stable adhesion

Modification of culture conditions,
addition of growth factors,
blocking antibodies, etc.

Incubation (several days) -
cytotrophoblast outgrowth - fixation and
sectioning or whole - mount analysis

plane of section

Figure 6 Diagram summarizing the steps for the establishment of explant cultures.

Protocol 9

Explant culture of first trimester placental tissue

Equipment and reagents

- Laminar flow hood
- CO_2 incubator
- Dissecting microscope
- 6-, 12-, or 24-well culture plates (Costar)
- Pastettes
- 9- and 5-cm Petri dishes
- Rat tail collagen type I (3 mg ml^{-1}; Becton Dickenson, distributed by Stratech)
- Matrigel (Becton Dickenson, distributed by Stratech)
- 10× DMEM (Sigma)
- 7.5% sodium bicarbonate (Gibco-BRL)
- Culture medium: 1:1 mixture of DMEM and Ham's F12
- 25-gauge needles
- 1-ml syringes
- 10-ml stripettes
- AAM (Sigma)
- FCS (Gibco-BRL)
- Growth factors
- Humidified box
- 70% ethanol
- Collagenase (Sigma type IA; 100 μg ml^{-1} in PBS)

A. Collagen gel[a]

1 For one multiwell plate, using the laminar flow hood, withdraw 1 ml of collagen from the manufacturer's container using a 1-ml syringe and needle. To prevent the introduction of air bubbles, discard the needle before transferring to a 1.5-ml microfuge tube.

2 Add 100 μl of 10× DMEM and mix carefully with a disposable pastette, taking care not to introduce bubbles (solution is a yellow colour).

3 Add approximately 200 μl of 7.5% sodium bicarbonate to neutralize the collagen and cause it to set (the colour changes from yellow to pink). Again, mix this carefully with a pastette as the solution will set from the bottom.

4 Use a pastette to transfer one drop (~80 μl) to the centre of each well of the plate. Avoid including air bubbles by careful pipetting. The culture plate can either be left in the hood or carefully transferred to a 5% CO_2 incubator. The collagen should take around 10 min to gel.[b]

5 Add culture medium to each well to cover the gel and incubate the plate in a humid box at 37 °C in a CO_2 incubator until required.

B. Matrigel[c]

1 Frozen (−20 °C) aliquots of Matrigel should be defrosted overnight at 4 °C. Pipette tips and culture plates should also be pre-cooled to 4 °C to prevent premature gelling of the Matrigel.

2 Mix 1 ml of Matrigel with 100 μl of 10× DMEM in a 1.5-ml microfuge tube and start the gelling process by adding 200 μl of 7.5% sodium bicarbonate, ensuring complete mixing before placing a drop of approximately 80 μl of Matrigel mix in the centre of each culture well.

3 Incubate the culture plates at 37 °C for 10–15 min and then cover the gel in 1 ml of culture medium. Store in a sealed humid box at 4 °C until required.

C. Selection of tissue and explant attachment (see *Figure 7*)

1 Using scissors, cut numerous terminal portions (~2–3 mm) of the villous tree (see *Protocol 8*) for explant culture. For optimal results, villi should either be trimmed from the placental periphery under a dissecting microscope or the pieces checked for suitability using the microscope. Optimal attachment is obtained when villous clusters include several branches. Observe the frond-like structure of the villi after gently teasing out with fine forceps. If the mesenchymal (terminal) villi appear elongated and stringy to the naked eye, then success of attachment and outgrowth tends to be low. Use a portion that includes an intermediate villus with several mesenchymal branches emanating from it.

2   Place selected portions in a drop of culture medium in a Petri dish until enough material has been collected.

3   Once the placental tissue has been dissected into appropriately sized pieces, drain the gels completely of medium; during incubation they will have contracted slightly and become firm, aiding tissue attachment.

4   Transfer one piece of villous tissue to each well using fine forceps and carefully tease out the branches over the gel surface, taking care not to score the gel. Ideally, the villous termini should be arranged radially over the surface of the gel drop. If correct placement does not occur, use a pastette and a minimal volume of medium to lift and reposition the villous tissue gently onto the drop. Alternatively, leave the medium in the well and add the selected villous tissue to the medium.

5   Following placement, cover each piece of tissue with 20 μl of pre-warmed culture medium and incubate at 37°C in 5% CO_2 in a humid box for 3–12 h. After this period, initial outgrowth is sometimes seen using a high magnification phase contrast objective. For maximum attachment, the villi are incubated overnight and attachment may be indicated by the appearance of radial stress lines of bundled collagen in the gel.

6   Following this initial period, gently immerse cultures in 1 ml of culture medium, directing a pipette to the side of the culture well to avoid detachment of the tissue from the gel. At this stage, growth factors, bromodeoxyuridine, or adhesion-modulating antibodies may be added to the explant in the culture medium at suitable concentrations (32).

7   Transfer the culture plate carefully back to the humid box in the incubator. After a further 12 h, initial outgrowth from viable attached terminal villi structures should be apparent. If tissue becomes detached when 1 ml of medium is added or during the overnight incubation that follows, then reattachment can be attempted by draining the well and repositioning the explant. However, repositioned tissue often has poor viability and is likely to detach again as well as exhibiting delayed growth characteristics.

8   Monitor daily cytotrophoblast column outgrowth and subsequent cell migration from gel-attached villi using low and high power whole-mount microscopy.[d]

D. Isolation of extravillous cytotrophoblast

1   Remove the explant conditioned medium, filter, and store it at −20°C, and wash the explant three times with PBS.

2   Remove the villous tissue from the collagen with fine forceps and incubate the remaining adherent cells[e] with collagenase solution[f] at 37°C for 10 min to release cytotrophoblast cells. Gentle agitation by pipetting can aid cell release. Trypsinization is also effective.

Protocol 9 continued

3   Pellet the isolated cytotrophoblast cells by centrifugation at 1500 r.p.m. for 10 min at 4 °C, and wash with PBS twice to remove all traces of collagenase.

[a] Collagen gels should be prepared the day before the experiment. Storing gels for a longer time period may result in deterioration, as indicated by a stringy appearance. For each 12-well plate, 1 ml of collagen type I is required. Ideally, the collagen concentration should be more than 3 mg ml^{-1}. Gels become less firm with storage time of the collagen, and more dilute stock solutions tend to have a shorter shelf life. If more than 3–4 ml of collagen are to be used, then it is recommended that it be prepared in separate 1–2 ml batches to ensure even mixing. For best results, use collagen, 10× medium, and sodium bicarbonate directly from the fridge.

[b] Alternatively, gels can be set using ammonia. This is accomplished by adding a few drops of ammonia onto a small Petri dish placed inside the box which is then sealed and incubated for 15–30 min at 37 °C. Care must be taken not to disturb the collagen droplets and cause them to spread—they will be more fluid than in the normal method—and not to let the gels dry out. After setting, the gels are incubated as before in a box free from ammonia vapour.

[c] Matrigel is an alternative to collagen gel and should be prepared at least 24 h prior to setting up explant cultures. Matrigel sets at room temperature; therefore, all preparation should be performed on ice or at 4 °C.

[d] For daily comparisons of column growth, low power image capture is recommended (*Figure 6*). The area of outgrowing cells is very variable, being highly dependent on the size of the initial villus–gel contact (32). Outgrowth is limited to the tips of the villi but cells from adjacent tips merge to form a shell-like structure surrounding the explant (35).

[e] Removal of tissue from the surface of a collagen gel after up to 7 days' culture leaves a pure population of extravillous cytotrophoblast at the surface.

[f] Collagenase should be reconstituted in diethylpyrocarbonate (DEPC)-treated sterile PBS if the isolated cytotrophoblasts are to be used for RNA extraction.

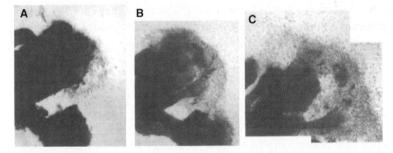

Figure 7 (**A**) Explant of villous tissue onto a gel showing outgrowth of cytotrophoblast cells from the upper villus after 1 day. (**B**) After 2 days, outgrowths from several independent villi have fused to form a 'shell' or sheet of cytotrophoblast cells. (**C**) After 4 days in culture, the shell is more extensive and single cells can also be seen migrating across the gel. The black bars show the radial increase in growth.

2.3.3 Staining protocols for placental explant cultures

The protocol given for whole-mount staining uses a single primary antibody with nuclear counterstaining, but it can be extended to double antibody staining.

Protocol 10

Whole-mount staining of placental explant cultures in multiwell plates

Equipment and reagents

- Inverted fluorescence microscope with long working distance objectives
- Dissecting microscope
- Platform shaker
- Methanol
- Blocking solution (4% BSA in PBS, with 0.02% sodium azide for longer storage)

- PBS
- Primary antibody diluted in 4% BSA/PBS
- Secondary antibody diluted in 4% BSA/PBS
- Nuclear counterstain (propidium iodide or 4′,6-diamidino-2-phenylindole dihydrochloride (DAPI) at 5 μg ml^{-1} in PBS; stocks prepared in water at 5 mg ml^{-1})

Method

1 Wash the cultures gently and thoroughly in PBS twice.

2 Fix in 1 ml of methanol. Replace the first solution immediately with a second aliquot. Incubate for 30 min.

3 Wash three times with PBS.

4 At this stage, cultures can be trimmed under the dissecting microscope with a small pair of scissors. Remove parts of the explant that are not participating in anchorage to the gel. Take care not to damage the gel surface. If floating villi are removed, there is less likelihood of damaging the culture during processing.

5 Incubate overnight (at least) at 4°C in blocking solution.

6 Incubate in primary antibody solution for 30 min at room temperature.[a] Ideally, the culture should be submerged. If necessary, the volume can be minimized (e.g. 100 μl) by repeatedly pipetting the solution over the explant, preventing it from drying out.

7 Wash in PBS, then overnight on a platform shaker in 4% BSA/PBS.

8 Incubate for 30 min–2 h in fluorescent-conjugated secondary antibody solution. Plates should be wrapped in foil.

9 Wash overnight in several changes of PBS using a platform shaker. Cultures can be inspected under the inverted microscope at any stage to monitor the removal of background fluorescence in the gel. Keep the plates in the dark throughout.

10 Counterstain nuclei for 30 min and wash further in PBS.

[a] A good positive control antibody is anti-cytokeratin 7 (OV-TL 12/30; Dako; use at 1/50). This identifies the outgrowing cells unequivocally as trophoblast. Anti-cytokeratin 8/18 can be used for marking cells (e.g. CAM5.2, Beckton-Dickenson) but beware that some placental fibroblasts express this marker so it is not specific for trophoblast.

Protocol 11

Whole-mount staining of placental explant cultures in tubes[a]

Equipment and reagents

- See *Protocol 10*
- Aqueous hardening mountant (Histotec, Serotec, UK)

Method

1 Fix as described in *Protocol 10* and rehydrate in PBS.

2 Trim the gel using a scalpel, retaining the central area with the explant.

3 Gently tease the remaining gel off the well surface using a small spatula.

4 Use a pastette to transfer it to a 0.5-ml microfuge tube.

5 Add blocking solution and proceed as in *Protocol 10*. Incubations can be carried out on a roller mixer.

6 After staining, transfer the gel fragment to a microscope slide and remove excess PBS by blotting with a tissue. Orient the fragment as during culture.

7 Encase the gel fragment in aqueous hardening mountant. Do not use a coverslip. Incubate at room temperature overnight in the dark.

8 The culture can now be inspected directly using water or oil immersion objectives and either upright or inverted optics.

[a] This method allows more economical use of antibody solution than that in *Protocol 10*.

Protocol 12

Cryosectioning of placental explant cultures[a]

Equipment and reagents

- Cryostat
- Cryotubes
- OCT compound
- Liquid nitrogen
- Poly-L-lysine (100 μg ml^{-1} in PBS)
- Acetone
- Protein blocking solution (Dako)

Method

1 Trim the edges of the gel with a scalpel if the culture is confined to one part of the collagen drop. Gently tease the gel carrying the culture from the floor of the well and transfer to a cryotube containing a drop of OCT compound. Place a further drop of OCT over the culture, then place the tube in liquid nitrogen.[b]

2 Pre-coat cleaned glass microscope slides with poly-L-lysine at 100 μg ml^{-1} for 10 min.

Protocol 12 continued

3   Mount the tissue on a stub using OCT as adhesive.

4   Cut 7 μm sections using a cryostat. Air-dry the sections then place in a tray or box, wrap in foil, and store at −80 °C. Sections should be used as soon as possible and certainly within 2 months.

5   Allow sections to warm to room temperature then fix with dry acetone for 10 min.

6   Rehydrate sections using PBS then incubate in protein blocking solution for 10 min.

7   Proceed using standard immunostaining protocols (e.g. *Protocol 10*).[c]

[a] Cryosections are required for immunostaining with a large panel of antibodies on the same culture.

[b] It is convenient to keep the tissue near the top of the cryotube. Liquid nitrogen-cooled isopentane may also be used for freezing. Tissue may be kept frozen for several weeks.

[c] Peroxidase- or fluorescent-conjugated antibodies may be used (14). Note that alkaline phosphatase conjugates should be used with caution as this enzyme is expressed by trophoblast.

Protocol 13

Resin embedding of placental explant cultures[a]

Equipment and reagents

- Dissecting microscope
- Glass vials with plastic snap-on tops.
- Oven/incubator at 48 °C and 60 °C
- Ultramicrotome
- Hotplate
- Silicone rubber moulds (rectangular, ~6 × 11 × 5 mm deep; Agar Scientific Ltd, Taab Laboratories Equipment Ltd)
- 25% glutaraldehyde (EM grade; Agar Scientific Ltd, Taab Laboratories Equipment Ltd)
- 2% aqueous osmium tetroxide (Agar Scientific Ltd, Taab Laboratories Equipment Ltd)
- Propylene oxide (use in a fume cupboard, wearing protective gloves)
- 0.1 M sodium cacodylate buffer, pH 7.3 ± 3 mM $CaCl_2$

- Epoxy resin (Epon, Araldite, or Taab embedding resin, or other epoxy resin mixtures, can be used; prepare and use resin in a fume cupboard, wearing gloves, and following the manufacturer's instructions; Agar Scientific Ltd, Taab Laboratories Equipment Ltd)
- Propylene oxide:resin mixture (1:1 and 1:3; prepare and use in a fume cupboard, wearing gloves; do not inhale vapour)
- 2.5% glutaraldehyde in 0.1 M sodium cacodylate buffer, pH 7.3
- 1% osmium tetroxide in 0.05 M sodium cacodylate buffer, pH 7.3 (mix equal parts 2% aqueous OsO_4 solution and 0.1 M buffer; prepare and use in a fume cupboard and wear protective gloves; do not inhale vapour)
- 1% toluidine blue in 1% borax

Method

1   Rinse the explant culture in PBS, then fix in freshly prepared 2.5% glutaraldehyde in cacodylate buffer without $CaCl_2$ for 2–3 h at room temperature.

Protocol 13 continued

2   Rinse in cacodylate buffer with $CaCl_2$ three times over 24 h and store at 4°C until ready for processing. Then, using a dissecting microscope and razor blade, trim the gel around the explant into a rectangular shape about 4×8 mm, ensuring that one short edge is cut at right angles across the explant growth. This will be the face that is sectioned, and the maximum area of outgrowth possible is desirable. Transfer to a glass vial for processing.

3   Post-fix in 1% osmium tetroxide at 4°C for 1 h in a closed glass vial. After fixation, decant the fix and rinse in cacodylate buffer.

4   Dehydrate in an ascending alcohol series: 15 min each in 50, 70, 95, and 100% ethanol.

5   Incubate in propylene oxide twice for 15 min.

6   Infiltrate in propylene oxide:resin 1:1 mixture for 1 h.

7   Leave overnight in closed vials at 4°C on a rotator in propylene oxide:resin 1:3 mixture.

8   Replace with three changes of freshly prepared resin at 48°C during the day and then embed in silicone rubber moulds. Place the gel so that the cut surface of the explant is up against the 6 mm edge of the mould which will be the cutting face. Fill the moulds to the top with resin. Polymerize at 60°C for 72 h.

9   With a razor blade or using an ultramicrotome trimming facility, trim into the block face until the outgrowth is reached. This can be seen as a fine black line and the cut villus surface as a round profile. Then cut 0.5 μm thick sections and mount on a glass slide. Leave on a hotplate (70–80°C) to dry and stain with 1% toluidine blue in 1% borax. Wash and examine.

10 Suitable areas can be photographed and then ultrathin sections cut with a glass or diamond knife, mounted onto copper grids, contrasted with uranyl acetate/ lead citrate, and examined in the electron microscope.

[a] Examination of semi-thin resin sections can show the pattern of cell growth across and within the gel (*Figure 8*). Electron microscopy enables cell–cell and cell–matrix interactions to be visualized.

2.4 Isolation and culture of placental fibroblasts (see *Figure 9*)

The following can be modified for use with term or first trimester tissue. We have transferred serum-free conditioned culture medium from first trimester placental fibroblasts at 8–10 passages onto trophoblast explants to demonstrate paracrine effects on cell migration (36, 37). The cells are also useful for studies of the composition and control of placental ECM production. Note that fibroblasts have only been seen very rarely using the explant method for trophoblast (*Protocol 9*). However, trophoblast may occasionally be obtained when using the following fibroblast protocols. Cell morphology is not a reliable method for cell

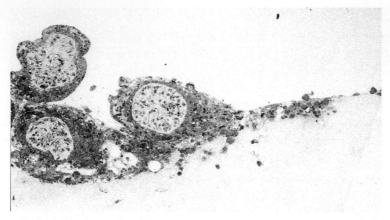

Figure 8 Semi-thin section through a resin-embedded explant culture showing three villi cut in cross-section and the outgrowth from them spreading across the surface of the gel. Toluidine blue stain.

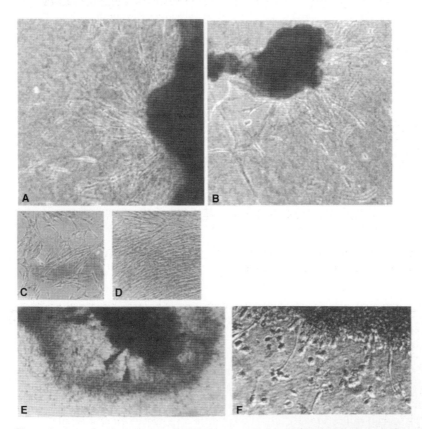

Figure 9 (**A** and **B**) Characteristic fibroblast outgrowth from an explant into a gel. (**C** and **D**) Fibroblasts grown on a plastic substrate show a typical morphology. (**E** and **F**) Cytotrophoblast cells after 4 days culture have formed a typical 'shell' with single, rounded cells at the periphery; at higher magnification (F), their morphology is distinctly different from that of the fibroblasts.

254

identification as cytotrophoblasts and fibroblasts can resemble one another. As a precaution, cells resulting from all the following protocols should be characterized further by antibody markers. Vimentin and α-smooth muscle actin are useful markers for fibroblasts because they are both absent from trophoblast, although α-smooth muscle actin generally is not expressed by all the fibroblasts in a culture population (23).

Protocol 14

Isolation and culture of placental fibroblasts from direct explants

Equipment and reagents

- Except where indicated, equipment and reagents are as in *Protocol 9*
- Culture medium: DMEM (Gibco-BRL) supplemented with 10% FBS (Gibco-BRL), penicillin (100 U ml^{-1}), streptomycin (100 μg ml^{-1}), glutamine (2 mM, 10 ml l^{-1}),

- non-essential amino acids (10 ml l^{-1}; Sigma), and 2-mecaptoethanol (8 ml l^{-1}). Note that fibroblasts do not grow well in DMEM:F12 as used for trophoblast in *Protocol 9*

Method

1 Mince placental villous tissue using small scissors and fine forceps into pieces of approximately 1 mm diameter and place in 12-well culture plates. Allow the tissue to air dry and attach to the plastic surface by leaving to stand at room temperature for approximately 10 min.

2 Add 1 ml of culture medium gently to each well. Incubate cultures at 37°C in 5% CO_2, 95% air. After outgrowth of cells from the explants (2–3 weeks), establish subcultures by selective trypsinization of the primary cultures.[a]

3 After growth to confluence and expansion, preserve cells by standard freezing protocols.

[a] A minority of these explants generate trophoblast; only colonies containing spindle-shaped cells should be chosen to ensure recovery of fibroblasts.

Protocol 15

Isolation and culture of fibroblasts from enzymatic dispersal of placental tissue[a]

Equipment and reagents

- Ringer-bicarbonate buffer (0.9% NaCl, 0.042% KCl, 0.02% NaHCO$_3$, 0.15% glucose)
- Trypsin-EDTA solution (0.25% in Ringer-bicarbonate buffer; Gibco-BRL, UK)
- 25-cm^2 tissue culture flask

- DNase I (10 U ml^{-1} in Ringer-bicarbonate buffer; Sigma)
- Culture medium (composition as in *Protocol 14*)

Protocol 15 continued

Method

1 Mince placental tissue (*Protocol 14*), transfer it to a centrifuge tube, and subject it to four or five 10-min sequential digestions with trypsin-EDTA solution and DNase I solution. In each fraction, pipette the tissue vigorously (but avoiding foaming) for 5 min.

2 Allow the tube to sit for 5 min undisturbed to let the large pieces of tissue settle.

3 Remove the supernatant to a fresh tube. Pool all supernatant fractions and spin at 1000 *g* for 10 mins.

4 Resuspend the pellet in 5 ml of culture medium and seed in a tissue culture flask.

5 After 1 h incubation to allow for cell adherence, wash the flask with medium to remove loosely adherent trophoblastic cells. Maintain cultures in culture medium, changing daily during the first week and then every other day.

[a] Under these conditions, small colonies of proliferating fibroblasts are visible within 7 days of culture.

Protocol 16

Isolation and culture of fibroblasts obtained via explant in collagen gel

Equipment and reagents

- As in *Protocol 9* for collagen gels
- Collagenase (Sigma Type IA-S, 5 mg ml^{-1} in water)
- Culture medium as in *Protocol 14*
- Trypsin-EDTA (Gibco-BRL; 1/10).

Method

1 Prepare collagen gels (*Protocol 9*).

2 Immediately after adjusting the pH and prior to setting, mix minced villous placental tissue (~1 mm pieces) with the collagen solution and place a drop containing one or more fragments in the centre of each well of a 12-well culture plate.

3 Add culture medium after incubating the gel for a few minutes to set.

4 Incubate at 37°C in 5% CO_2 with daily medium changes during the first week and then every other day.[a]

5 After 6–8 weeks, release fibroblasts which should have grown out in large numbers by collagenase treatment and seed into flasks for growth to confluence and passage in trypsin-EDTA.

[a] Under these conditions, initial outgrowth of fibroblast cells may be seen after a few days.

References

1. Aplin, J. D. (1997). *Rev. Reprod.*, **2**, 84.

2. Bergh, P. A. and Navot, D. (1992). *Fertil. Steril.*, **58**, 537.

3. Aplin, J. D. (2000). *The cell biological basis of human implantation. Bailliere's Best Practice Research Clinical Obstetrics and Gynaecology*, **14**, 757.

4. Simón, C., Piquette, G. N., Frances, A., and Polan, M. L. (1993). *J. Clin. Endocrinol. Metab.*, **77**, 549.

5. Simón, C., Piquette, G. N., Frances, A., El-Danasouri, I., and Polan, M. L. (1994). *J. Clin. Endocrinol. Metab.*, **78**, 675.

6. Simón, C., Gimeno, M. J., Mercader, A., O'Connor, J. E., Remohí, J., Polan, M. L., *et al.* (1997). *J. Clin. Endocrinol. Metab.*, **82**, 2607.

7. Simón, C., Mercader, A., Garcia Velasco, J., Remohí, J., and Pellicer, A. (1999). *J. Clin. Endocrinol. Metab.*, **84**, 2638.

8. Pellicer, A., Valbuena, D., Cano, F., Remohí, J., and Simón, C.(1996). *Fertil. Steril.*, **65**, 1190.

9. Meseguer, M., Aplin, J. D., Caballero-Campo, P., O'Connor, J.E., Martin, J.C., Remohi, J., *et al.* (2001). *Biol. Reprod.*, **64**, 590.

10. Glasser, S. R., Julian, J., Decker, G. L., Tang, J. P., and Carson, D. D. (1988). *J. Cell Biol.*, **107**, 2409.

11. Jacobs, A. L., Decker, G. L., Glasser, S. R., Julian, J., and Carson, D. D. (1990). *Endocrinology*, **126**, 2125.

12. Yee, G. M. and Kennedy, T. G. (1991). *Biol. Reprod.*, **45**, 163.

13. McCormack, S. A. and Glasser, S. R. (1980). *Endocrinology*, **106**, 1634.

14. Kennedy, T. G., Ross, H. E., Barbe, G. J., Shu, M. A., and Zhang, X. (1998). *Mol. Reprod. Dev.*, **49**, 268.

15. Pratt, H. P. M. (1987). In *Mammalian development: a practical approach* (ed. M. Monk), p. 13. IRL Press, Oxford.

16. Lawitts, J. A. and Biggers, J. D. (1991). *J. Reprod. Fertil.*, **91**, 543.

17. Lawitts, J. A. and Biggers, J. D. (1993). In *Methods in enzymology* (ed. P. M. Wassarman and M. L. DePamphilis). Vol. 225, p. 153. Academic Press, London.

18. Ho, Y., Wigglesworth, K., Eppig, J. J., and Schultz, R. M. (1995). *Mol. Reprod. Dev.*, **41**, 232.

19. Aplin, J. D. (1991). *J. Cell Sci.*, **99**, 681.

20. Kohnen, G., Kertschanska, S., Demir, R., and Kaufmann, P. (1996). *Histochem. Cell Biol.*, **105**, 415.

21. Kliman, H. J., Nestler, J. E., Sermasi, E., Sanger, J. M., and Strauss, J. F. III (1986). *Endocrinology*, **118**, 1567.

22. Jones, C. J. P. and Fox, H. (1991). *Electron Microsc. Rev.*, **4**, 129.

23. Haigh, T., Chen, C.-P., Jones, C. J. P., and Aplin, J. D. (1999). *Placenta*, **20**, 615.

24. Frank, H.-G., Morrish, D. W., Potgens, A., Genbacer, O., Kumpel, B., and Caniggia, I. (2001). *Placenta*, **22**, Suppl. A, S107.

25. Tanaka, S., Kunath, T., Hadjantonakis, A. K., Nagy, A., and Rossant, J. (1998). *Science*, **282**, 2072.

26. Damsky, C. H., Librach, C., Lim, K.-H., Fitzgerald, M. L., McMaster, M.T., Janatpour, M., *et al.* (1994). *Development*, **120**, 3657.

27. Irving, J. A. and Lala, P. K. (1995). *Exp. Cell Res.*, **217**, 419.

28. Vicovac, L. and Aplin, J. D. (1996). *Acta Anat.*, **156**, 202.

29. Choy, M. Y. and Manyonda, I. T. (1998). *Hum. Reprod.*, **3**, 2941.

30. Aplin, J. D., Sattar, A. and Mould, A. P. (1992). *J. Cell Sci.*, **103**, 435.

31. Church, H. J., Richards, A. J., and Aplin, J. D. (1997). *Trophoblast Res.*, **10**, 143.

32. Aplin, J. D., Haigh, T., Jones, C. J. P., Church, H. J., and Vicovac, L. (1999). *Biol. Reprod.*, **60**, 828.

33. Jones, C. J. P., Haigh, T., Vicovac, L., and Aplin, J. D. (1998). *Reprod. Hum. Horm.*, **11**, 402.

34. Pijnenborg, R., Dixon, G., Robertson, W. B., and Brosens, I. (1980). *Placenta*, **1**, 3.

35. Aplin, J. D., Jones, C. J. P., Haigh, T., Church, H. J., and Vicovac, L. (1998). *Hum. Fertil.*, **1**, 75.

36. Aplin, J. D., Lacey, H., Haigh, T. Jones, C. J. P, Chen, C.-P., and Westwood, M. (1999). *Biochem. Soc. Trans.*, **28**, 199.

37. Aplin, J. D., Haigh, T., Lacey, H., Chen, C.-P., and Jones, C. J. P. (2000). *J. Reprod. Fertil.*, Suppl. **55**, 57.

List of suppliers

Abbott Laboratories Ltd. Queensborough, Kent ME11 5EL, UK
Tel: 01795 580099

Advanced Biotechnologies Inc. Rivers Park II, 9108 Guilford Road, Columbia, MD 21046-2701, USA

Agar Scientific Ltd 66a Cambridge Road, Stansted, Essex CM24 8DA, UK

Alpha Innotech Corp. PO Box 2499, Cannock, Staffordshire WS12 5UN, UK
Alpha Innotech Corp., 14743 Catalina Street, San Leandro, CA 94577, USA

Amersham Pharmacia Biotech (UK) Ltd
Amersham Place, Little Chalfont, Buckinghamshire HP7 9NA, UK (see also Nycomed Amersham Imaging UK; Pharmacia)
Tel: 0800 515313   Fax: 0800 616927
URL: http//www.apbiotech.com

Anachem Ltd 20 Charles Street, Luton, Bedfordshire LU2 0EB, UK

Anderman and Co. Ltd 145 London Road, Kingston-upon-Thames, Surrey KT2 6NH, UK
Tel: 0181 5410035   Fax: 0181 5410623

Applied Biosystems 7 Kingsland Grange, Woolston Warrington, Cheshire WA1 7SR, UK

Balzers Ltd Bradbourne Drive, Tilbrook, Milton Keynes MK7 8AZ, UK
Balzers, Inc., 495 Commerce Drive, Amherst, NY 14228, USA

Barnstead/Thermolyne 2555 Kerper Boulevard, Dubuque, IA 52001-1478, USA

BASF Corp. PO Box 4, Earl Road, Cheadle Hulme, Cheadle, Cheshire SK8 6QG, UK

BDH now Merck Ltd, Merke House, Poole, Dorset, BH15 1TD, UK

BD PharMinogen 10975 Torreyana Road, San Diego, CA 92121, USA

Beckman Coulter (UK) Ltd Oakley Court, Kingsmead Business Park, London Road, High Wycombe, Buckinghamshire HP11 1JU, UK
Tel: 01494 441181
Fax: 01494 447558
URL: http://www.beckman.com
Beckman Coulter Inc., 4300 North Harbor Boulevard, PO Box 3100, Fullerton, CA 92834-3100, USA
Tel: 001 714 8714848
Fax: 001 714 7738283
URL: http://www.beckman.com

Becton Dickinson and Co. 21 Between Towns Road, Cowley, Oxford OX4 3LY, UK
Tel: 01865 748844
Fax: 01865 781627
URL: http://www.bd.com
Becton Dickinson and Co., 1 Becton Drive, Franklin Lakes, NJ 07417-1883, USA
Tel: 001 201 8476800
URL: http://www.bd.com

Bibby Sterilin Ltd Tilling Drive Stone, Staffordshire, ST15 0SA, UK

Billups-Rothenburg PO Box 997, Del Mar, CA 92014-0977, USA

Bio 101 Inc.
Bio 101 Inc., c/o Anachem Ltd, Anachem House, 20 Charles Street, Luton, Bedfordshire LU2 0EB, UK
Tel: 01582 456666   Fax: 01582 391768
URL: http://www.anachem.co.uk
Bio 101 Inc., PO Box 2284, La Jolla, CA 92038-2284, USA
Tel: 001 760 5987299   Fax: 001 760 5980116
URL: http://www.bio101.com

Biosoft Quantiscan Biosoft Quantiscan, c/o Team Data Support A/S, Sverigesvej 20, DK-4200 Slagelse, Denmark

Bio-Rad Laboratories Ltd Bio-Rad House, Maylands Avenue, Hemel Hempstead, Hertfordshire HP2 7TD, UK
Tel: 0181 3282000   Fax: 0181 3282550
URL: http://www.bio-rad.com
Bio-Rad Laboratories Ltd, Division Headquarters, 1000 Alfred Noble Drive, Hercules, CA 94547, USA
Tel: 001 510 7247000   Fax: 001 510 7415817
URL: http://www.bio-rad.com

Biowhittaker UK Ltd 1 Ashville Way, Wokingham, Berkshire RG41 2PL

BOC Edwards Manor Royal, Crawley, West Sussex RH10 2LW, UK
BOC Edwards, 301 Ballardvale Street, Wilmington, MA 01887, USA

Boehringer-Mannheim (see Roche)

Branson Sonic Power Co. 41 Eagle Road, PO Box 1961, Danbury, CT 06813-1961, USA

Calbiochem-Novabiochem Corp. PO Box 12087, La Jolla, CA 92039-2087, USA

Cedarlane Laboratories Ltd 5516 – 8th Line, R.R.#2, Hornby, Ontario L0P 1E0, Canada

Cellular Products Inc. 872 Main Street, Buffalo, NY 14202, USA

Chance Propper Ltd PO Box 53, Spon Lane South, Smethwick, West Midlands B66 1NZ, UK

City University now known as Citifluor Ltd, c/o Newby Castleman Ltd, Chartered Accountants, 110 Regent Road, Leicester LE1 7LT, UK

Clark Electromedical contact Harvard Apparatus Ltd, Fircroft Way, Edenbridge, Kent TN8 6HE, UK

Coopers Needle Works 261–265 Aston Lane, Perry Barr, Birmingham B20 3HS, UK

Corning Costar GmbH Am Kuemmerling 21–25, D-55294 Bodenheim, Germany

CP Instrument Co. Ltd PO Box 22, Bishop Stortford, Hertfordshire CM23 3DX, UK
Tel: 01279 757711
Fax: 01279 755785
URL: http//:www.cpinstrument.co.uk

Dako Ltd Denmark House, Angel Drove, Ely, Cambridgeshire CB7 4ET, UK
DAKO Corporation, 6392 Via Real, Carpinteria, CA 93013, USA

Dupont (UK) Ltd Industrial Products Division, Wedgwood Way, Stevenage, Hertfordshire SG1 4QN, UK
Tel: 01438 734000   Fax: 01438 734382
URL: http://www.dupont.com
Dupont Co. (Biotechnology Systems Division), PO Box 80024, Wilmington, DE 19880-002, USA
Tel: 001 302 7741000
Fax: 001 302 7747321
URL: http://www.dupont.com

Dynal Biotech UK 11 Bassendale Road, Croft Business Park, Bromborough, Wirral CH62 3QL, UK

Eastman Chemical Co. 100 North Eastman Road, PO Box 511, Kingsport, TN 37662-5075, USA   Tel: 001 423 2292000
URL: http//:www.eastman.com

Electron Microscopy Sciences PO Box 251, 321 Morris Road, Fort Washington, PA 19034, USA

ELF Atochem

Eli Lilly and Company Lilly Corporate Center, Indianapolis, IN 46285, USA
Eli Lilly Holdings Limited, Erl Wood Manor, Sunninghill Road, Windlesham, Surrey GU20 6PH, UK

Eppendorf AG 10 Signet Court, Swann Road, Cambridge CB5 8LA, UK

Fisher Scientific (UK) Ltd Bishop Meadow Road, Loughborough, Leicestershire LE11 5RG, UK
Tel: 01509 231166   Fax: 01509 231893
URL: http://www.fisher.co.uk
Fisher Scientific, Fisher Research, 2761 Walnut Avenue, Tustin, CA 92780, USA
Tel: 001 714 6694600
Fax: 001 714 6691613
URL: http://www.fishersci.com

Forma PO Box 649, Marietta, OH 45750, USA

Fluka PO Box 2060, Milwaukee, WI 53201, USA
Tel: 001 414 2735013
Fax: 001 414 2734979
URL: http://www.sigma-aldrich.com
Fluka Chemical Co. Ltd, PO Box 260, CH-9471, Buchs, Switzerland
Tel: 0041 81 7452828
Fax: 0041 81 7565449
URL: http://www.sigma-aldrich.com

Gelman Sciences Arheiliger Weg 6a, D-64380 Rossdorf, Germany

Gibco-BRL (see Life Technologies)

Glenwoods Ltd Unit 3, Hadstock Road, Linton, Cambridge CB1 6NR, UK

Greiner Labortechnik Ltd Brunel Way, Stonehouse, Gloucestershire GL10 3SX, UK
Greiner Labortechnik GmbH, Maybachstraße 2, D-72636 Frickenhausen, Germany

Hoechst Foundation now Aventis Foundation, Altes Schloss Höchst, Höchster Schlossplatz 16, D-65929 Frankfurt am Main, Germany

Hoeffer Scientific 654 Minnesota Street, Box 77387, San Francisco, CA 94107, USA

Hook and Tucker Instruments Ltd Vulcon Way, New Addington, Croydon, Surrey CR0 9UG, UK

Hybaid Ltd Action Court, Ashford Road, Ashford, Middlesex TW15 1XB, UK
Tel: 01784 425000   Fax: 01784 248085
URL: http://www.hybaid.com
Hybaid US, 8 East Forge Parkway, Franklin, MA 02038, USA
Tel: 001 508 5416918
Fax: 001 508 5413041
URL: http://www.hybaid.com

HyClone Laboratories 1725 South HyClone Road, Logan, UT 84321, USA
Tel: 001 435 7534584
Fax: 001 435 7534589
URL: http//:www.hyclone.com

Hygenic Corp. 1245 Home Avenue, Akron, OH 44310, USA

ICN Ltd Cedarwood, Chineham Business Park, Crockford Lane, Basingstoke, Hampshire RG24 8WG, UK

Intervet (UK) Ltd Walton Manor, Walton, Milton Keynes, Buckinghamshire MK7 7AJ, UK

Invitrogen Corp. 1600 Faraday Avenue, Carlsbad, CA 92008, USA
Tel: 001 760 6037200   Fax: 001 760 6037201
URL: http://www.invitrogen.com
Invitrogen BV, PO Box 2312, 9704 CH Groningen, The Netherlands
Tel: 00800 53455345   Fax: 00800 78907890
URL: http://www.invitrogen.com

Jackson Laboratories 600 Main Street, Bar Harbor, ME 04609, USA

Laser Photonics Inc. 12351 Research Parkway, Orlando, FL 32826, USA

Leica Microsystems Davy Avenue, Knowlhill, Milton Keynes MK5 8LB, UK

Life Technologies Ltd PO Box 35, 3 Free Fountain Drive, Inchinnan Business Park, Paisley PA4 9RF, UK
Tel: 0800 269210   Fax: 0800 243485
URL: http://www.lifetech.com/
Life Technologies Inc., 9800 Medical Center Drive, Rockville, MD 20850, USA
Tel: 001 301 6108000
URL: http://www.lifetech.com

Medical Systems Corp. 1 Plaza Road, Greenvale, NY 11548, USA

Merck Sharp & Dohme Research Laboratories Neuroscience Research Centre, Terlings Park, Harlow, Essex CM20 2QR, UK
URL: http://www.msd-nrc.co.uk
MSD Sharp and Dohme GmbH, Lindenplatz 1, D-85540, Haar, Germany
URL: http://www.msd-deutschland.com

Millipore (UK) Ltd The Boulevard, Blackmoor Lane, Watford, Hertfordshire WD1 8YW, UK
Tel: 01923 816375   Fax: 01923 818297
URL: http://www.millipore.com/local/UKhtm

Millipore Corp., 80 Ashby Road, Bedford, MA 01730, USA
Tel: 001 800 6455476   Fax: 001 800 6455439
URL: http://www.millipore.com

MJ Research Inc. 590 Lincoln Street, Waltham, MA 02451, USA

Molecular Probes Inc. 4849 Pitchford Avenue, PO Box 22010, Eugene, OR 97402, USA

Narishige (USA) Inc. 1 Plaza Road, Greenvale, NY 11548, USA

NASCO 901 Janescille Avenue, Fort Atkinson, WI 53538, USA

New England Biolabs 32 Tozer Road, Beverley, MA 01915-5510, USA
Tel: 001 978 9275054

Nikon Inc. 1300 Walt Whitman Road, Melville, NY 11747-3064, USA
Tel: 001 516 5474200   Fax: 001 516 5470299
URL: http://www.nikonusa.com
Nikon Corp., Fuji Building, 2–3, 3-chome, Marunouchi, Chiyoda-ku, Tokyo 100, Japan
Tel: 00813 32145311   Fax: 00813 32015856
URL: http://www.nikon.co.jp/main/index_e.htm

Novagen Inc. 565 Science Drive, Madison, WI 53711, USA

Nunc Inc. 2000 North Aurora Rosd, Naperville, IL 60563-1796, USA

Nycomed Amersham Imaging Amersham Laboratoriess, White Lion Road, Amersham, Buckinghamshire HP7 9LL, UK
Tel: 0800 558822 (or 01494 544000)
Fax: 0800 669933 (or 01494 542266)
URL: http//:www.amersham.co.uk
Nycomed Amersham, 101 Carnegie Center, Princeton, NJ 08540, USA
Tel: 001 609 5146000
URL: http://www.amersham.co.uk

Olympus Optical Company (UK) Ltd 2–8 Honduras Street, London EC1Y 0TX, UK

Organon Laboratories Ltd Cambridge Science Park, Milton Road, Cambridge CB4 4FL, UK Organon Inc., 375 Mount Pleasant Avenue, West Orange, NJ 07052, USA

Oxoid Ltd Wade Road, Basingstoke, Hampshire RG24 8PW, UK

Perkin Elmer Ltd Post Office Lane, Beaconsfield, Buckinghamshire HP9 1QA, UK
Tel: 01494 676161
URL: http//:www.perkin-elmer.com

Pierce Chemical Company 3747 N. Meridian Road, Rockford, IL 61105, USA

Pharmacia Davy Avenue, Knowlhill, Milton Keynes, Buckinghamshire MK5 8PH, UK (also see Amersham Pharmacia Biotech)
Tel: 01908 661101   Fax: 01908 690091
URL: http//www.eu.pnu.com

Polysciences Inc. 400 Valley Road, Warrington, PA 18976, USA

Promega (UK) Ltd Delta House, Chilworth Research Centre, Southampton SO16 7NS, UK
Tel: 0800 378994   Fax: 0800 181037
URL: http://www.promega.com
Promega Corp., 2800 Woods Hollow Road, Madison, WI 53711-5399, USA
Tel: 001 608 2744330   Fax: 001 608 2772516
URL: http://www.promega.com

Qiagen (UK) Ltd Boundary Court, Gatwick Road, Crawley, West Sussex RH10 2AX, UK
Tel: 01293 422911   Fax: 01293 422922
URL: http://www.qiagen.com
Qiagen Inc., 28159 Avenue Stanford, Valencia, CA 91355, USA
Tel: 001 800 4268157   Fax: 001 800 7182056
URL: http://www.qiagen.com

Research Organics 4353 East 49th Street, Cleveland, OH 44125, USA

Roche Diagnostics Ltd Bell Lane, Lewes, East Sussex BN7 1LG, UK
Tel: 0808 1009998 (or 01273 480044)
Fax: 0808 1001920 (01273 480266)
URL: http://www.roche.com
Roche Diagnostics Corp., 9115 Hague Road, PO Box 50457, Indianapolis, IN 46256, USA
Tel: 001 317 8452358
Fax: 001 317 5762126
URL: http://www.roche.com
Roche Diagnostics GmbH, Sandhoferstrasse 116, D-68305 Mannheim, Germany
Tel: 0049 621 7594747
Fax: 0049 621 7594002
URL: http://www.roche.com

RS Unit 6, Flanders Park, Hedge End, Southampton, Hampshire SO30 2FZ, UK

Scandinavian IVF Science now Vitrolife Fertility Systems, UK Distributor: Hunter Scientific Ltd, Unit 3a, Priors Hall, Widdington, Saffron Walden, Essex CB11 3SB, UK

Schleicher and Schuell Inc. 10 Optical Avenue, Keene, NH 03431A, USA
Tel: 001 603 3572398

Serono Laboratories, Inc. 100 Longwater Circle, Norwell, MA 02061, USA

Serotec Ltd 22 Bankside, Station Approach, Kidlington, Oxford OX5 1JE, UK

Shandon Scientific Ltd 93–96 Chadwick Road Astmoor, Runcorn, Cheshire WA7 1PR, UK
Tel: 01928 566611
URL: http//www.shandon.com

Sigma–Aldrich Co. Ltd The Old Brickyard, New Road, Gillingham, Dorset SP8 4XT, UK
Tel: 0800 717181 (or 01747 822211)
Fax: 0800 378538 (or 01747 823779)
URL: http://www.sigma-aldrich.com

Sigma Chemical Co., PO Box 14508, St Louis, MO 63178, USA
Tel: 001 314 7715765
Fax: 001 314 7715757
URL: http://www.sigma-aldrich.com

Sigma-Genosys London Road, Pampisford, Cambridgeshire CB2 4EF, UK
Sigma-Genosys, 1442 Lake Front Circle, The Woodlands, TX 77380-3600, USA

Specialty Media Division of Cell and Molecular Technologies, Inc., 580 Marshall Street, Phillipsburg, NJ 08865, USA

Spectronics Corporation 3190 North East Expressway, Suite 220, Atlanta, GA 30341, USA

Spectrum-Microgon 23022 La Cadena Drive, Suite 100, Laguna Hills, CA 92653, USA

Stratagene Inc. 11011 North Torrey Pines Road, La Jolla, CA 92037, USA
Tel: 001 858 5355400
URL: http://www.stratagene.com
Stratagene Europe, Gebouw California, Hogehilweg 15, 1101 CB Amsterdam Zuidoost, The Netherlands
Tel: 00800 91009100
URL: http://www.stratagene.com

Stratech Scientific Ltd 61–63 Dudley Street, Luton, Bedfordshire LUZ QNP, UK

Sutter Instrument Co. 51 Digital Drive, Novato, CA 94949, USA

Taab Laboratory Equipment Ltd. 3 Minerva House, Calleva Park, Aldermaston, Berkshire RG7 8NA, UK

Ted Pella Inc. PO Box 492477, Redding, CA 96049-2477, USA

United States Biochemical (USB) PO Box 22400, Cleveland, OH 44122, USA
Tel: 001 216 4649277

Vector Laboratories, Inc. 30 Ingold Road, Burlingame, CA 94010, USA
Vector Laboratories, Ltd, 3 Accent Park, Bakewell Road, Orton Southgate, Peterborough PE2 6XS, UK

VH Bio Ltd. PO Box 7, Gosforth, Newcastle upon Tyne NE3 4DB, UK

Whatman International Ltd. Whatman House, 20 St Leonard's Road, Maidstone, Kent ME16 0LS, UK

World Precision Instruments, Inc. 175 Sarasota Center Boulevard, Sarasota, FL 34240, USA

Worthington Biochemical Corp. 730 Vassar Avenue, Lakewood, NJ 08701, USA

Xenopus Express 5 Gerbera Court, Homosassa, FL 34446, USA

Index

ablation of cells 137, 142, 148–9
actin cytoskeleton 2, 12, 15, 21, 59, 85, 169, 255
α-actinin 24
adherens junctions 2, 3, 37, 40, 203
affinity chromatography 108, 115
agglutinins 39
Alexa dyes 48, 56, 77, 134
angiopoietin 39
angiotensin-converting enzyme 39
antisense methods 66–9, 162, 194–8, 244
antral follicles 177–80, 182–3, 185, 189, 193
apoptosis 205, 222–4
aromatase 193
Armadillo family proteins 24; see also catenins
avian retinal ganglion cells 49–51

baculovirus 165–7, 172
biocytin 48
biotinylated dextran (BDA) 134, 136
blastocoel cavity 203–5, 233–4
blastocyst 186, 203–5, 214, 218–9, 223–225, 232–5, 241
blastula 154, 156
BODIPY® dyes 85
bromodeoxyuridine (BrdU) 134–136, 248

CaCo-2 cells 168
cadherins
    biochemical analysis 24–9
    desmosomal cadherins (desmoglein, desmocollin) 2, 26

E-cadherin 1-3, 11–12, 14–6, 18, 20, 24–5, 34, 71, 75, 87, 238
    functional assays 20–4
    immunocytochemistry 14–20
    N-cadherin 16
    P-cadherin 2,12, 16, 20, 25
    phosphorylation assays 30–5
    VE-cadherin 37, 39, 41
    Xenopus 154
calcium
    cytosolic 88–9
    chelation 6
    epithelial cell culture 3–6, 12–13, 20, 23, 28, 71–4, 78, 87, 154, 205
    signalling 87
    switching, see epithelial cell culture
calmodulin 71, 87
carbocyanin 127–130; see also DiI
Cascade blue 52, 56
catenins 23–30, 32, 35, 157
CD3 97–8
CD8 126
CD11 94–5, 105; see also integrins
CD18 94–5, 99; see also integrins
CD36 39
CD44 2, 12, 22
CD62 96; see also selectins
CD68 231
cell adhesion
    endothelial cells 37–45
    calcium 3–6, 12–13, 20, 23, 28, 154
    functional assays 20–4, 111–2
    implantation 229–42
    keratinocytes 1–35
    leukocytes 93–118
    placental development 243–56

preimplantation mammalian embryo 203–5, 214–20, 222
cell cycle 205–9
cell settlement 53
choriocarcinoma cells 243
cholera toxin 5
cingulin 71, 87, 153, 157, 162, 174, 204
claudin 37, 71, 80
cobalt 127
collagen 72, 96, 114, 181–2, 244–9, 256
collagenase 180, 182, 188, 230, 246, 248, 249, 256
compaction 206, 209, 233
confocal microscopy 19–20, 50–2, 55–6, 58–9, 85–6, 156, 197, 204–5, 207, 209, 214–16, 222
connexins 47, 50, 54–63, 179, 193–4
Coomassie blue 105, 170–2
COS cells 108–10
cumulus cells 179, 184–98; see also granulosa cells, oocyte-granulosa cell complexes
cyclic adenosine monophosphate (cAMP) 185
cycloheximide 185
Cy dyes 56, 77, 134
cysteine 157
cytokeratin 231, 239, 243, 250
cytotrophoblast 243–4, 246, 248–9, 253–4

DAPI 53, 250
deciduum 244–5
desmocollin (DSC) 210, 219

desmosomes 2, 12, 203, 218, 238
dextran-fluorochrome conjugates 51-2, 83
diaminobenzidine (DAB) 129, 131-4, 158
diethylstilbestrol 192
differential nuclear labelling 223
DiI 53, 127-30
diptheria toxin 140
dissections 121-3, 237
dissociation constant (K_d) 171-3
Drosophila
central nervous system 119-51
dissection 121-3
dye injections 126-30
immunocytochemistry 130-4
lineage tracing 135-7
manipulation of cells 137-49
reporter genes 123-6
dye injections 126-30
dye transfer 48-53

ectopic gene activation 137-9
electrical potential 83-4
electron-dense tracers 78-9
electron microscopy 77-82, 153, 156-7, 252-3
embryos
chick 67-9
Drosophila 130-3, 141-2, 146
human 229-35
mammalian pre-implantation 203-28, 240-2
Xenopus 153-64, 173
endometrium 229-32, 234-5;
see also uterine epithelium
endothelial cells 4, 25, 37-44, 94, 96
enhancer trapping 124, 139-40
enzyme-linked immunosorbent assay (ELISA) 112-13
epidermal growth factor 5
epidermis; *see* keratinocytes
epithelial cells 1-36, 71-91, 153-76, 203-28, 229-58
expression vectors 5, 9, 108, 162, 165-7
extracellular matrix (ECM) 177,

189, 230-2, 244, 253; *see also* collagen, laminin
factor VIII-related antigen 39
FAD medium 5, 6, 22
Fc chimeric proteins 108-12
ferritin 78
fertilization 155-6, 183-4, 186-7, 231, 233, 235
fetuin 186
fibrinogen 94
fibroblasts 5, 39, 243, 250, 253-6
fibronectin 42-4, 96, 243
flippase recombinase target recombination system (FLP-FRT) 140-1, 143-7
fluorescence microscopy 48, 156, 158, 197, 207, 223, 224, 250;
see also immunocytochemistry; confocal microscopy
follicle stimulating hormone (FSH) 155, 186, 189, 193, 233
freeze fracture 80-2, 153
freeze substitution 58
Fura-2 88

GAL4 124, 139-40, 142-7, 149
GAL80 146-7
β-galactosidase; *see LacZ*
gall bladder 72, 83
gap junctions 47-69, 178-80, 192-4
gastrula 156, 162
germinal vesicle breakdown (GVBD) 184, 187
giant cells 243
glial cells 123-37
glucose transporters 40
glutathione S-transferase (GST) fusion proteins 10, 106-7, 116, 160, 164-73
glyceraldehyde 3-phosphate dehydrogenase 59
Graafian follicles 198
granulocytes 96
granulosa cells 177-201; *see also* oocyte-granulosa cell complexes
green fluorescent protein (GFP) 125-6, 143, 145-7

GTP-binding proteins 88
HaCat cells 5, 9
haemaglobin 78
halothane 64
hatching of blastocyst 234, 241-2
heat shock promoter 138-40, 143, 145-7
high performance liquid chromatography (HPLC) 194
histamine 42
Hoechst dye 53, 204, 223
horseradish peroxidase 43, 48, 77, 113, 127, 134, 136
HPRT 227
human chorionic gonadotrophin (hCG) 155, 187, 208-9, 233, 236, 243
human leukocyte antigen 243
HUVECs 38-44
hyaluronic acid 189
hyaluronidase 188
hydocortisone 5

imaginal discs 136
immunoblotting 27-9, 32, 54, 68, 101, 157, 160-2, 164-71, 174, 215
immunocytochemistry
antibodies 16-19, 54-8
detergent solubility 14-16
Drosophila 130-4
embryo-endometrial interactions 231, 238
endothelial cells 43-4
epithelial cell cultures 76-7
fixation 54
gap junctions 48, 54-8, 68
keratinocytes 14-20
oocytes 196-7
permeabilisation 54
placental development 249-52
preimplantation mammalian embryos 214-18
Xenopus development 156-8, 162
see also confocal microscopy
immunoelectron microscopy 156
immunoglobulin superfamily adhesion molecules 93, 94, 96
see also intercellular cell adhesion molecules (ICAMS)

immuno-PCR 216
immunoprecipitation 25-9, 31, 99-100, 103, 160, 165, 171, 173-4, 215
immunosurgery 204, 218-20, 223
implantation
  embryo-endometrial interactions 229-42
  human placental development 243-56
inhibin 194
inner cell mass (ICM) 203-5, 210, 214, 218-20, 222, 233
insect cell expression 153, 165-7, 171, 172
in situ hybridisation 54, 61-3, 219-22
insulin 5, 230
integrins 2, 12, 22, 39, 93, 94, 96, 97, 104-7, 111, 113-16, 157, 243
  see also CD11, CD18
interleukin 231
intestine 72, 83
intercellular cell adhesion molecules (ICAMS) 39, 94, 95, 97, 102, 105, 111-14
inulin 83
involucrin 2
junction adhesion molecule 2-3, 41

keratinocytes
  biochemical analysis 24-30
  culture 5-6
  expression vectors 5, 9
  immunocytochemistry 14-20
  microinjection 7-14
  orthophosphate labelling 30-1
  protein phosphorylation assays 30-5
kinesin 126
Kit receptor 179

LacZ 124, 125, 139, 163
laminin 96, 114, 239
Landsteiner-Wiener antigen 94
latex beads 22-4

laser gene activation 140-3
leukocytes
  activation of adhesion 97-8
  adhesion molecules 93-6, 98-118
  placental development 243
  see also lymphocyte T-cells
leukocyte adhesion deficiency 96
lipid diffusion 86
low density lipoprotein 39
Lucifer yellow 48-50, 52, 193
luteinization 179, 193, 198
luteinizing hormone (LH) 184, 187, 189, 192
lymphocytes T-cells 97-104
L-lysine 54-6

macrophages 94, 231
mannitol flux 78, 82-4
Matrigel 244-7, 249
Madin-Darby canine kidney (MDCK) cells 24, 73-4, 77-8, 80, 82, 85, 87-8, 168, 203, 235
meiosis 183, 184-5, 188
membrane conductance 73, 80, 82
microelectrodes 48-52, 65, 78, 84, 153
microinjection 7-14, 20, 162-4, 195-8
mitomycin C 5
monocytes 94
mosaic analysis 143-8
multidrug resistance protein 40
myometrium 244

Na/K-ATPase 87, 203
nervous system 119-51
neurobiotin 48-52, 127
neuroblasts 130
neutrophils 94
Normarski interference contrast microscopy 50, 51
Northern blotting 69

occludin 41, 71, 75, 80, 85, 87, 157, 159, 160, 204

oestradiol 187, 192, 193, 233
oocytectomy 189-91
oocyte-cumulus cell complexes; see cumulus cells, oocyte-granulosa cell complexes
oocyte-granulosa cell complexes
  antisense strategies 194-8
  gap junctions 193-4
  granulosa cell effects 188
  in vitro maturation 183-8
  in vivo systems 198
  isolation and culture 179-83
  oocyte effects 188-93
  oocyte growth 177-9
ouabain 87
ovulation 229

paracellin-1 80
paracellular flux 82-4
patch electrodes 50-1
peptides 65
phage display 113-16
pharmacological agents 64, 71
phorbol esters 72, 97, 98, 101
phosphate labelling 101-4
phosphoamino acid analysis 103-4
phospholipase C 71, 87
phosphopeptide mapping 33, 103-4
phosphorylation of proteins 30-5, 40, 71-2, 87-8, 97, 101-4, 157
phosphatase inhibitors 30
photoreceptors 48
plakoglobin; see catenins
platelet endothelial cell adhesion molecule-1 (PECAM-1) 37, 39, 41
platelets 94, 96
Pluronic gel 67-9
polar body 184, 187, 233
polyacrylamide gel electrophoresis (PAGE) 29, 33, 97-100, 102-3, 105, 107, 160-1, 164-72, 174
polymerase chain reaction (PCR) 108-10, 116, 225
  see also reverse transcriptase-PCR
polyornithine 72
potassium channels 73

pre-antral follicles 177, 178, 179–83, 188, 192, 193
preimplantation mammalian embryos
cell-cell interactions 203–5
cell lineage segregation 218–27
culture 229–35, 240–2
gene expression 210–14, 220–2
protein expression 214–18
sexing 223, 225–7
synchronous embryos and cell clusters 206–9
pre-ovulatory follicles 178, 192
pregnant mares serum (PMS/G) 185, 187, 188, 236
primordial follicles 177, 178, 183
primordial germ cells 177
progesterone 198
pronucleus 184, 233
propidium iodide 57, 223, 250
protein kinase C 71, 87, 97, 101
reticulocyte lysates 153, 169–71
reporter genes 123–6, 130, 163
see also green fluorescent protein, immunocyto-chemistry, LacZ
reverse transcriptase-polymerase chain reaction (RT-PCR)
gap junction genes 59–61
preimplantation mammalian embryos 210–14, 218, 219, 223
RHO GTPases
cdc-42 2, 7, 8
Rac 2, 7, 8, 13, 14
Ras 7
Rho 2, 7, 8, 13, 14
ricin 140
RNase protection 69
ruthenium red 77, 78

scanning electron microscopy 159–60, 231
scrape loading 51–2
selectins 39, 93, 95, 96

sexing embryos 225–7
sharp electrodes 48–50, 65
SRY 227
stroma 229, 231, 232, 235, 244
talin 23
tau 124, 126
T cells; see lymphocytes
testis 72
tetanus toxin 124, 140
Texas red 134
theca cells 179, 180, 192
thin layer chromatography 34, 102
thrombin 41, 42
thrombomodulin 39
tight junctions
biochemical analysis 160–161, 164–175
calcium signalling 87
conductance 73, 82
electron microscopy 77–82, 159–160
electron dense tracers 78–79
endothelial cells 37
epithelial cells 71–91, 164–75, 203–5, 218, 219, 238
freeze-fracture 80–2
immunocytochemistry 76–7, 156–8
immunoprecipitation 173–5
lipid diffusion 86
paracellular flux 82–4
phosphorylation of proteins 87–8
preimplantation embryos 203–5, 218, 219
protein-protein interactions 164–75
Xenopus development 153–64
see also cingulin, claudin, occludin, junction adhesion molecule, ZO-1, ZO-2, ZO-3
transepithelial electrical resistance 71, 74, 78, 83–5, 87, 238–40
transforming growth factor-β 244
trophectoderm 203–5, 210, 214, 218–20, 222–3, 229

trophoblast 233, 243, 244–53, 255, 256
see also cytotrophoblast, giant cells
tumour necrosis factor-α (TNF-α) 41
TUNEL labelling 223–4
tyrosine phosphorylation 101
umbilical cord 38, 245
uterine epithelium 235–40, 241–2; see also endometrium

vascular cell adhesion molecule-1 (VCAM-1) 39, 96
vascular endothelial growth factor (VEGF) 39, 41
vascular permeability 40–3
vimentin 231, 255
vinculin 24
vitelline membrane 131, 156, 157, 159, 160, 163

Weibel-Palade bodies 96
Western blotting; see immunoblotting

Xenopus development 153–64

ZFY 227
ZO-1 37, 41, 71, 87, 157, 159, 174, 204, 219
ZO-2 71, 87, 174
ZO-3 71
zona pellucida 186, 195–7, 206, 208–9, 216–17, 219, 223, 234, 242
ZP3 194